P9-CFD-242

THE EARTH GAZERS

BY THE SAME AUTHOR

You Are Here

How to Make a Human Being

THE EARTH GAZERS

GAZERS

ON SEEING OURSELVES

CHRISTOPHER POTTER

PEGASUS BOOKS
NEW YORK LONDON

THE EARTH GAZERS

Pegasus Books Ltd
148 West 37th Street, 13th Floor
New York, NY 10018

Copyright © 2018 by Christopher Potter

First Pegasus Books hardcover edition February 2018

All rights reserved. No part of this book may be reproduced in whole or in part
without written permission from the publisher, except by reviewers who may quote brief
excerpts in connection with a review in a newspaper, magazine, or electronic publication; nor may
any part of this book be reproduced, stored in a retrieval system, or transmitted in any
form or by any means electronic, mechanical, photocopying, recording, or other,
without written permission from the publisher.

ISBN: 978-1-68177-636-1

10 9 8 7 6 5 4 3 2 1

Printed in the United States of America
Distributed by W. W. Norton & Company, Inc.

And whan that he was slayn in this manere,
His lighte goost ful blisfully is went
Up to the holughnesse of the eighthe spere,
In convers letying everich element;
And ther he saugh, with ful avysement,
The erratik sterres, herkenyng armonye
With sownes ful of hevenyssh medolie.

And down from thennes faste he gan avyse
This litel spot of erthe, that with the se
Embraced is, and fully gan despise
This wrecched world, and held al vanite
To respect of the pleyn felicite
That is in hevene above; . . .

—*Troilus and Criseyde,* CHAUCER

Once a photograph of the Earth, taken from
the outside, is available . . . a new idea as
powerful as any in history will be let loose.

—FRED HOYLE, 1948

The stature of man in prodigious
confrontation with the cosmos emerges
immensely small and immensely large.

—POPE PAUL VI, 1968

For my father, who woke me up to watch the first moonwalk. I wish I had shown more enthusiasm at the time.

PRELUDE

On 21 December 1968, at around 10.30am Eastern Standard Time, three men saw what no human being had ever seen before, the Earth as a sphere in space. Only 24 human beings have seen the Earth from the outside: the Apollo astronauts who went on the nine manned missions to the moon that took place between 1968 and 1972.

On 20 December 1968, the day before the launch of Apollo 8, the first of the manned missions, Charles Lindbergh and his wife Anne were given a tour of Cape Kennedy, a 'city' of 30 launch towers built in the heart of a nature reserve. They were shown around the Flight Crew Training Building where the astronauts practised flight and landing manoeuvres in Apollo and Lunar Module simulators. They walked through the Vehicle Assembly Room where, like some exceptionally complex puzzle, the rockets were assembled out of parts that had been made elsewhere by thousands of companies all across America. It was then the largest building in the world: 'The lower bay alone is the size of the United Nations Building,' their guide informed them. They were taken, too, on a tour of what felt like a city within a city, a region of Kennedy where the many NASA administrators, scientists, engineers and technicians were making the final preparations for tomorrow's launch. As they moved between one region of the vast launch site and another, dotted here and there they saw various vintage rockets repainted to preserve them against the salt air. Anne wrote later that already these post-Second World War rockets looked as dated as Civil War cannon come upon in a country graveyard.

With a sense of relief they arrived at the astronauts' quarters, the last stop on the tour. A notice on the door warned anyone with a cold, or symptoms of a cold, not to pass beyond that point. On the other side of the door was a small reception room decorated with an artificial Christmas tree. A real one would have been a fire risk, they were told. In an adjoining room Charles and Anne surprised a group of astronauts and geologists bent over celestial maps. After some animated talk, they were asked if they would stay to lunch.

On the walls of the otherwise unornamented, bleak room that served as a dining room was a coloured photograph of a Greek temple, and a view of the White House showing the Washington Monument. They were about fifteen to lunch, seated around a single rectangular table; among the party was the crew of Apollo 8, whose last lunch this would be before the launch.

During lunch Charles told the astronauts about his youthful experiences of wing-walking. They told him about their experiences of walking in space. Charles wondered how much fuel the rocket would burn on takeoff. Someone said around 20 tons in the first second. Ten times the amount of fuel, Lindbergh said, that had taken *The Spirit of St Louis* all the way from New York to Paris in 1927. He told them about a conversation he had had with the plane's designer, Donald Hall, about the size of the fuel tanks. Hall had said to him, 'Say, how far is it to Paris?' And Lindbergh had realized that he didn't know for sure. He thought they could get a good estimate of the planned route by scaling off a globe. 'Do you know where there is one?' he asked. 'At the public library,' Hall said. 'It only takes a few minutes to drive there. My car's right outside.' At the library Lindbergh rummaged in his pockets and found a piece of white grocery string. He placed one end on New York, ran the string up the coast to Newfoundland, from where he would begin his Atlantic crossing, and across to Paris. He pulled the string tight and measured it: 3,600 miles, they calculated. The estimate would turn out to be surprisingly accurate.

Charles told his attentive audience about how, just two years after his pioneering solo crossing of the Atlantic, he met Robert Goddard, America's first – and at the time *only* – rocket engineer. Goddard believed, when few others had, that a rocket might one day take human beings to the moon. Goddard told him during their first conversation that though he thought it was possible to send men to the moon, he worried that the venture might prove to be too expensive. It might cost as much as a million dollars, he said.

'Think,' one of the Apollo 8 astronauts at the table said, 'it's hard to believe, this time tomorrow we'll be on our way to the moon.' There was something boyish about these young men, Anne later wrote. She was reminded perhaps of her young hus-band when she first met him, soon after his triumphal tour of every American state. He was in his mid-twenties and she 21, but both childlike; both, too, intelligent and brave. Now, they had grown up into messy, complex adult lives.

After a long day, Charles suggested that he and Anne go and look at the Apollo rocket on its launch pad: the tallest, heaviest and most powerful rocket ever built – the masterpiece of the German–American engineer Wernher von Braun. It was mid-night and they would need to be up at 4.30am for the launch, but she readily agreed.

Anne thought that distance and night had simplified the rocket 'into the sheer pure shape of flight, into beauty'. She was reminded of Henry Adams' reaction when the historian had first walked into the 'great hall of dynamos' at the Great Exposition of

1900. He had stood before the giant machine half in awe and half repelled; so shaken that he had begun to pray to this 'silent and infinite force'. Here at Cape Kennedy, Anne acknowledged the presence of another such 'dynamo'.

A few hours before takeoff, a helicopter flew around the site in an attempt to persuade as many birds as possible temporarily to leave the area; the Cape was on their migratory route. Around the launch sites, 50,000 acres had been preserved as a wildlife sanctuary.

The three Apollo 8 astronauts were in their eyrie capsule. Now, with only moments to go before takeoff, the nearest other human being was at least 3 miles away.

With the gaze of hundreds of thousands of spectators fixed on it, the rocket began to rise slowly: 'as in a dream,' Anne wrote, 'so slowly it seemed to hang suspended on the cloud of fire and smoke'. A number of recalcitrant birds took flight. Instinctively, Anne turned to watch them, only to find that when she turned back the rocket had left the launch tower. Nearly 40 years after Goddard and Lindbergh had first met, and 23 years after Goddard's death, the world's first manned journey to the moon had begun. Three men were being transported to the moon on top of a multistage liquid-fuelled rocket just as Goddard had predicted.

On Christmas Eve 1968, the crew of Apollo 8 saw the Earth rise over the horizon of the moon, and took photographs to bring back for the rest of us. We, stuck here on our home planet, know intellectually that the Earth is a sphere falling through space, but to know it is one thing, to see it – even in a photograph – is something else. Later that same day, during their final broadcast to Earth, each member of the Apollo crew read in turn from the Book of Genesis. The next day, Christmas Day, the *New York Times* reported that here in the manned missions to the moon was an opportunity to bring the sacred and secular together into some new kind of alliance suitable to the modern age.

The photographs that came back from the Apollo missions immediately catalyzed the rapid growth of what were then nascent fields of ecology and environmentalism. The Genesis reading, however, had been quickly protested by Madalyn Murray O'Hair, the founder of American Atheists, and at the time the country's most powerful defender of the separation of church and state. Because of her objections, future Apollo astronauts were warned against making any kind of religious observance from space. As a result of O'Hair's challenge, familiar battle lines would be drawn up between fundamentalists on both sides of the religious divide. Was the opportunity to acknowledge the numinous in a secular way lost, or has the battle, even now, hardly yet begun?

PART ONE

CHAPTER ONE

According to family legend, when Charles Lindbergh's paternal grandfather, August Lindbergh, lost an arm as the result of an accident at the local sawmill, he asked that it be buried in its own pine coffin. He had apparently addressed his limb in farewell: 'You have been a good friend to me for fifty years, but you can't be with me any more. So good bye. Good bye, my friend.' Even making due allowance for the magnifying, coarsening and mythologizing effects of time, August Lindbergh's life was clearly the stuff of legend.

August Lindbergh had once been Ola Månsson. He was born in Sweden in 1808. Despite the lack of any formal education, he became a parliamentarian known for his brilliant rhetoric. He was a farmer who spoke on behalf of farmers, a liberal who defended women's and children's rights, and those of Jews. He was for land reform, the lessening of trade restrictions and the expansion of railways. He argued that the Lutheran Church's sway was too great. He was friendly with the King. He was the director of a bank. At the height of his fame, he was brought down by his enemies on a largely trumped-up charge of embezzlement. Then in his early fifties, Ola Månsson fled Sweden for America, leaving behind his wife and their seven legitimate children but taking with him his mistress Lovisa – a waitress almost 30 years his junior – and their illegitimate son, Karl; a gold

Ola Månsson c. 1850

medal he had been given by his constituents; a gold watch, Lovisa's only heirloom; and very little else. In America, as many immigrants did who were starting over, they changed their names. They were now the Lindbergh family: father August, mother Louisa and son Charles August, father of the future aviator. In Minnesota, in typical pioneering fashion, August built his family a log cabin. The gold medal was traded for a plough and August was a farmer once more. There was by now already a second child, and soon a third.

It was a couple of years after they had arrived in Minnesota that August accidentally fell against the blade at the sawmill. His arm was mutilated and his chest cut through. His beating heart, as well as part of a lung, could be seen through the wound. The doctor took three days to arrive. There was nothing to be done except cut the arm off at the shoulder, an operation that was performed without anaesthetic. August apparently didn't so much as groan. Soon he was back working on the farm, swinging a scythe he had adapted for one-armed use.

Then, five years after he had abandoned his wife, news came that she had died. August married Louisa. Two of his sons from the first marriage came to join the Lindberghs in Minnesota. And still the family grew. August was to have seven children by each wife, a curious prefiguring of the famous aviator's outsized life to come. The log cabin over the years grew to be one of the largest properties in the area.

Charles August Lindbergh trained as a lawyer and became a significant figure in the Little Falls community. He married Mary La Frond in 1887. She died in 1898, a few days before her thirty-first birthday, of complications following what should have been a routine operation. Two children survived the marriage: daughters, Lillian and Eva. A third child, Edith, had died aged 10 months.

Charles August's second marriage was to Evangeline Lodge Land, who now became the even more gloriously named Evangeline Lodge Land Lindbergh. The Lodges and the Lands were patrician families proud of their ancestors, among whom were numbered the first European settlers of America. Lands fought for George III in the American Revolution. Lodges came over in the *Mayflower*. Evangeline's father, Charles Henry Land, was a dentist and inventor of the jacket porcelain crown, patented in 1889. Their only child, Charles, was born in 1902.

Like his father, Charles August was a politician and a farmer. He was a congressman for a decade from 1907. He had demanded regulation of the railroads, changes to the democratic process, conservation measures and various economic reforms. He was one of only 50 representatives who voted against America joining the First World War. In 1918 he tried for the Senate and failed, coming a poor third in the election. He had been followed by mobs during the campaign, arrested on charges of conspiracy, dragged from the podium during one speech, escaped another meeting amid a volley of shots and even hanged in effigy.

Charles August and Evangeline split up when Charles was seven. Charles lived with his mother, but his father visited frequently. His half-sister Eva said that her father and Evangeline were not suited to each other. Evangeline was apparently emotionally volatile, Charles August austere and difficult to know. Eva had been left in the care of her stepmother after the separation but ran away from home aged 14, when her half-brother was five years old. Her father wrote to reassure her: 'I couldn't live with her, and you don't have to either.' In later life Eva said that her stepmother mocked other women in the town to their faces, said she 'was a cruel and crazy woman'.

Looking back from the perspective of his adult self, Charles Lindbergh considered his childhood to have been one of idyllic freedom. His maternal grandfather, Charles Henry Land, gave him a .22-calibre rifle when he was six: 'Father thought six was young for a rifle, but the next year he gave me a Savage repeater; and the year after that, a Winchester 12-gauge automatic shotgun; and he loaned me the Smith and Weston revolver that he'd shot a burglar with.' Charles said that his father shot the intruder as he had tried to make his escape, and that there had been blood on the window-ledge to prove it. This half-sister Eva remembered the story differently, as is often the way with family history. She said that her father had not been able to bring himself to fire at the burglar, even though the burglar had been armed. And so out of such competing anecdotes do family legends fight for precedence. During his lifetime Charles would write about his ancestors on a number of occasions, but as his biographer Scott Berg points out, 'despite his fascination with detail, [Lindbergh] never examined his family history closely enough to see that it included malfeasance, flight from justice, bigamy, illegitimacy, melancholia, manic-depression, alcoholism, grievous generational conflicts, and wanton abandonment of families'.

When Charles was ten, his mother took him on a trip to Panama. In that same year, 1912, his father bought a car, the soon to be ubiquitous Model T Ford. Neither parent drove with any confidence, he said. It was years before his father was at all competent, and his mother was always a timid and alarming driver. Charles was the designated driver. No license was needed in those early days of motoring. It seemed dangerous, he wrote, but only at first. By the age of 12, Charles spent the summer exploring Minnesota by car; he gives the distinct impression that he went on his own. He was engrossed as mechanics disassembled and reassembled the engine. A seasoned driver by the age of 14, he bought a Saxon Six and drove his mother across country from

Charles (right) and his father   Evangeline Lindbergh

their home in Little Falls to California. It took weeks to get there – the weather was often atrocious, the going slow and hazardous. And then he drove her back. The car is still used in town parades to this day.

Charles was a crack shot. He and a friend took it in turns to shoot 25-cent pieces out of each other's fingers from a distance of 50 feet. He once shot 50 bull's eyes in a row. Like many an adventurous teenager he bought a motorcycle, an Excelsior. It wasn't the speed, he said, that gave him the greatest thrill, it was the mastering of his machine. The Excelsior came to feel as if it were an extension of his body. He rode thousands of miles on his own, exploring the surrounding countryside. He was a loner for much of his young life. His mother once revealed – which puts a curious cast on Charles's assertion that his childhood had been idyllic – that she had at times paid local children to play with him.

Charles August was not a particularly avid, nor competent, farmer. When war came, Charles had little difficulty persuading his father to entrust the farm to him. Various legal exemptions were made for farmers during the war years: for example, they might take their children out of school if they were needed on the land. Charles gave up on school and became a teenage farmer. Already in his short life Charles had taken many risks, but here on the farm he said he felt 'death brush past several times'. One day a ploughshare shattered and a piece of it flew by his head like a missile. He estimated that it had missed him by inches. He learned then, he said, that danger was a part of life, to be confronted not turned away from – that life is risky no matter what we do. Risk could not be avoided, he said, but it could be assessed.

After the war, now in his late teens, Charles moved to Madison, Wisconsin, to the university there. His mother – who had taught chemistry at the local school – uprooted herself, and moved into rented rooms to be near him. Charles was an inattentive student and left before he'd finished his sophomore year. He had by now, however, had his first experience of flying, and wanted more. As with the motorcycle, it was not the thrill nor the danger he remembered, but the experience of being taken beyond danger, 'beyond mortality'. To rise above the planet was astonishing. It was as if he were leaving behind his body and the dimensions of earth for another state of being. In the air, from a god-like perspective, he wrote that he was 'never more aware of all existence', never less aware of himself.

He moved again, this time to Lincoln, Nebraska, to take flying lessons. His teacher left after he'd had only eight hours of training, but he knew now where his future lay. To earn money Lindbergh had taken up barnstorming. He and a crewmate would fly from town to town coming in low as one of them stood on the wing in an attempt to attract a crowd. Sometimes one of them would stand on his head, or tie himself in a standing position to the top wing as the plane looped the loop. There was something of the circus, of showmanship, about those early days of aviation. A hero among pilots was Roscoe Turner, who flew with a waxed moustache and a pet lion named Gilmore.

Lindbergh was 6 feet 2½ inches tall. He said it was as safe there on the wing as it was in the cockpit. He claimed flying wasn't as dangerous as the public imagined, that most serious accidents were caused – clearly meaning to exclude himself – by inexperienced pilots who took ill-judged risks. Danger 'lay coiled in the hidden, in the subtle, not the obvious'. They made their money by taking the braver spectators up for a short flight.

Charles saved to buy his own plane. His father also contributed. When Lindbergh arrived at the airfield to collect and pay for it, he simply handed over the money and the plane was his. No license was required in those early days of aviation. The plane was a wartime training plane, a Curtiss JN4-D, affectionately known as the Jenny. Now that war was over, they were being sold off by the government in large numbers. Fitted out with a new engine, it cost him $500. Though he had barnstormed for months, he had previously only flown a Jenny for a few minutes. He still had only those eight hours of formal training behind him, and that was six months ago now. He had had no experience of flying alone. Despite his best efforts he could not get the Jenny more than a few feet off the ground. Embarrassment being the better part of valour, he brought the plane to earth again. A young pilot by the name of Henderson – what pilot wasn't young then? – took pity on him. 'I expect you're just a bit rusty,' he said, and offered to go up with him a few times until he'd got the hang of it. Later that day Lindbergh took off on his own and flew to 4,500 feet. He landed safely, if not elegantly.

The underpowered Jenny was hard to master. It had 'to be wished up over low trees'. It required almost instinctive skill – the ability to synchronize the movement of all the controls at once, and that just to keep the craft in level flight. The Jenny came down hard, often splintering the undercarriage. A pilot had to know more than how to fly the plane; had to be a technician, know how to take apart an engine and put it back together, even had to 'know how to lock-stitch, how to bind the ends of a rubber rope, how to lap a propeller hub to its shaft. There were hundreds of details you had to learn ... you were your own helper, rigger, and mechanic.' A needle and thread was needed as often as a spanner. A surprising flying hazard was cattle, not because they got in the way during an emergency landing, but because cows seemed to enjoy the taste of the dope-soaked wing fabric and might strip the wings in a matter of hours if given the chance.

Both parents were encouraging of Charles's flying ambitions. Charles persuaded his father, who was at first somewhat reluctant, to join him in the air. They dropped leaflets to help promote his father's senatorial ambitions. His mother made a number of flights with her son from the start, once joining him on a ten-day barnstorming tour. She was a less nervous passenger than his father, Charles said.

If you could fly a Jenny you could fly just about anything, and Lindbergh wanted to fly everything. He wanted to fly more modern and more powerful planes, and he knew the only way he'd get to do that was if he joined the army (in those days the air force was still part of the army) and train as a pilot. To win his wings, he would have to go through a year of rigorous training at the United States Army flying school.

Lindbergh graduated from his class with the highest marks. Out of an intake of 104, 33 passed the first stage of training and just 18 got their wings at the end of the course. He was now a Second Lieutenant in the Air Reserve Corps. He enlisted in the 110th Observation Squadron of the 35th Division Missouri National Guard and was soon commissioned First Lieutenant, and then a few months later, Captain.

He went back to barnstorming while he looked for a job. The Robertson Aircraft Corporation promised him the position of chief pilot if they won their bid for the mail route between St Louis and Chicago. America's Air Mail service first began on

15 May 1918 under the auspices of the United States Army Air Service, flying six Jennys that were modified to carry mail. The Post Office took over the service in October and began to employ civilian pilots. In 1925 Congress decided that the business should be put out to private tender. The Robertson Corporation won the St Louis–Chicago route that same year, and Lindbergh got the job.

The life expectancy of a pilot was short, about 800 flying hours. Flying the mail was particularly hazardous. The Robertson Corporation flew modified Jennys and a modified de Havilland biplane salvaged from the army, the DH-4. In the army it had been known as the 'flaming coffin'. The planes had to be flown from the rear – where the navigator would have sat when the plane was in army service. The mail went up front, which meant that there was no forward window. To navigate, the pilot had to look to the side.

Lindbergh assessed the risks: 'How tightly should one hold on to life? How loosely give it rein?' How much risk was he prepared to take? Somewhere in *The Iliad* we are told that Achilles was offered the choice of the long life of a pastoral farmer or the short life of a warrior hero. Lindbergh gave himself a similar choice: 'Of course I would like to have become a centenarian, but I decided that ten years spent as the pilot of an airplane was in value worth more than an ordinary lifetime.' Too much security brings life to a standstill. Without adventure he might as well be a stone as a living human being. There were ways of reducing the risks. One was to know your machine intimately. And then there was the parachute, recently introduced. It drastically reduced the mortality rate of pilots and changed the experience of flying. Now all was not lost if the engine stopped or the plane fell apart in the air. If forced to it, in most types of flying emergency there would be time to think and to take action, the final step being to bail out.

The parachute was still something of an innovation. When Lindbergh attempted his first parachute jump – a double jump, made with two chutes tied together, one to open after the other – he almost died. The wrong type of string had been used, rotten grocery string that could be rubbed apart in the fingers. The second chute opened 'as a useless wad of fabric', which only by chance spread out in time. Lindbergh was unaware until he was told afterwards how close he had come to death.

Parachuting brought with it a new sensation, different again from flying. Lindbergh understood now why the Earth is mother Earth. Dropping, weightless, through the air, the sensation was not of falling but of being held. And though he might have disobeyed her laws, strayed too far from his rightful realm, still he was welcomed back to Earth as a frightened child might run to his mother's arms. First there was fear, but beyond the fear 'life rose to a higher level, to a sort of exhilarated calmness'. He felt now that he was living 'on a higher plane than the skeptics on the ground'.

The first time a parachute saved Lindbergh's life, his plane collided with another in mid-air. The planes locked together, milled around, the wing wires whistling. He unbuckled his belt and climbed onto the damaged wing, pushed himself away from the ship. A second or two of thought, long enough to assess another danger, that the wreckage might fall on him. He jumped. 'How safe the rushing air . . . seemed when I cleared those planes – like a feather bolster supporting me.' He waited until he had fallen what he guessed might be a few hundred feet before pulling the ripcord. 'Next I turned my attention to locating a landing place.' Both pilots landed safely. It was the first time anyone had survived the collision of planes in the air. Both automatically were members of the Caterpillar Club, founded in 1922 by the inventor of the free-fall parachute, Leslie Irvin. Club motto: *Life depends on a silken thread.* Charles Dawson McAllister, the other pilot, was member number 12, Lindbergh number 13.

The second time a parachute saved his life, the plane Lindbergh was test-piloting – a commercial four-seater OXX-6 Plywood Special – went into a sudden spin. On this occasion, because he was so much nearer to the ground when he abandoned the plane, he had to pull the ripcord immediately. The plane fell past him, missing him by only 25 feet or so. He saw it crash in a grain field. 'Then I turned my attention to landing.'

The third time a parachute saved his life, Lindbergh was flying the mail route. He was not far outside Chicago in fog and at night, the twin horrors of early aviation. There was nothing to be done except hope that he could fly out of it, or that the fog cleared. The fog neither cleared nor came to an end. The engine sputtered and died. At 5,000 feet he abandoned his craft. Lindbergh said in later life that he had confronted fear that night at its most pure, as if it were outside him.

Falling through the dark alongside an abandoned plane might be thought terrifying enough, but then suddenly the plane's engine came back to life. There must have been some residual fuel in the tank that sloshed back into the fuel line. He could hear the plane heading his way, and then sensed it pass close by. He could not see a thing. When should he pull his ripcord? Was there a bottom to this bank of fog? He must not wait too long. He guessed that he might be 1,000 feet above the Earth, and yet still he was in fog. Nothing to be done except pull the cord, keep his legs together and hope for the best. The air was turbulent and there was sheet lightning. The chute became so heavily soaked from the saturated air it kept collapsing, only to refill with air once more when a new gust caught it. He landed in a field. The plane narrowly missed crashing into a farmhouse.

The fourth time Lindbergh's life was saved by a parachute, he was once again in fog at night not far outside Chicago, and again he ran out of fuel. This time he landed on a barbed-wire fence. His thick flying suit protected him.

Neither of the mail planes he had abandoned burst into flames when they crashed. There was too little fuel left. The mail could be saved. A telegram was sent ahead, a car arrived, a train was met. Most nights, somewhere in America, a mail plane would come down, usually because of bad weather. Lindbergh was the only member of the Caterpillar Club who qualified twice over, three times over, four times over.

Mail pilots risked their lives not for the mail itself, 'but in obedience to orders which ennoble the sacks of mail once they were on board ship'. The words are not Lindbergh's, though he would have agreed with the sentiment, but were written by his French doppelgänger, Antoine de Saint-Exupéry, the author of the children's classic *Le Petit Prince* (1943), writing here of his own uncannily similar experiences as a mail pilot working for the Latécoère Company, later Air France. Across the world pilots understood – connected perhaps by that silken thread – that there was something sacred about their task, even when, as was often the case, the sacks weighed more than the letters they contained. Until very recently – until the advent of the Internet, which turned physical mail into something ethereal – mail delivery had a mystique about it. The more incomplete the address, the more isolated the destination, the more sacred the duty of delivery.

There was not much demand for airmail in the earliest years of aviation. Who cared if a letter sent from New York to San Francisco might arrive in 36 hours rather than the four days that the journey by train took? But the time was coming when people would care. Lindbergh had seen it. By the mid-1920s there were 2.5 million miles of airmail routes crisscrossing America delivering 14 million letters a year. In 1926 Dwight Morrow, Lindbergh's future father-in-law, was appointed by President Coolidge to chair a board set up to recommend national aviation policy. Lindbergh predicted that in a few years the United States would also be covered by a network of passenger routes, a vision he would help realize. There had been a very few passenger planes from as early as 1913 but only the most intrepid travellers would have been prepared to take the risk.

Pilots were bound together like chivalric knights under some sworn oath. 'A novice taking orders could appreciate this ascension towards the essence of things,' Saint-Exupéry wrote in *Wind, Sand and Stars* (1939), 'since his profession too is one of renunciation: he renounces the world; he renounces riches; he renounces the love of woman. And he renounces his hidden god.' Fanciful perhaps, given that Saint-Exupéry had both wife and mistress. In this band of knights, if Saint-Exupéry was mystical but worldly Lancelot, virginal Lindbergh was Percival.

More practically, out of intense camaraderie, one pilot's experience might save another's life. The animated recounting of, typically, some near-miss was called 'ground flying'. 'We were over strange territory on a dark night and with a rapidly diminishing fuel supply...' In those early days of flying, aviators rarely lost touch with each other altogether, unless, of course, separated by death. Wherever and whenever they met, airmen took up conversations that might have been interrupted by years of silence.

And yet the fellowship served only to underline the essential solitariness of being a pilot. Flying alone above the clouds at night, does a human being ever escape the bonds of the world so completely? What could be more magical, Saint-Exupéry asked, than flying on a clear starlit night, 'its serenity, its few hours of sovereignty'? Lindbergh said that he never saw the Earth so clearly – he meant metaphorically as well as literally – as he did in those early days of flight. 'I feel aloof and unattached in the solitude of space. Why return to that moss, why submerge myself in brick-walled human problems when all the crystal universe is mine?' Lindbergh said that when he came in to land it was as if he were leaving a better life behind: 'Sometimes I circled to delay my landing ... I became conscious of a relativity of time that escaped my mind and senses in ordinary moments. My airplane was my world to me: the world itself was quite unessential. I entered a core of timelessness in a turbulence of time, like the eye of a tornado. Permanence lay only in the instant. Outside all was fleeting ... Riding the wings of power, I realized the fragility of power exposed to the dynamic elements of time.' Saint-Exupéry once annoyed colleagues on the ground when, coming in to land, he circled the landing strip many times while he finished the novel he was reading. Pilots were frontiersmen in search of a homeland they had not yet found. Up there, Saint-Exupéry wrote, was 'a silence even more absolute than the clouds, a peace even more final'. The clouds were a frontier 'between the real and the unreal, between the known and the unknowable'.

To be distracted by the view was to take a risk. 'One can't be following a satellite's orbit and watching these dials at the same time,' Lindbergh wrote of looking, from high above the clouds, at the moon: 'I return abruptly to the problems of temperature, oil pressure, and rpm.' It was a risk then, and it would be a risk still, decades later, to astronauts struggling not to gaze out the window at the mesmerizing view of a receding Earth. They called it 'Earthgazing'. It was addictive, but a flying machine requires a great deal of attention. Dreams and machines do not mix easily.

Those early aviators were masters of their machines. For most of us our tools remain forever separate from ourselves, something out there to be manipulated as best we can; for them – as perhaps, say, a violin becomes for a great violinist – the tool, instrument, machine becomes an extension of the self. 'It was as though the wings, nose, and tail were a part of me,' Lindbergh wrote. 'They followed my wishes just as did my arms and legs ... With practice, the handling of a plane becomes instinctive. You move without thinking because you have no time for thought. So long as you have to think to make your plane take action, you have not become its master and its complement.'

Many years later, during one of the regular Saturday recreational flights he took with his youngest daughter Reeve, the engine cut out. 'What I noticed was my father's sudden alertness, as if he had opened a million eyes and ears in every direction,' Reeve said. 'Are we going to crash?' she had asked, not out of fear but conversationally. It hadn't occurred to her to be scared, so confident was she of her father's abilities as a pilot. She said he coaxed and willed 'the plane to do what he wanted it to do . . . He could feel its every movement, just as if it were part of his own body. My father wasn't flying the airplane he was *being* the airplane . . . Now I knew.' Afterwards no one could work out how he'd managed to land in such an enclosed space. The plane had to be taken apart to get it out of the field.

There was something magical, too, about the mere fact of flight. 'There seems to be no reason whatever to keep you from plummeting earthward like a rock,' Lindbergh wrote. 'It's not until you put your arm outside, and press hard against the slipstream, that you sense the power and speed of flight. The air takes on the quality of weight and substance.'

It was to a heightened experience of wakefulness that these aviators were in thrall. 'It is not danger I love,' Saint-Exupéry wrote, 'I know what I love; it is life.' Flight then had something of the visionary about it; as if consciousness for a moment opened out. Even from the ground and to a non-aviator, there was something other-worldly about man-made flight, at least in those early days when all was new. When Marcel Proust, as the narrator of his long novel, saw a plane for the first time he burst into tears. He felt that for the pilot, and through the pilot for himself too, 'there lay open . . . all the routes in space, in life itself'. He saw the plane glide for a few moments over the sea, before the pilot quietly made up his mind, seemed 'to yield to some attraction that was the reverse of gravity', and returned to 'his native element'.

Planes were machines capable of 'annihilating time and space', Saint-Exupéry wrote, making them tools in our central struggle to understand one another. But it is early days, he warned; we are 'young barbarians still marveling at our new toys'. Goethe wrote that the human being is the most precise machine that can exist. Perhaps, but which is elevated: the machine or human beings? As the centuries roll on and our tools, instruments and measuring devices become more and more refined, we might be tempted to suppose, at some point, that our early days are behind us. And yet we should be careful. Science, as the philosopher David Deutsch has said, is at the 'beginning of infinity'. Should our days turn out even to be numberless, we might, as judged by the elusive and infinite universe, be young barbarians forever.

CHAPTER TWO

He had flown a mail plane through winter nights, fog and storms. He had abandoned a plane mid-air four times over, and survived. Lindbergh wondered if attempting to fly between New York and Paris could be any more hazardous.

In 1919 the French-born American hotelier Raymond Orteig offered prize money of $25,000 to whoever was first to fly between Paris and New York, in either direction. The prize in part was meant to promote good relations between America and France, relations that had become strained during and after the negotiation of the peace settlement at the end of the war. France had wanted Germany permanently weakened and to make the country pay, literally and metaphorically. William Bullitt, a member of the US delegation to the talks, later wrote of the Treaty of Versailles that it was 'one of the stupidest documents ever penned by the hand of man'. Many were sure even then that the humiliation of Germany would make another war inevitable.

More substantial prize money was to be had elsewhere and for less of a challenge. The London *Daily Mail* was offering $50,000 to the first crew to fly across the Atlantic, starting and finishing points unspecified. British aviators John Alcock and Arthur Brown made the first crossing and claimed the *Mail*'s prize money only a few weeks after the Orteig prize was first established. They flew from Newfoundland to Ireland, nearly 2,000 miles, in 16 hours and 12 minutes. John Alcock died later that same year when his plane came down in fog.

Alcock and Brown's 'flying crate' – a modified version of the Vimy IV, a British bomber biplane – laden with 865 gallons of fuel had struggled to take off. No aircraft of the time could have carried enough fuel to sweep the more than 3,500-mile arc between New York and Paris. Even by 1923 the long-distance record stood at 2,500 miles, for a flight undertaken between New York and San Diego.

After five years, without a single contender stepping forward, the Orteig prize expired, just at the moment that technology had become advanced enough to make such a challenge achievable, at least in principle: planes had become more streamlined, more fuel-efficient.

Orteig decided to re-establish the prize just weeks after it expired, and almost immediately a number of competitors declared themselves.

The French ace René Fonck and his American co-pilot Lieutenant Lawrence Curtin made an early attempt at the crossing. On 21 September 1926 their Sikorsky plane crashed on takeoff in New York, killing two of the four-man crew, the radio operator Charles Clavier and the mechanic Jacob Islamoff. A few months later Fonck announced that a new plane was being built for him, and that he would make a second attempt. He never did. A number of other teams were by now also in the running.

Lindbergh wrote to his mother – his father had died of cancer in 1924 – to tell her of his decision to try for the Orteig prize. Evangeline wrote back saying that he must lead his own life: 'I mustn't hold you back. Only I can't see the time when we'll be together much again.' Their correspondence was frequent, their letters intimate, at times reading more like love letters than those between a mother and her only child. She was right: they were not to be together much again.

Lindbergh decided that his best shot at the prize would be in a single-engined monoplane, flying alone. He spent months looking for backers, but without success. No one was prepared to fund that combination of risks. Most of the other contenders had opted to fly tri-motor planes, and no one else was preparing to make the flight alone. What if he became exhausted? What if the engine failed? He argued that on his own he would have only himself to blame or fall out with, and that monoplanes were known to travel relatively faster when heavily loaded with fuel than less streamlined planes. Monoplanes were what the army had used on their multistage round-the-world flights. A single-engined plane would also be cheaper; and the clincher, Fonck's plane had had three engines and hadn't even got into the air. In any case three engines did not reduce the risk, rather increased it: the chance of any single engine failing would be correspondingly higher simply because there were three of them. No tri-motored plane of the time would have been able to make it across the Atlantic on two engines.

Lindbergh thought about raising the money by public subscription, but there wasn't time. He decided to try to split the investment between a number of businessmen in his home town of St Louis. The first to come in on the venture was Major Albert Bond Lambert, the city's first licensed pilot, and an avid balloonist and keen golfer (he won a silver medal at the 1904 Olympics); second was Harold M. Bixby, a banker and the head of the St Louis Chamber of Commerce. Bixby had wanted to stump up the rest of the money but his wife was against the idea. She was sure Lindbergh would be killed and that her husband would then appear culpable. Bixby decided to approach fellow members of the St Louis Racquette Club. Seven agreed to contribute, making a consortium of nine in all. It was Bixby, too, who came up with a name for the plane, *The Spirit of St Louis*, not just in honour of their city but because King Louis IX of France became St Louis, the patron saint of Paris. The name seemed propitious, must surely appeal to the French, and suited the ambition of the prize.

Lindbergh now had the modest funds he needed to compete – a fraction of the money his rivals had each raised – but he didn't yet have a plane. By the time he heard about Ryan Airlines, a company based in San Diego that made monoplanes, he had pretty much decided in his own mind that it might well be too late for the Orteig, but he decided he'd make an attempt anyway, and if not the Orteig he could still try for some kind of endurance record. He sent a wire: 'CAN YOU CONSTRUCT WHIRLWIND ENGINE PLANE CAPABLE FLYING NONSTOP BETWEEN NEW YORK AND PARIS.' With the nonchalance that seemed to characterize the age, the founder, Claude Ryan, wrote back: 'CAN BUILD PLANE SIMILAR M-I BUT LARGER WINGS ... DELIVERY ABOUT THREE MONTHS.' Lindbergh asked, could they build it in two months? The owner of the company, Benjamin Franklin Mahoney, only in his twenties himself, wired to say two months would be long enough. Nothing ever fazed him, an employee recalled.

Lindbergh was due to fly out and meet the Ryan team but changed his plans when he got a telegram from the aircraft builder Giuseppe Bellanca offering to build him a plane for $15,000. Lindbergh went to New York to discuss the deal only to discover that the offer was conditional on the company using their own pilot. Lindbergh was disgusted. He finalized the deal with Ryan.

The order went in on 28 February 1927. It was for a single-seater monoplane, custom-built around a Wright J5C 'Whirlwind' engine. In addition he would need Pioneer navigating instruments including an Earth Inductor Compass. The vibrations of a plane cause the needle of an ordinary compass to jump about too much to give accurate readings; something more substantial was required. To save time, the plane was to be modelled on their

Ryan M-2, a mail plane that had gone into production only the year before. Estimated cost, $10,580.

When Lindbergh turned up at the factory, he was immediately reassured by the familiar banana smell of dope drying on wing fabric. Just a few days earlier the company had appointed a new chief engineer, Donald Hall, aged 29. 'How far is it to Paris?' Hall asked Lindbergh. And so history began to be made.

Hall and his team worked around the clock, seven days a week. Hall, whose task was to redesign the M-2 according to Lindbergh's needs, was at one time discovered to have been at his drafting table for 36 consecutive hours, longer than the time it would take Lindbergh to make his historic crossing.

Out of lengths of steel tubing, spruce wood and piano wire the plane began to take shape. It was calculated that 450 gallons of fuel should do it; enough for a flight of 4,000 miles; a decent margin of error, assuming the plane was not blown too far off course. Though 450 gallons is not much more than half the amount of fuel Alcock and Brown carried, that was still 2,700 lbs of cargo, far more than the plane itself would weigh. The fuel was to be stored in five tanks, three of them in the wings. Lindbergh asked that the two largest tanks go in the front of the plane. It would be more balanced that way, and in the event of a crash he would be less likely to get crushed (though presumably more likely to be incinerated). The disadvantage was that there would be nowhere to put a front window screen. But Lindbergh was used to navigating by looking to the side, and *The Spirit of St Louis* would have what the mail planes he flew did not have, the luxury of a periscope so that he might avoid high buildings and trees at low altitudes. He decided against taking a radio – too heavy. Extra weight meant greater fuel consumption – and anyway they were too unreliable, too difficult to use while flying, and only useful when within range of a signal, which mostly he would not be. The engine would have the distinct advantage of being self-lubricating.

It must run like a human heart, without interruption. It was built to run for 9,000 hours, but that of course was no guarantee that it would not fail at any given moment. Lindbergh understood what designers of intricate machinery have always known, that detail is key. If every smallest part is made precisely, there is a greater chance that the whole will work too. Lindbergh was apt to quote Longfellow to make his point:

> In the elder days of Art,
> Builders wrought with greatest care
> Each minute and unseen part
> For the Gods see everywhere.

The craft was built to Lindbergh's needs and specifications, but even so the cockpit was a mere 36 × 32 × 51 inches in size. Lindbergh was slender – 'Slim' was one of his nicknames – but he was also tall: there would not be anything like enough room to stretch out his legs. (The first astronauts would not have to bend double as Lindbergh did, but they would fly in capsules not that much more spacious.) His flying seat was made of wicker – light, and deliberately uncomfortable so that he would be less likely to fall asleep. Staying awake, Lindbergh reckoned, would be a major hurdle.

Only 60 days after the order had first been placed, the plane was complete. The total cost came to $14,000; a modest overrun. *The Spirit of St Louis* was 28 feet long with a span of 46 feet. Empty it weighed 2,150 lbs, laden with fuel (a little more than the original estimate) 5,135 lbs. Lindbergh in his flying suit weighed 160 lbs. In addition there would be 40 lbs of accessories:

Water and tinned food in case of an emergency landing
2 flashlights
1 ball of string
1 ball of cord

1 hunting knife

4 red flares, sealed in rubber tubes

1 match safe with matches

1 large needle

1 canteen, 4 quarts

1 canteen, 1 quart

1 Armbrust cup [worn over the face, it converts
   condensation in the air into drinking water and comes
   with helpful instructions printed on it: unscrew here
   for pure water, drain saliva here]

1 air raft with a pump and repair kit

5 cans of army emergency rations

2 air cushions

1 hacksaw blade

A map, with the top and bottom cut off to gain an extra
   ounce or so

A pair of boots, made by Lindbergh himself out of light
   material

No parachute: Lindbergh decided to take the risk that he'd make it to the Atlantic without needing one. After that it wouldn't have made any difference if he had one or not. List-making and stripping away anything that was not essential would be guiding principles of Lindbergh's life. By contrast, Fonck's plane had a leather interior, two radios, a four-man crew, and the means to produce a hot dinner on arrival.

Under the heading 'Results', Lindbergh itemized just two possible outcomes:

1. Successful completion, winning $25,000 prize to cover
   expense.
2. Complete failure.

The test flight was a great success, due in part both to the dedication of the small team and to Lindbergh's own vital input. He knew what he wanted, and he had witnessed the machine he was to fly being built from scratch.

One of Lindbergh's rivals for the Orteig was Commander Richard E. Byrd, already a national hero. Just the year before, he and fellow aviator Floyd Bennett were lionized in America for being the first humans to fly over the North Pole. Lindbergh had applied to be Byrd's co-pilot but his application had arrived too late. The European press was more sceptical of Byrd and Bennett's achievement. After his death it was discovered that Byrd's diary entries for the relevant period had been erased and rewritten, and his official report of the flight probably contained false sextant readings. On 16 April 1927, a week or so before Lindbergh took delivery of *The Spirit of St Louis*, Byrd announced that he was ready to make the test flight of his $100,000 Fokker C-2 monoplane, *America*. The plane made a bad landing, all four of the crew were injured, one seriously, and the plane was structurally damaged. Another $100,000 bid came to grief on 26 April when Lieutenant Commander Noel Davis' Keystone Pathfinder plane, making its last trial flight, crashed on takeoff, killing both Davis and his co-pilot Lieutenant Stanton Hall Wooster. They had planned to make an attempt on the Orteig prize the following week. Now the main rival was the team of French ace Captain Charles Nungesser and navigator Lieutenant Coli, 'the famous one-eyed airman'. They meant to try for the prize soon, flying from Paris to New York in a single-engined plane carrying 800 gallons of fuel. Not only was time running out for Lindbergh, each week seemed to bring some new addition to the record books. On 25 April, Clarence Chamberlin and Bert Acosta circled New York City for 51 hours, 11 minutes and 25 seconds in their $25,000 Bellanca WB-2 monoplane,

The Spirit of St Louis

covering a distance of 4,100 miles. Not the most glamorous of records but a new endurance record nevertheless. Now the team had its sights set on the Orteig prize too.

On 8 May, Nungesser and Coli took off from Paris. By now *The Spirit of St Louis* had been put through 23 test flights from between five minutes long to more than an hour. The *Spirit* proved itself to be one of the most aerodynamic planes of its time, taking off in just over six seconds, but it was less stable than Lindbergh had imagined it would be. As with the Jenny, it took a lot of effort even to keep the plane flying level. Nungesser and Coli were due to land in New York on 9 May. If they did, Lindbergh would need to think about what other records he might challenge. In France an early edition of *Le Presse* announced Nungesser and Coli's success under the headline: 'THE ATLANTIC IS CONQUERED, THE GOLDEN HOUR OF FRENCH AVIATION'. The report said that every ship and boat in New York sounded its horn at the news of the landing. French and American flags flew together from windows in every skyscraper. Nungesser's mother was reported as saying that luck had always been with her son. Crowds poured onto the streets. People in cinemas and theatres burst into tears when the news was announced. All of it a fabrication, so determined was the paper to be first with the news. Once the truth came out that the crew hadn't in fact arrived there were demonstrations at the newspaper offices. The backlash resulted in the paper going bankrupt.

On 10 May, as Lindbergh took off from San Diego in *The Spirit of St Louis* on his way to New York via St Louis, the Nungesser team had still not arrived. There had been a number of sightings, but the reports were inconsistent. Lindbergh decided to press on and see where fate took him. These first two long flights to New York were an opportunity to really get to know his craft. 'A pilot doesn't feel at home in a plane until he's flown it for thousands of miles. At first it's like moving into a new house. The key doesn't slip in the door smoothly.'

Lindbergh with *The Spirit of St Louis*

Now he really understood why he had insisted on making this attempt alone; only on his own could he really get to understand the idiosyncrasies, as if of a living creature, of his machine. Now he really appreciated what his father meant when he had said to him years ago not to rely too much on others. If this endeavour was to prove successful, it would have been because of both the cooperative endeavour of his team, and, crucially, the final authority of Lindbergh as solo pilot.

Lindbergh took off from San Diego at 3.55pm, giving him three hours of daylight in case of problems with the plane. He flew over Arizona at 5,000 feet, needing to gain further altitude to cross the Continental Divide; but at 8,000 feet his engine began to sputter and the plane to vibrate. He wondered if he might have to abort the flight. He brought the plane down to a lower altitude and the engine stopped misfiring. He guessed that the problem was to do with the carburettor getting too cold at higher altitudes. He would need to install a heater when he got to New York. For now, he changed the fuel mix and hoped that that would be enough. He slowly climbed to 13,000 feet, urging the engine not to stall. From time to time, the engine continued to sputter and misfire, but he decided to take the risk and hope that sputtering was the worst of it.

Then his Earth Inductor Compass stopped working. He would need to order a new one. And yet, to his amazement, when he was able finally to check his location by daylight – this was the first time he had flown through an entire night – he was only 50 miles off course.

Lindbergh landed at the Lambert Field in St Louis – named for his backer Albert Bond Lambert who owned the land – at 6.20am, 14 hours and 25 minutes after leaving San Diego, a new record for a non-stop flight of that distance. The next day he took off again, on the second leg of his flight across the continent. After flying for seven hours and 20 minutes, he landed at the

Curtiss Airfield in New York. From San Diego to New York, he had been in the air for 21 hours 20 minutes, five and a half hours shorter than the previous transcontinental record. Lindbergh was already making history. By the time he landed in New York, he was somewhat famous.

There were journalists swarming everywhere. With four men killed, two missing (surely Nungesser and Coli must now be presumed lost), three injured and three planes seemingly near ready to take off at short notice, coverage was reaching fever pitch. Chamberlin's team, having secured the world endurance record, were ready to try their luck at the Orteig prize even as Lindbergh was arriving in New York, but the owner of Chamberlin's Bellanca plane then got into an acrimonious dispute with the co-pilot Lloyd W. Bertaud about the route, and whether or not to take a radio. Their planned takeoff was delayed.

Byrd had arrived in New York in a newly rebuilt *America*, the first time the plane had made an appearance since the crash. His team was now ready to go, though observers noted that Byrd seemed not to have fully recovered his spirits. The team was testing new equipment and instruments in preparation for a quick takeoff when Byrd's financial backer, Rodman Wanamaker, insisted that they wait until the fate of the French team was known. Lindbergh was warned that if he planned to go ahead regardless, he had better be prepared for an unpleasant reception in France.

On 15 May Lindbergh completed further test flights. He had decided to put aside the risk of French opprobrium; all he was waiting on was the weather. The Weather Bureau at that time was a network across America of 210 fixed-land weather stations each with a staff of up to 15, together with thousands of other sub-stations with part-time employees. The reports that came in over the next days told of dense fog over Nova Scotia and storms over the Atlantic. Not until the evening of 19 May did a report indicate that the low-pressure system over Newfoundland was clearing and moving out over the Atlantic. There was still no news of Nungesser and Coli.

That night, Lindbergh was in Manhattan about to see a show at the Ziegfeld Theatre, *Rio Rita*. A journalist got a lucky break when he overheard the Lindbergh party talking about the favourable weather report. A mere three hours later the story was headlines in the *Mirror* newspaper. Technically, if Lindbergh took off now he was not eligible for the Orteig prize since the prize's stipulated 60-day notice period had not yet passed. He asked his backers what he should do. To hell with the prize they said, just go.

Lindbergh's party made their way back to the airfield from Manhattan, stopping off at a small restaurant at Queensboro Plaza, across the 59th Street Bridge in Queens, to have a meal. At a nearby drugstore Lindbergh bought five sandwiches for the flight, two ham, two beef, one hard-boiled egg.

Lindbergh was told that he should be back on the field in two and a half hours' time. In the past Lindbergh had felt the benefit of even an hour's sleep, but tonight sleep eluded him. And then, just as he was finally nodding off, there were loud knocks on the door. Standing there was George Strumpf, a young man who had been sent by the St Louis Chamber of Commerce to serve as Lindbergh's aide-de-camp. Lindbergh wondered what the problem could be. Has the weather changed? Is there trouble at the field? 'Slim, what am I going to do when you're gone?' the young man asked. Lindbergh told him that he did not know, that there were other problems he had to think about. George Strumpf left the room, and Lindbergh to wakefulness. What had got into George Strumpf? Was he being literal- and simple-minded? Selfish? Or was he perhaps smitten? George Strumpf, person from Porlock, had unwittingly changed the nature of the ordeal facing Lindbergh. He had always suspected that the greatest challenge for him would be how to stay awake for the 30 or 40 hours – who knew how long? – it would take to reach Paris; now it looked as if he was about to set off already deprived of sleep for 24 hours. If not for that knock, would those other visitors have turned up, those shadowy presences about whom Lindbergh would remain silent for over 20 years?

At 3am Lindbergh abandoned any further attempts to try to sleep. He went out onto the airfield and into the damp, dark night. An engineer, Ken Lane, said to him, 'Did my message get through, Slim?' 'What was it?' Lindbergh asked. 'I said to let you sleep until daybreak, and we'd have everything set for you to take off.' For whatever reason, the message had not got through. Another opportunity for sleep had been lost.

After a number of last-minute inspections they were ready to attempt takeoff, not from Curtiss Airfield where he had landed, which was too short for the fully fuelled plane, but from Roosevelt Field, which Byrd generously allowed Lindbergh to use. Byrd had also shared his weather reports, but now – perhaps because he had been stymied by his backer – he behaved pettily, perhaps petulantly too. He kept Lindbergh waiting for two hours while he made a test flight that surely he did not have to make at that hour of the day.

Harry Guggenheim, a flying enthusiast associated with the Byrd team, came over to wish Lindbergh luck. They posed for photographs together. He told Lindbergh to look him up when he got back. They would later become close friends, but Guggenheim said that at the time he was sure the flight was doomed.

The Spirit of St Louis was trundled out onto the waterlogged field. The overloaded plane lurched from side to side.

It appears completely incapable of flight – shrouded, lashed and dripping. Escorted by motorcycle police, pressmen, aviators, and a handful of onlookers, the slow, wet trip begins. It's more like a funeral procession than the beginning of a flight to Paris.

Despite the early hour, word of an imminent takeoff had leaked out and a crowd of some 500 people was assembled on the field (a fact at variance with Lindbergh's memory that there had been just 'a handful of onlookers'). A schoolteacher, Katie Butler, persuaded a policeman to give a medallion of St Christopher to Lindbergh, who took it from the officer and put it in his pocket distractedly. At 7.40am the plane's engine was started up and the aircraft began slowly to plough its way through the field. The tyres cut deep into the mud as if they were the wheels of a heavy truck, not of a plane made of tubing, wood, wire and canvas. Every breath of wind seemed to press the plane more firmly into the ground. Only Lindbergh could decide if now was the moment to proceed. No one would think the worse of him – rather the opposite – for abandoning takeoff. There were still a few minutes left to him during which he might turn back. And then, all of a sudden, the conviction came over him that the time had come, that the plane would make it over the distant trees. The decision was made.

The end of the field seemed to be approaching, nearer and nearer, and still he was not airborne, but at 7.52am – some accounts say 7.54am – on 20 May 1927, the wheels of *The Spirit of St Louis* lifted out of the mud; 5,000 lbs 'balanced on a breath of air'. It was a slow climb; the plane barely cleared the trees at the end of the airfield.

Lindbergh was aware that there were a number of other planes in the air. They were following him with cameramen on board. Film footage of the takeoff would be shown all around the world. It was so popular that a company was formed to feed the demand for almost live filmed news coverage: Fox Movietone News. Lindbergh had by now come to despise the popular press. Journalists had so worried his mother about the risk he was taking that she travelled to New York from Detroit to see for herself. She stayed only a few hours and left days before takeoff, careful to show support without getting in Charles's way. She declined to be photographed kissing her son: 'I wouldn't mind if we were used to that,' she said, 'but we come from an undemonstrative Nordic race.' She did admit, however, that for the first time in her life she realized 'that Columbus also had a mother'.

The press had pursued Lindbergh everywhere these last weeks, had forged a photograph of him embracing his mother, had written about him with such inaccuracy that he had come to despair of journalism. No matter what they might offer him, he had decided that he would not write about his flight for a Hearst-owned paper. 'The excesses are what bother me – the silly stories, the constant photographing, the composite pictures, the cheap values . . .'

Soon the last of the press planes turned back, and Lindbergh was alone. Later that morning his mother went to school as usual, leaving instructions that she was not to be disturbed by reporters. She had taught at the Cass Technical High School in Detroit since 1924, and except for one year she would spend teaching in Turkey, it would be the only job she ever had. She wore a hat at school at all times, and during staff meetings sat alone reading a book. She would retire suddenly, giving no reason, in 1942.

The greatest hazard during these first hours of the flight was the new experience of controlling the plane as it was now, carrying 1,000 lbs more weight than it had ever carried before. And here was something else novel: now, this was the first time Lindbergh had crossed over such a large body of water as Long Island Sound. The air, as it often is crossing between land and water, was turbulent. Lindbergh sensed that The Spirit of St Louis could not stand much more stress. The craft's wires felt taut, stretched almost to breaking point. He had known there would be turbulence, but now that he was experiencing it in a plane at its most vulnerable he wondered if the flight was doomed even in its first moments. But the wires did not snap and within a minute or so the air became completely still.

During the next hours Lindbergh made precarious progress

up the Eastern seaboard, carefully steering his overladen craft, flying low. He had set off during the day precisely so that these first difficult hours could be faced in sunlight.

Flying over a wood in Nova Scotia, the air was so clear and he was flying so close to the surface of the earth that he could see the moss on stones. He had been in the air for four hours and had already burned through 400 lbs of fuel, out of almost 3,000 lbs.

Higher now, he saw thousands of ducks, 'like a cloud's shadow drifting over land and water', but a cloud seen from above, as if reflected in a lake, rather than from below. 'No man before me had commanded such freedom of movement over earth,' he wrote. From this exalted perspective he felt as separated from the country below him as though he were 'looking through a giant telescope at the surface of another planet'.

He made a slight detour to St Johns, flying low so that he could be seen. He decided he owed it to his backers that word get back that he had made it so far.

He had brought along plastic side windows but decided not to insert them, preferring to feel the elements directly. It brought him into 'a sensual contact with the geography below'. He needed to feel part of the outside, not separate from it, but he was annoyed at the additional weight. He could see a lump of mud still attached to the plane, thrown up from takeoff; even that little bit of unwanted cargo irked him. Then came a moment when he almost regretted his decision not to put in the windows: a sudden breeze snatched the map out of his fingers. He grabbed it back just before it was lost to the world outside. Such a small item to lose, but without it he would have lost more than just his way.

He had expected to battle tiredness, he just hadn't expected to battle it so soon. There were times even during these first few hours of the flight when he was so tired that he brought the plane down to within a man's height of the surface of the water so that he could feel the sea-spray on his face. Touch the wing accidentally to the water and he knew that he would come down forever, as if a spell had been broken. How ignominious if tiredness were to have brought his venture to so early an end.

Lindbergh again mapped the shape he saw from the air to the shape he saw on his chart. This was it, the place where he was to turn away from the land and head out over the ocean. He had been in the air about 11 hours. (He was charting a course familiar to all commercial airlines that fly between New York and London or Paris today, known until 1957 as the Lindbergh Line, and since then as North Atlantic Track 101.) An hour later the stars began to appear in the sky above, but below there was a bank of fog, which was quickly rising. In order to try and avoid it he had to increase his altitude even more quickly. He could see nothing below except fog, only stars above. He took the *Spirit* up from 800 to 7,500 feet. By 9pm he had climbed to 10,000 feet. In addition to the fog there was now a thunderhead looming. Ice began to form on the wings. To make matters worse he seemed to have entered a magnetic storm, a local disturbance of the Earth's magnetic field caused by solar winds. The air above the Atlantic was uncharted territory. Very little was known about the Earth's atmosphere at that time, or of the weather at high altitudes. Until Lindbergh entered this magnetic storm it was not known for sure that such a phenomenon actually existed, and certainly not the reason why. And yet this had not been planned as a scientific expedition. For all his interest in science, this was primarily an adventure.

As a result of the magnetic disturbance, both his inductor and liquid compasses were temporarily of no use to him. Lindbergh had to plot his course by dead reckoning only. He had been using this method anyway as a double check; every hour he judged his position by the speed of the plane and the wind direction. Sometimes he would have to come down to just above sea level to work out the wind direction from the way the foam blew off the crest of the waves.

Now, in the storm, he decided he had no choice but to begin to make a large loop back from where he had come and try to fly around the thunderhead. Fortunately, after a few minutes the ice started to thin and gaps began to open up in the storm clouds. The moon appeared.

Each stage of the journey brought its own terrors, but it was the coming dawn that Lindbergh feared most of all. Pilots who have flown through the night know that with dawn come feelings of great fatigue. Around 17 hours into the flight, before it was 3am in New York, the sky began to lighten at the horizon. The pain of being cramped into the tiny cockpit was as nothing compared to the mental pain of trying to keep himself awake. He could think of nothing except sleep. He tried to talk himself into wakefulness, to scare himself into wakefulness, but nothing worked. There were times when he realized that he had been asleep with his eyes open; that he had been flying as if without the need of his ego. He sensed that the barrier between living and dreaming was breaking down in him. He looked at the time but his flight had become disconnected from clock time: 'I'm flying in a plane over the Atlantic Ocean, but I'm also living in years now far away.'

Lindbergh was involved in an epic struggle like no other he had ever faced. He would never blame anyone for his tiredness, not even George Strumpf.

Back in America the popular columnist Will Rogers wrote mawkishly, inaccurately and illiterately, that he would make no attempts at jokes that day, because a 'slim, tall, bashful, smiling American boy is somewhere over the middle of the Atlantic ocean, where no lone human being has ever ventured before. He is being prayed for to every kind of Supreme Being that has a following. If he is lost it will be the most universally regretted loss we ever had.'

After 21 hours in the air, Lindbergh was drifting in and out of sleep. After 24 hours, he had visitors.

The fuselage behind was filled with ghostly beings, 'vaguely outlined forms, transparent, moving, riding weightless with me in the plane'. He was not surprised at their arrival, nor did it feel as if they had come suddenly; they were simply there. He turned his head and looked at them as he might have done at anything that had fallen within his normal field of vision. It was as if his whole skull had become 'one great eye, seeing everywhere at once', and he heard the visitors as if his 'entire being were an ear'. The benign presences were 'neither intruders nor strangers...more like a gathering of family and friends after years of separation'. They were human in outline but without rigid bodies; 'transparent forms', 'intangible as air'. Some of the spirits were more prominent than others. Some whispered in his ear, helping him to steer his course.

Like many who have been through such an experience, Lindbergh felt as if he was standing on some threshold between worlds. He was 'caught in the field of gravitation between two planets', acted on by forces he could not control.

And then, somehow without his being aware of it, the visitors had gone. Afterwards, he would not be able to recall a single word any of them had said to him. He would not write about the experience for two decades.

For the next few hours, Lindbergh faced a new and unexpected danger. It was as if he had once again become a mere novice flyer. He could fly only with great concentration. The ability that had been worked into his body had disappeared.

And then, for the first time since he had left Newfoundland, he saw a human being. On the sea below a man stuck his head out of the porthole of a boat. He looked disembodied, as if he were only his head. Lindbergh flew down low and called out, 'Which way is Ireland?' The man did not react. Lindbergh wondered if he was still in the land of faery.

After a while Lindbergh saw coastline, and after careful

scrutiny of his map was astonished to recognize the shape of the south-western coast of Ireland. He was less than 3 miles – 3 miles after 3,000 miles! – off course, and two hours ahead of schedule. Had his weeks spent studying navigation charts and navigational techniques paid off? He had learned spherical mathematics. He had learned the art of plotting a route. But these techniques, no matter how meticulously they might be studied, are inaccurate. And what of the magnetic storm that had rendered his compass useless? Emerging out of that storm he had had no idea whether or not he was still on course. Arriving here, so precisely where he was meant to be, looked like something more than good fortune. 'Before I made this flight, I would have said carelessly that it was luck. Now, luck seems far too trivial a word, a term to be used only by those who've never seen the curtain drawn or looked on life from far away.'

Lindbergh was exultant. He saw people in the harbour at Dingle Bay looking up. (Decades later, on a visit to Ireland he would overhear a man say that he had been one of those in the harbour. Lindbergh would choose not to reveal himself.) News of Lindbergh's arrival would soon travel from the crowd in the bay to Paris, back to New York and across the world. There would be several more sightings of him across Ireland, then at Plymouth and again at Cherbourg.

It was mid-afternoon in Ireland (late morning in New York). Lindbergh might with luck be able to make it to Paris before dark. He increased his speed. By early evening he saw the coast of England. A couple of hours later the sun had set. Paris was still another 200 miles away. He no longer cared about tiredness or the darkness. He was almost there. In Paris he flew once around the Eiffel Tower. Now he had to work out how to find the landing field at Le Bourget.

As he began his approach, he was alarmed that he had still not regained his 'feel' for flying. He had never tried to land a plane without feel before. And there was yet another danger. From the air he could see streams of lights along the roads radiating from the unlit airfield. It took him a while to realize that these were car headlights. As he came down, he saw that the landing field was a sea of people.

CHAPTER THREE

Nungesser and Coli had not been found, and were never found. It seems most likely that they disappeared somewhere in Newfoundland, where there had been a number of last sightings. The predicted hostility towards Lindbergh never materialized, indeed something about Lindbergh – his youth, reticence and diplomacy were presumably all ingredients – captured the imagination of the French. Almost immediately, he asked if he might call on Mme Nungesser. She wept and hugged him. He told reporters that Nungesser and Coli's flight had been the harder for going west. He was embraced by Louis Bleriot, who, in 1909, had become the first person to fly across the Channel. Lindbergh called him his hero. Here was naturalness and youthful vitality, commentators observed at the time; but here, too, was a man who knew his own mind. At a lunch held in his honour the following day his speech lasted barely two minutes:

Gentlemen, 132 years ago Benjamin Franklin was asked: 'What good is your balloon? What will it accomplish?' He replied: 'What good is a new born child?' Less than twenty years ago when I was not far advanced from infancy M. Bleriot flew across the English Channel and was asked: 'What good is your aeroplane? What will it accomplish?' Today those same skeptics might ask me what good has been my flight from

New York to Paris. My answer is that I believe it is the fore-runner of a great air service from America to France, America to Europe, to bring our people nearer together in understanding and in friendship than they have ever been.

The press praised him for the brevity, modesty and tactfulness of his speeches. The American Ambassador Myron T. Herrick hoped that Lindbergh's achievement might help further good relations between America and France. His best expectations were exceeded.

A member of the Diplomatic Corps in Paris wrote that he saw the cult being created right before his eyes, even though, he said, his eyes couldn't see it. In those first few days, 'Lindbergh's personality was reaching out and winning the French just as surely as his flight had reached out and found their city.' Ambassador Herrick saw it too: 'I am not a religious man,' he said, 'but I believe that there are certain things that happen in life which can only be described as the intervention of a Divine Act.' The Arctic explorer Fitzhugh Green wrote that there was something artificial about the first newspaper reports from the American side (one of the more curious headlines in the New York Times the day after Lindbergh landed ran: 'ATE ONLY ONE AND A HALF OF HIS FIVE SANDWICHES'). It was in France, he said, that the Lindbergh cult had its origins. From our perspective today, we see the first long crossing of the Atlantic from New York to Paris as a clear historical milestone, but at the time it was not obvious that the achievement was going to capture the public imagination in quite the way it did.

Lindbergh had never been abroad before and was excited about exploring Europe. He planned to spend a few weeks travelling from airfield to airfield chatting to pilots about the future of flight. Lindbergh told the American Ambassador in London that he intended to fly back to America – in The Spirit of St Louis of course – via Europe and Asia, then Alaska and Canada. The Ambassador told him that the President had other plans. Inflamed by his European fame, and rumours that he planned an extended stay abroad, the press in America was agitating, a week after he had landed, to have Lindbergh back. President Coolidge ordered him home and sent a naval destroyer. Lindbergh thought it was insulting to his plane that it had to suffer the ignominy of travelling back home by ship. *The Spirit of St Louis* was a part of himself. When he talked of flying Lindbergh talked of 'we' – not a royal we, he meant my plane and I.

In Washington – having been escorted up the Potomac by a fleet of warships and military aircraft – Lindbergh and *The Spirit of St Louis* were met by the President. On the same day the US Postal Service issued a 10-cent stamp bearing his name, the first in American history to bear the name of a living person.

Coolidge said that he had been told that more than a hundred companies had provided parts for the construction of *The Spirit of St Louis*. Lindbergh claimed his accomplishment not as the act of a single pilot but as 'the culmination of twenty years of aeronautical research and the assembling together of all that was practicable and best in American aviation. It represented American industry.' The French press had claimed Lindbergh's success for the world, and Lindbergh as a citizen of the world. Back home his achievement was quickly reframed as an American, not a global one.

Lindbergh was promoted from Captain to Colonel. After seemingly unrivalled celebrations in Washington, he was greeted in

The US Postal Service 10-cent commemoration stamp

New York by the largest tickertape parade in history. Around 2,000 tons of paper descended onto the streets. The next day the *New York Times* devoted its first 16 pages to coverage of Lindbergh. The celebrations in New York went on for four days. 'The greatest torrent of mass emotion ever witnessed in human history' had been let loose, wrote Fitzroy Hugh, an adventurer in his own right, and Lindbergh's editor at Putnam. Ridiculous hyperbole, of course. And yet, measured in column inches – surely as plausible a measure of celebrity as any other – Lindbergh was the most famous person alive. The only man who came close was the Prince of Wales, the future Edward VIII, whose exploits were avidly followed by the press worldwide.

Between July and August that year, Lindbergh flew over 22,000 miles, visited every state, gave 142 speeches in 82 cities. He was 260 hours in the air. No one in history had ever experienced America as he now experienced it. No one had ever seen so much of America, nor seen it from his perspective, and 30 million people turned out to see him and *The Spirit of St Louis*. There were reports of hard-bitten reporters – though aren't the so-called hard-bitten always the most sentimental? – breaking down in tears. He received over 2 million letters and hundreds of thousands of telegrams. It is said that 5,000 poems were written about his flight across the Atlantic. A town was called Lindbergh. Babies were named after him. He was awarded the Medal of Honor by act of Congress. He was advised to enter politics, told that there was a good chance he could become president. 'He has lifted us into the freer and upper air that is his home,' wrote one commentator. 'He has displaced everything that is petty; that is sordid; that is vulgar. What is money in the presence of Charles A. Lindbergh?' Within 18 months of landing in Paris he had earned $1 million, and could have earned a great deal more. He was offered $50,000 to endorse a brand of cigarettes even though he didn't smoke. Randolph Hearst offered him $250,000 to appear in a movie, and was amused when Lindbergh declined. 'Never was America prouder of a son,' an editorial in the *New York Times* proclaimed.

Putnam commissioned an account of his flight, ghostwritten, to be brought out as soon as possible. Lindbergh was so horrified by the ghostwriter's hyperbole and inaccuracy that he insisted on doing the job himself. Harry Guggenheim had been back in touch and offered him a safe haven in which to write. In typical Lindbergh fashion, he worked out how many words he would need to write each day in order to get the book written in the short time available. He thought 30,000 would be the minimum he could get away with, and he gave himself three weeks. The resulting book, inevitably titled *We*, was a bestseller. Ironically, given the title, he did not mention the spirits who had visited him. He didn't even tell his mother.

A collector wrote to the publishers and offered them $30,000 for the manuscript. The publishers offered to split the money with the author. Lindbergh wrote in reply that if the manuscript was theirs then they should have the money, and if it was his, then he should have it. It was, of course, his. For Lindbergh it was never about the money but always about principle, what he believed was the right thing to do.

One of his last long flights in *The Spirit of St Louis* (no one else ever piloted it) was to Mexico at the behest of the American Ambassador there, Dwight Morrow. Ambassador Herrick had exploited Lindbergh's charm to help improve relations between America and France; now Morrow hoped that he could do the same for those between America and Mexico. Lindbergh set off for Mexico City from Washington. It was to be the first ever non-stop flight between those cities. Around 150,000 people turned out to welcome him. A festival put on in his honour lasted six days.

At the Ambassador's home, Lindbergh identified the woman he decided would make the ideal wife, one of Dwight Morrow's daughters. He and Anne were well suited, both shy and solitary, both liked nature, both had the same ideas about service and integrity. Charles had made up his mind. Now all he had to do was tell Anne.

Anne had been secretly in love with Charles before they met. 'He is the only saint before whom I light a candle,' she wrote in her diary, in French. She described him as being the last of the gods. 'He is unbelievable and it is exhilarating to believe in the unbelievable.' Anticipating Lindbergh's visit, she confided to her diary that she was not sure she could bear it: not just because she would be in the physical presence of someone with whom she had become obsessed, but because of her conviction that Charles would naturally prefer her older sister Elizabeth. Everyone always preferred Elizabeth. She was smarter, more attractive, more socially at ease. She imagined him falling for Elizabeth and herself fading into the background, 'feeling in the way, stupid, useless', but hoping nevertheless that there had been a mistake and that she would be missed when he left. When it became clear to the press that something was afoot, they too assumed that his interest was in Elizabeth. 'It wasn't Elizabeth, it was Anne – isn't it funny?' the Ambassador's wife told the embassy staff, not very diplomatically. 'Unlike most brides to be,' Anne later wrote, 'it was I who was congratulated, not he.'

'I think she loves him,' Anne's mother wrote in her diary. 'I don't expect to be happy,' Anne told an ex-boyfriend, 'but it's gotten beyond that, somehow.' One of Anne's friends said of Charles that it wasn't that he was cold, more that he was immature.

On 27 May 1929 they married, 18 months after they had first met. At the wedding Charles had trouble cutting the cake, sawing at it rather than slicing through it. His mother-in-law said she would get him a sharper knife. He grabbed her wrist in a way she found threatening. It was a moment, a small detail, that she never forgot.

It was not long before Anne learned how to fly, taught by Charles of course. She also learned how to be a navigator, and was the first woman in America to earn a glider's license.

Lindbergh had decided that he would devote his life to bringing into reality a new era of flight. In this he was as assured and determined as in other ways he was uncertain and naive. It was a heady combination of character traits. Lindbergh had used his tour of America to promote an image of the future in which there would be a large airport situated close to every large city in America. He envisioned air routes 'radiating in every direction', connecting up the whole continent. On 7 July 1929 the first transcontinental service was launched by Transcontinental Air Transport (TAT). Lindbergh was pilot for part of the inaugural trip. He was on TAT's board, and would be on the board of every emerging American airline. In September that year one of

Anne Morrow Lindbergh, looking strikingly like Lindbergh's mother at the same age

TAT's planes crashed in mountains. Charles and Anne joined in the search party. There were no survivors. They would visit two other crash sites in the next 18 months. Their presence may have helped restore a measure of calm and confidence to the industry, but every crash undermined what was still a tentative business. Doubters said TAT stood for 'Take a Train'. Certainly flying was not for everyone. It was both expensive and hazardous. The first passengers were an elite, not at all like the frontiersmen of the past, though perhaps as brave as they had been.

Even then, at the end of the 1920s, Lindbergh was imagining what would come next. He thought of rockets, and 'thinking of rockets extended my vision from air to space'. But he could find no one else who shared his interest. Experts at DuPont told him that no rocket could ever produce enough thrust to escape the Earth's atmosphere. Even the smallest engine would require 400 lbs of black powder and a firebrick combustion chamber. The machine would be so heavy that every time you tried to increase the thrust, the extra weight of the rocket would cancel out the advantage.

One day when Lindbergh was staying with Harry and Carol Guggenheim, Carol began to read aloud a semi-humorous article in *Popular Science* magazine about the rocket pioneer Robert Goddard. The article was headed: 'AIMS ROCKET AT ROOF OF SKY. GODDARD TESTS NEW MISSILE TO EXPLORE UPPER AIR FOR SCIENCE'. Lindbergh was intrigued. Some of the naysayers' doubts appeared to be have been addressed by Goddard. He was using a thin metal called Duralumin, not firebrick, for his combustion chamber, and liquid fuel – gasoline and liquid oxygen – not black powder to produce the thrust.

Lindbergh approached MIT to make sure Goddard's work was bona fide. Satisfied, he telephoned Goddard, asking if he might come and visit. Robert Goddard answered the telephone himself and thought at first it must be a hoax call.

CHAPTER FOUR

Robert Goddard was born in 1882 in Worcester, Massachusetts. He was, according to his first biographer, a sickly and frail child who suffered from stomach problems, pleurisy and bronchitis. It is possible that this was in part later myth-making. His mother may have become oversensitive to Robert's health after the death, before the age of one, of a younger son. Robert was kept at home for much of his early childhood, falling two years behind at school. His father bought him a microscope, a telescope, a subscription to *Scientific American* and showed his son how to generate electricity from the carpet. Robert became an avid reader, borrowed books on science from the local library and started to conduct experiments at home, at least one of which resulted in a large explosion. By the time he was a teenager he was interested in flight, experimenting with kites and then balloons. At the age of 16 he constructed a balloon made out of aluminium and filled with hydrogen, but it remained earthbound. After reading *The War of the Worlds* by H. G. Wells, and *Edison's Conquest of Mars* by Garret P. Serviss, both as serials in the *Boston Post*, he dreamed of a spacecraft that could fly to Mars. But the formative moment of his life came the following year.

Goddard climbed a cherry tree in order to cut off dead branches. Looking 'towards the fields at the east', he imagined 'how wonderful it would be to make some device which had the *possibility* of ascending to Mars'. In that moment it came to him what such a device might look like: 'It seemed to me then that a weight whirling around a horizontal shaft, moving more rapidly above than below, could furnish lift by virtue of the greater centrifugal force at the top of the path.' He said he was a different boy when he descended the tree from when he had ascended it: 'Existence at last seemed purposive.' He took photographs of the tree and of the ladder. He celebrated the day, 19 October 1899, every year for the rest of his life as his *miraculosa die*. He even venerated the saw that cut the branch off.

Goddard contracted tuberculosis in 1913 and was not expected to survive. Against the odds, he was much recovered within a year. During that year he took out his first rocket patents. US Patent 1102653 describes a multistage rocket and US Patent 1103503 the kinds of fuel that might propel a rocket, either powder or, crucially for the future development of rockets, liquid fuel. Both patents were registered in 1914.

Why is rocket science famously so difficult? A rocket is basically an elaboration of a spear or an arrow. Distant ancestors discovered that a straight stick weighted at the end and thrown with enough force takes flight in a controlled manner. If hurled hard enough a rocket can be put into orbit around the Earth, or even ejected from the Earth's gravitational field and sent flying into outer space. How hard can it be to get an object – any object, even the smallest object – into space? The British astronomer Fred Hoyle once remarked that space isn't remote at all, 'It's only an hour's drive away if your car could go straight up.' And yet until the twentieth century no physical object had ever escaped the pull of the Earth's gravitational field. Everything and everyone for all of history had been bound to the Earth. Hoyle was being deliberately disingenuous. Even if you could drive straight upwards you wouldn't get very far. The grip of the Earth's gravitational field is too insistent. To escape the pull of the Earth you would have to accelerate your car/rocket to 25,020 mph: the Earth's so-called escape velocity. At speeds less than 17,500 mph the car/rocket would be gradually bent into a trajectory that inevitably would bring it back to the surface of the Earth. At speeds between 17,500 mph and escape velocity you might be able to put your projectile into orbit around the Earth. Most scientists of Goddard's time thought, as Lindbergh had found out, that space rockets were not only far beyond the technology of the day but that they were impossible in principle. Goddard's genius was to have worked out how, in theory, the Earth's gravitational pull might be overcome. His brilliant concept was the multistage rocket propelled by liquid fuel. The first stage would fire up and launch the rocket, then, at some point, the first stage would fall away, and a second stage would fire up; what was left of the rocket would be accelerated further because what remained was that much lighter. The rocket that eventually took the first humans to the moon was a three-stage rocket fuelled by liquid oxygen and liquid hydrogen.

What a rocket transports is called its payload. If the payload is a warhead the rocket becomes a missile. The payload might also be scientific equipment, in which case the rocket is called a 'sounding' rocket. In less than half a century after Goddard took out his patents, a rocket would safely transport a human being as part of its payload.

At Clark University, where he taught physics, Goddard first worked on solid-fuelled rockets, liquid fuel at that time being too difficult to handle. He developed a nozzle that focused the expulsion of exhaust gases and increased both the rocket's propulsion and the efficiency of the fuel from 2 per cent to 64 per cent. It was the first major practical development in rocket science. During the early years of the First World War, Goddard tried to sell the idea of rockets to the military, but even solid-fuelled rockets were still largely impractical. In any case his rockets were, at the time, 3 inches long. The army rebuffed him. They wanted hardware not ideas.

In 1916 Goddard wrote a letter to the Smithsonian that summarized his work on solid-fuel rockets as a means of exploring the Earth's atmosphere. The Smithsonian came up with five years' worth of funding. The head of Clark's physics department urged Goddard to publish the letter, but Goddard was slow to agree. The letter, by now a paper titled 'Results on a Method of Reaching High Altitudes', was eventually published in December 1919. In the paper, Goddard presented the mathematics of rocket flight: the theory mastered decades before the engineering problems were solved. In America Goddard's work was unrivalled.

His paper immediately made him famous. In a final section, just eight lines in a paper 69 pages long, Goddard wrote about a future in which it might be possible to launch a rocket that escaped the Earth's gravitational pull. In a thought experiment, he calculated how large a rocket would have to be in order to send a small object to the moon. He reckoned that a rocket of 3.2 tons would be required to get a payload weighing 10.7 lbs to the moon. If the payload was a flare that ignited once it reached the moon, perhaps we would be able to see the evidence from Earth. Those eight lines brought him overnight acclaim. A number of provocative headlines appeared: 'NEW ROCKET DEVISED BY ROBERT GODDARD MAY HIT FACE OF THE MOON' (*Boston Herald*), 'AIMS TO REACH MOON WITH NEW ROCKET' (*The New York Times*), 'SAVANT INVENTS ROCKET WHICH WILL HIT MOON' (*San Francisco Examiner*), and 'SCIENCE TO TRY SHOOTING MOON WITH ROCKET' (*Chicago Tribune*). To describe these as jumping the gun is an understatement. In the *New York Times* of 12 January 1920, a press release that Goddard had sanctioned resulted in the headline: 'BELIEVES ROCKET CAN REACH MOON'. The following day it ran an unsigned article headed 'A SEVERE STRAIN ON CREDULITY' that, in typical newspaper fashion, attempted to temporize its own sensationalism. The article accused Goddard of knowing less about science than a schoolchild. How could a rocket move through space? In the vacuum of space there would be nothing for the rocket's propulsive gases to react against. Goddard did not even understand Newton's Third Law, that for every action there is an equal and opposite reaction. He laughed the article off; at least at first. He had proved as early as 1915 that a rocket would work in a vacuum. 'Every vision is a joke,' he wrote, 'until the first man accomplishes it; once realized, it becomes commonplace.'

Goddard wrote a measured response, a statement that was released to the Associated Press: 'Too much attention has been concentrated on the proposed flash powder experiment, and too little on the exploration of the atmosphere... Whatever interesting possibilities there may be of the method that has been proposed, other than the purpose for which it was intended, no one of them could be undertaken without first exploring the atmosphere.' He was right; rockets would be used first of all to explore the atmosphere, but that hardly made for good copy. Over the years that followed he would continue to send out provocative press releases and the *Times* and other papers would continue to respond to them. In early 1921 Goddard claimed that a 'workable model' of the moon rocket was almost ready to be test-launched. The *Times* duly ran the story under the headline: 'MOON ROCKET READY SOON'. When the deadline passed Goddard said it was for lack of funding.

In 1920 Goddard sent a report to the Smithsonian that put forward the possibility of using both liquid oxygen and liquid hydrogen as propellants. The groundbreaking idea was almost lost in a paper full of more fanciful speculation, about ion propulsion,

Goddard launched this rocket on 16 March 1926, the world's first liquid-fuelled rocket. The bulkiest part of the rocket can be seen in the lower portion of the launch frame that supports it.

solar mirrors and the problems of talking to extraterrestrials. But by September 1921 he was taking the idea of using liquid oxygen seriously and beginning to work with it practically. There are major difficulties in working with oxygen. It turns into a gas at temperatures higher than minus 297 degrees Fahrenheit, and it violently combusts when it comes into contact with many organic substances. (Liquid hydrogen is even harder to manage, boiling at minus 423 degrees Fahrenheit.)

When Goddard's grant from the Smithsonian ran out, he was given a small top-up to allow him to continue his studies on liquid oxygen. It was suggested to Charles Abbot, the director, that here was a chance for some publicity. 'I think the less publicity given to Dr Goddard's work the better,' Abbot said, 'as it results in a flood of correspondence and in newspaper interviews, which only give trouble and consume time.' The additional funding was given on the condition that Goddard stop promoting the idea of a moon rocket. By the end of 1923 Goddard was testing the world's first engines to be powered by liquid oxygen.

In March 1926, at Auburn, Massachusetts, Goddard launched the world's first liquid-fuelled rocket. It was airborne for two and a half seconds, reached a top speed of 60 mph, an altitude of 41 feet, and landed 184 feet from the launch frame in a cabbage field. In 1966 the field was designated a National Historic Monument. No flight to the moon, but nevertheless a historic maiden flight to be set alongside that of the Wright Brothers' *Kitty Hawk* on 17 December 1903, a flight that had itself lasted only 12 seconds. The 12-horsepower aircraft flew a distance of 120 feet at an altitude of at most 10 feet before its bicycle wheels touched the ground once more.

In 1927 Goddard couldn't resist writing another of his incendiary press releases. He said that he was near to completing a new moon rocket. The *Boston Herald* ran an article headed: 'WANT TO BE FIRST TO VISIT THE MOON? Apply to Robert Goddard, Clark

University'. Around 100 people offered themselves as volunteers to ride in the theoretical rocket. Captain Claude Collins of the New York Police Department even offered to fly to Mars. Goddard issued a more sober press release claiming, quite accurately, that 'No one in the history of heat engines has ever increased the efficiency of an engine in the same degree as Goddard.' Needless to say, that press release attracted no attention at all.

Goddard was delighted when, during that year, he learned that an exhibition in Moscow was featuring his work, but was less delighted when he also learned that the elderly Russian rocket scientist Konstantin Tsiolkovsky was getting greater attention than he was. Robert Goddard had become known as the father of rocket science, but now there was another claimant. Goddard was the first person to have launched a liquid-fuelled rocket, but Tsiolkovsky had laid out the mathematics of space flight, in particular the trajectory needed to get a rocket to the moon in an article he had written in 1903, anticipating Goddard by more than a decade. The paper – 'The Exploration of Space by Means of Reactive Propelled Devices' – was published in a Russian science journal but went unregarded, even in Russia, for decades.

The article that Carol Guggenheim read aloud to Lindbergh described a launch that took place on 17 July 1929. An 11-foot-long rocket carrying an aneroid barometer, thermometer and camera flew for 18.5 seconds, travelled for 171 feet and reached an altitude of 90 feet. It was the first time a rocket ever carried scientific instruments. The Weather Bureau had become interested in Goddard's work earlier that year, but not so interested that they came up with any funding. All the instruments functioned during the flight, even the camera, which was triggered to take a photograph of the moment when the parachute opened. Reporting the launch, a local paper ran the headline: 'MOON ROCKET MISSES TARGET BY 238,799 ½ MILES'. Charles Lindbergh recognized a fellow sufferer at the hands of the press. Anne Morrow

Lindbergh wrote that 'the lurid publicity about the "moon-rocket" man disgusted the scholarly professor, but it brought him a new supporter'. What they did not know was that Goddard's wounds were largely self-inflicted.

Goddard was undoubtedly eccentric. In August 1928 the cartoon 'Buck Rogers in the 25th Century' had first appeared in *Amazing Stories* comic, then ran as a newspaper strip cartoon from 1929 to 1967. Buck Rogers was aided and abetted by Dr Huer, the bald, moustached, ever-optimistic scientist genius. It was said that Dr Huer was modelled on Robert Goddard. Students recalled Goddard coming in from the rain absent-mindedly walking the corridors with his umbrella still held aloft. One visitor remembered being driven for 15 miles in first gear, Goddard all the while talking non-stop. Everyone liked him. He was always making jokes (though – according to his biographer – after he died no one could remember any of them). He was a good public speaker and frequently gave talks in order to generate publicity and enthusiasm for his work.

Lindbergh and Goddard met for the first time on Saturday 23 November 1929, at Goddard's office at Clark University in Worcester, Connecticut. For Goddard the meeting was particularly timely. During that year he had again run out of funding. After a tour of his laboratory, Goddard took Lindbergh home to meet his wife, Esther, who was waiting with hot chocolate and cake. Lindbergh had his arm in a sling. Esther asked if it was the result of a flying accident. No, Lindbergh told her, he had wrenched it trying to persuade a puppy to come out from under a bed.

Lindbergh was taken with Goddard's work and told him he wanted to help him find funding. What did he need? Goddard said $25,000 a year for four years would allow him to build a proper launch tower.

Lindbergh knew that the obvious person to approach was Harry Guggenheim. Harry was President of his father's charity, the Daniel Guggenheim Fund for the Promotion of Aeronautics. Founded in 1926, and dedicated 'to the country that had given them a freedom of action Jewish peoples were denied in Europe', the charity had sponsored Lindbergh's 48-state tour of America. But Lindbergh was also reluctant to ask a favour from a good friend. He approached the chemical company DuPont but they said no. The Carnegie Institution for Science came up with $5,000. All other avenues proved fruitless. As a last resort Lindbergh went to Harry, and Harry invited him to talk to his father in person. After less than 10 minutes' conversation, Guggenheim Sr agreed to the entire funding from his own pocket, and stumped up an initial payment of $50,000.

After careful consideration of local weather conditions, Goddard chose to move his new operation to Eden Valley, north-west of Roswell in New Mexico. All his staff made the move with him from Worcester. He launched his first rocket from the new site in December 1931. Those new rockets were encased and had fins; they began to resemble our idea of what a rocket should look like.

A rocket is an intricate array of pipes, pumps and electric wiring housed together inside a metal sheath. A rocket needs a guidance system. A rocket has to be stable. Rockets are inherently unstable. Rocketeers do all they can to prevent a rocket from

Goddard and colleagues carrying a liquid-fuelled rocket in 1932

fulfilling what appears to be the rocket's desire to blow itself to pieces. Fuel and oxidant must be mixed very carefully if the rocket is not to explode; the resulting combustion has to be carefully controlled if the rocket is not to explode; the combustion products have to be carefully directed and expelled if the rocket is not to explode. Nature abhors a rocket. But then Nature abhors all attempts, it could be argued, to constrain her. All experimental science is hard, but rocket science is particularly challenging.

In April 1932 Goddard launched a rocket to which he had added vanes attached to the exhaust and controlled by a gyroscope. After a brief ascent, the rocket crashed. Nevertheless, Goddard was ahead of the rest of the world – if only just.

Daniel Guggenheim, Goddard's main backer, died in 1930. This might not otherwise have affected the agreed funding had not his death coincided with the Great Depression. Harry agreed to continue supporting Goddard, but there was to be a hiatus while he sorted out his father's financial affairs. By 1932 the first tranche of Goddard's funding ran out. Goddard had no choice but to suspend his Roswell operation and return, along with his team, to Clark University in Massachusetts.

By 1934 Harry Guggenheim had restructured his father's finances and was in a position to offer Goddard further funding. After a two-year interruption, Goddard and his team made the journey back to the launch site and workshop at the Mescalero Ranch in Eden Valley, Roswell, where they would remain for the next seven years. Goddard inspired great loyalty. His long-time assistant Charles Mansur had started out as the janitor at Clark University. Goddard offered him $100 to come and work with him, telling him to leave when the money ran out. He stayed until Goddard's death, and went on to have his own distinguished future career as a rocket engineer.

Just two days after the Goddards had returned to the desert, the Lindberghs made an unexpected visit. Anne described the scene that greeted them: 'The telemetry in Roswell, from behind a stout wall shelter, consisted of a pair of binoculars, an old alarm clock to drive a recording drum, and, of course Esther Goddard's faithful movie camera. Esther Goddard was not only photographer, she was also secretary, and seamstress of parachutes in her husband's enterprise.' The launch tower was a converted windmill.

Goddard was working on a series of what would be 14 rockets, A-1 to A-14. The rockets were all known affectionately as Nell, after one of the crew had been heard singing a line from a Cole Porter song, 'They ain't done right by our Nell'. The first two launches took place early the following year and were both failures. The third A rocket went straight up into the air with great force and then there was a bang. The fourth, launched in March 1935, reached an altitude of 1,000 feet. Goddard claimed that the fifth rocket reached an altitude of over a mile. Other accounts suggest that a more modest altitude was achieved and that Goddard had mismeasured. The sixth and seventh flights were failures. And so on. Goddard didn't know it at the time, but he was no longer leading the world in rocket science; at least not as judged by power and altitude.

On 22 September 1935, Lindbergh made what would be his last visit to Roswell before emigrating to Europe. He had hoped to see a successful rocket launch before he left, but he was to be disappointed. Goddard's eleventh and twelfth A rockets were damp squibs. Neither rocket even left the launch pad.

Goddard was a great inventor but he wasn't much of an engineer. 'He'd cobble up some of the craziest looking monstrosities,' Charles Mansur later observed: 'nothing against him, he was a wonderful man – but he couldn't solder, he couldn't weld, he couldn't run a machine, but he did all of it though. He'd get a big chunk of a thing set together and then it would all fall to pieces ... He liked to do things himself. In other words he was a hobbyist.'

The naval rocketeer Robert C. Truax made a similar observation: 'If you review the record of the tests he made, it is simply astonishing in light of present practice how he would take one component, which maybe had only one successful test on it; he would combine it with a dozen other components, some of which had only one successful test and some of which had never been tested; and he would put them all together in a rocket and try to fly it ... And the most remarkable thing of all is that sometimes it worked.'

Given the nature of the work and his lack of engineering finesse, it seems amazing that no one on Goddard's team was ever killed. A colleague remembered that once, after a rocket had exploded, as it frequently did, either on the launch pad or in the air, Goddard said, 'We learned something today. We won't make this mistake again. We'll correct it.'

Both Lindbergh and Guggenheim pressed Goddard to send them a film of a launch, but no film ever materialized; nor did Goddard ever submit the progress reports that Guggenheim requested, and on which his funding was supposedly contingent. Lindbergh wrote to tell Goddard that 'the morale of everyone concerned would be greatly increased if you would find it possible to obtain a record-breaking flight'. Lindbergh knew about the importance of breaking records. And yet, despite their admonitions, Lindbergh and Guggenheim remained loyal to Goddard. Lindbergh reassured Goddard that rocketry was still in its infancy, where aviation had been in 1912. And Guggenheim continued to send the cheques.

One of the most significant American rocket societies was formed at the California Institute of Technology (Caltech) in

1936, under the direction of the famous aerospace engineer Theodore von Kármán. The Guggenheim Aeronautical Laboratory at the California Institute of Technology (GALCIT) was named for the Daniel Guggenheim Fund for the Promotion of Aeronautics which provided the finance. GALCIT was carefully named to avoid using the word rocket, which was thought to have fantastical associations, perhaps as a result of the kind of publicity that Goddard had generated. The Air Army Corps had sneeringly said that Caltech could have the Buck Rogers work. GALCIT called themselves the Suicide Squad, for obvious reasons. Later it would become the Jet Propulsion Laboratory, the designers and manufacturers of America's first satellite. Almost immediately Harry Guggenheim tried to bring his two rocket enterprises together. Frank Malina, a graduate student working at GALCIT, was sent to discuss von Kármán's work with Goddard, but Goddard refused to talk about his own research, apparently treating Malina as if he were a spy. 'Naturally we at Caltech,' von Kármán later wrote, 'wanted as much information as we could get from Goddard for our mutual benefit. But Goddard believed in secrecy...The trouble with secrecy is that one can easily go in the wrong direction and never know it.' Nevertheless, Malina himself was enthusiastic about the visit and said that despite his secrecy, Goddard was friendly, and Malina apparently learned enough that afterwards they made changes at GALCIT to their own liquid-fuelled rocket.

Towards the end of 1939 Guggenheim tried once again to bring his two operations together. This time he got von Kármán and Goddard in the same room. At the meeting, eye to eye with von Kármán, Goddard agreed to a number of concessions. One of them was that GALCIT would take on the design of the combustion chamber. Yet the moment he was back in the desert Goddard decided he couldn't after all offer any concessions and wrote to Guggenheim to say so. Goddard was pig-headed, ignored all advice, the advice of his two main supporters included. He would continue to make up his research as he went along, and keep it to himself.

Over the years Goddard became even more jealous of his American rivals than his rivals abroad. He became reclusive where he had once been outgoing and open. His overweening protectiveness proved to be his undoing. Goddard was sometimes so secretive that a number of his discoveries, for which he might have taken due credit, were rediscovered independently by others, leaving Goddard hopelessly fuming when others, not surprisingly, claimed the innovations as their own. For most of his life he had been irrepressibly optimistic, but now in his later years he became increasingly bitter – and a heavy smoker and drinker.

CHAPTER FIVE

For most of the 1920s Robert Goddard had had the rocket-building field to himself. From the late 1920s rocket clubs began to appear across America and across the world – particularly in Russia and Germany. They were more than just places for swapping ideas; they were ambitious groups that attempted (and sometimes they succeeded) to build and launch their own rockets. The world's first rocket club – the confidently named Verein für Raumschiffahrt (VfR; Society for Space Travel) – was formed in Germany in 1927 by Johann Winkler, Max Valier and Willy Ley. Almost from its inception it focused its attention on liquid-fuelled rockets. The three had worked together as consultants to Fritz Lang on his 1929 science fiction film *Frau im Mond* (The Woman in the Moon), based on a novel written by Lang's then wife Thea von Harbou, which had in turn been inspired by a 92-page book titled *Die Rakete zu den Planetenräumen* (*By Rocket into Interplanetary Space*) written by Hermann Oberth, also a member of the VfR.

By Rocket into Interplanetary Space had started out as a doctoral thesis, but was rejected for being too 'Utopian'. It was later published privately as a short book in 1923. It opens with the claim that it was already possible – by means of a multistage liquid-fuelled rocket – to take some kind of craft beyond the pull of the Earth's gravitational field. Oberth speculated that within a few decades humans would travel into space, and that the construction of space rockets would by then be commercially profitable. The rest of the work is apparently unreadably dry, but Max Valier, one of the VfR's founder members, wrote a popular version, *Der Vorstoß in den Weltenraum* (Advance into Space), which was an instant bestseller.

Oberth came across Goddard's paper 'A Method of Reaching Extreme Altitudes' as his own book was going to press. He wrote to Goddard and asked if he might discuss the paper in an appendix. Goddard wrote back warmly saying yes. When Goddard discovered what Oberth wrote he was appalled. It was clear to him that Oberth's criticisms of his work were an attempt to claim precedence. Oberth wrote that Goddard didn't have the vision to see beyond the exploration of the Earth's atmosphere, to see the significance of rockets in the future exploration of outer space. Here he touched on the rawest of nerves. Goddard had begun to play up rockets as the means for exploring the atmosphere precisely because he had previously over-emphasized the rocket as the future means of exploring interplanetary space. When *Nature* gave Oberth's book a favourable review, Goddard wrote to the journal claiming that Oberth had stolen his ideas. He also attempted to establish, ostensibly on behalf of his country, his own precedence: 'I have read carefully the books that have been written in Germany recently on the application of the rocket method to the problem of interplanetary flight ... and in every book disparagement is made of America's contribution to the subject. I believe that, unless I can present the case in the proper way, when the time comes, my own ... part in the problem will be put in an unfavorable light.' When Oberth claimed that he had first considered hydrogen and oxygen as liquid fuel for rockets in 1912, Goddard countered that he had had the idea as early as 9 June 1909. Goddard dismissed Oberth, whose theoretical ideas were arguably ahead of his own, as a mere *theorist*. Oberth claimed, 30 years later, that he had never heard of Goddard and that his major influence had been Jules Verne, whom he held up as the true father of rocket science.

In Russia, Konstantin Tsiolkovsky was at it too, invading Goddard's territory. When Tsiolkovsky learned of the publication of Oberth's *The Rocket into Interplanetary Space*, he rushed out a revised version of his 1903 paper as a book titled *The Rocket into Cosmic Space*. He sent copies to Goddard and Oberth. The paper had originally been ignored but now the book made a significant impact, at least in Russia.

Oberth and other members of the VfR had planned to launch a rocket to coincide with the release of Lang's film in October 1929. The launch failed, but Oberth and his team built and fired their first liquid-fuelled rocket on 25 January 1930. The flight lasted five minutes. It was a significant achievement. They were not far behind Goddard.

In September 1930 the VfR was given permission by the military to use a larger launch site, Raketenflugplatz Berlin (Rocket Flight Field Berlin). They were joined that same month by a precocious 18-year-old student named Wernher von Braun.

Wernher Magnus Maximilian von Braun was born in 1912, with Junker ancestors on both sides. His father was Baron von Braun, his mother a von Quistorp. They knew who they were back to the thirteenth century: knights, generals, diplomats. His mother was a descendant of the Scottish king Robert III and the English king Edward III. More recent ancestors had numbered Immanuel Kant among their friends. As a child Wernher had piano lessons from Paul Hindemith, played piano duets with his mother, played the cello too, and composed. When he was ten, his mother asked him what he wanted to be when he grew up. He said: 'I want to work on the wheel of progress.' She remembered because it had been such an odd thing to say. She said that he soaked up knowledge, that she never succeeded in being cross with him, that if he were ever badly behaved it was because of his 'exuberant joy of life'. His father gave up trying to admonish him, it had no effect. His son matured so exorbitantly, he said, that he was glad he had taken this position. He said his son didn't inherit his interest in science from him, that came from his mother. When he was 12, Wernher constructed a rocket-propelled wagon by attaching fireworks to it. His mother said that he certainly had a lot of fun as a child, going around junkyards with other boys, collecting bits of old cars to make new cars 'with and without rocket propulsion'. He had his first plane ride when he was 13, in an open-cockpit Junker F13. From then on one of his ambitions was to become a pilot.

When he was 15, von Braun read and loved Kurd Lasswitz's 1897 novel *Auf zwei Planeten* (*On Two Planets*). He was impressed by Lasswitz's description of the precise trajectory the spacecraft – powered by anti-gravity – took between Earth and Mars. Wanting to know more, he sent off for a copy of Oberth's *The Rocket into Interplanetary Space*. He was disappointed to find the book was densely written and full of mathematical equations. Mathematics and physics were two subjects that he had no interest in. It was at that moment that Wernher realized that his passion for rockets exceeded his loathing of maths and physics. As Lindbergh had done, he quickly turned himself from academic underachiever into prodigy. While he was still in his teens von Braun wrote a partial paper, 'On the Theory of the Long-range Rocket', trying to work out in mathematical terms the kind of orbital trajectories Lasswitz had described in his novel.

Von Braun was working in a Berlin machine factory when he decided to sign up to study with Hermann Oberth at the Technische Hochschule Berlin (Berlin Institute of Technology). It was an unusual choice for an aristocrat to have made. Invited to join the von Brauns at dinner, a student friend of Wernher's was dismayed to discover 'that every day the whole family spoke another language, French, Italian, English, Spanish, I think even Portuguese . . . I knew just a little French.' After the ordeal at the dinner table, the friend described the relief of retiring to Wernher's room to lie on the floor and talk about the thermodynamics of rocket propulsion.

At the Institute, and at the VfR, von Braun assisted Oberth in liquid-fuelled rocket motor tests. Oberth sent him out to raise funds for their research. Von Braun manned a display in a Berlin department store from where he delivered his pitch to passing Berlin housewives. He was Teutonic, tall, blonde, handsome (in a cartoonish sort of way – big-featured, square-jawed), and his charm was evident even then. Years later, he could still recall part of his patter: 'I bet you,' he told the bemused shoppers, 'that the first man to walk on the moon is alive today somewhere on this Earth.' And he was right. In 1930 moonwalkers had been born across America. Neil Armstrong and Buzz Aldrin were both born that year. Alan Shepard, the fifth and oldest of the Apollo astronauts to walk on the moon, was seven years old. Soon after von Braun joined the VfR in 1930, Oberth left Germany to teach in Romania.

In 1931 von Braun travelled to Zurich to be at the celebrations following Auguste Picard's record ascent. From the late eighteenth century, hot-air balloonists first saw the Earth from a vantage point that was not part of the Earth itself, not from the top of a mountain or a cathedral spire, but from the perspective of another medium. Over the next century, balloonists rose higher and higher above the Earth's surface. 'No man can have a just estimation of the insignificance of his species,' wrote the painter Benjamin Robert Haydon in 1828, 'unless he has been in an air-balloon.' On 10 June 1867 the French balloonist Camille Flammarion, brother to the famous publisher Ernest, took himself 10,827 feet above the River Loire, slightly higher, he noted, than Mount Olympus. He said the silence was so oppressive he could not help wondering if he were still alive. One early aeronaut used the word 'hilarity' to describe the sensation of being suspended, silently, in a gondola high above the ground. Another early balloonist described the Earth as looking like 'a giant organism, mysteriously patterned and unfolding like a living creature'. In 1909 H. G. Wells wrote that 'to be alone in a balloon at a height of fourteen or fifteen thousand feet is like nothing else in human experience. It is one of the supreme things possible to man.' On 21 May 1931 Auguste Picard ascended to a record height of 51,775 feet, almost 10 miles. The Jenny, Lindbergh's first plane, could reach a maximum altitude of around 21,000 feet. Even today most commercial planes are not permitted to fly higher than 45,000 feet. Von Braun introduced himself to Picard: 'You know, I plan on travelling to the moon some time,' he said. Picard was apparently encouraging.

Word of the VfR's experiments with rockets made its way back to the army. The group had launched two types of liquid-fuelled rockets: the *Mirak*, which had not been a success, and the *Repulsor*, named after the spacecraft in Lasswitz's novel. Several *Repulsor* rockets had reached 1,000 feet or more; one had crash-landed onto a local police barracks. The army sent along several observers who were mildly impressed by what they saw and gave the group a 1,000-Mark contract to make improvements. A junior officer, Captain Walter Dornberger, then aged 35, was sufficiently intrigued by von Braun that he offered to act as go-between. Dornberger had special responsibility for rocketry within Army Ordnance. He later wrote that he had been struck during his casual visits 'by the energy and shrewdness with which this fair, tall, young student with the broad massive chin went to work, and by his astonishing theoretical knowledge'. It was the beginning of a long working relationship.

The modest contract brought up the question of whether or not the VfR should accept military funding, and led to the group breaking apart the following year. A new grouping emerged under the leadership of von Braun. Through Dornberger's mediation, the group was offered the test-range site in Kummersdorf, 16 miles south of Berlin, as their new base. Von Braun was told to report to Dornberger and to keep the project secret. They began work on the first of a series of A rockets, A for *Aggregat* (the German word for aggregate) because of the number of parts that had to work together. The A-1 was designed by von Braun, stood 4 feet 7 inches tall, and was fuelled by a mixture of alcohol and liquid oxygen that provided a thrust of 660 lbs. It had a gyroscope in the nose meant to give it stability. The rocket, however, proved to be unstable and blew up on the launch pad.

Before he had reached the age of 22, von Braun was awarded a doctorate for his thesis 'Design, Theoretical and Experimental Contributions to the Problem of the Liquid-Fuelled Rocket'. The manuscript was sent to the army for reasons of security. He had also taken up flying and was now a qualified pilot.

After 18 months' work refining the A-1, von Braun was designing the A-2. The gyroscope was moved to the centre of the rocket in an attempt to solve the problem of instability. Two A-2 rockets were built, each 5 feet 3 inches tall, one nicknamed Max the other Moritz after famous cartoon characters. They were launched in December 1934 from Borkum (a German island in the North Sea), army dignitaries in attendance. One rocket reached an altitude of almost 1½ miles, the other over 2 miles – a new record if they had but known. This was better than anything Robert Goddard would ever achieve in America. One of the army officers in attendance asked von Braun if the rocket could be used as a military weapon and carry a warhead. 'Probably,' said von Braun, 'but what would be the point?' In August that year, immediately following the death of President Hindenburg, the

Chancellor – Adolf Hitler – had styled himself Führer. News of the successful launch of the A-2s attracted the attention of the Luftwaffe. Von Braun was summoned to make a presentation. He had only got halfway through before he was cut off, to be told that they, too, wanted to fund von Braun's group. They offered them a staggering 5 million Reichsmarks (then around $2 million in US dollars), 'so that you can get the ball rolling'. Not to be outdone, and in order to maintain overall control of the project, Dornberger's superior officer, Colonel Karl Becker, a scientist and army chief of ballistics and ammunition, increased the army's funding to 6 million Reichsmarks (they had recently offered half a million after witnessing a static firing test of an A-3 engine). Before he had turned 24 years old, von Braun was in charge of a budget of 11 million Reichsmarks. It was funding beyond all Goddard's dreams.

Von Braun's sole ambition was to build a space rocket. As a young man he once turned up at a fancy dress party dressed as himself in his seventies: he said he was the famous von Braun who had been to Mars and back. 'We always considered the development of rockets for military purposes as a roundabout way to get into space,' von Braun wrote. He felt, naively or not, that in the mid-1930s he and his group still retained the power balance. They were exploiting the army and air force to their own ends as the pioneers of aviation had done, or so he believed. All I ever wanted, he once said, was a rich uncle. And now he had two rich uncles, the army and the Luftwaffe.

Part of his 11 million Reichsmarks was earmarked to build a new launch pad. When von Braun told his mother that they were looking for a suitable location for a new launch site, she suggested Peenemünde at the northern tip of the island of Usedom, a beautiful area of dense forests, marshes, and sandy beaches, where her ancestors had had estates. 'Your grandfather used to go duck-hunting up there,' she said.

Two large operations were constructed on the site: Peenemünde East, Werk Ost, under the control of the army, and Peenemünde West, Werk West, under the control of the Luftwaffe. There would be great rivalry between the two divisions. Von Braun's group moved to Werk Ost during April and May 1937.

At the age of 25, looking 18 – he had trouble getting served in bars – von Braun was now technical director of the operations at Werk Ost in charge of a team of 350. He was 'the heart and soul of the place', wrote his biographer Michael Neufeld. It was here that von Braun's real genius for management and engineering flourished. Despite his patrician Prussian bearing he was never aloof. He was witty, diplomatic, always polite and enthusiastic. He was both a good listener and a good talker, and, famously, he had charm. For a bet von Braun once drove three times around a Berlin church in the wrong direction. The police officers were so disarmed by him that not only did they not press charges, but accepted his invitation to go for a drink. A colleague who had been there at the time wrote later that it was that experience that made him realize that he would go through fire for such a man.

The army, von Braun claimed, took great care to keep the Nazi Party at bay, at least at first. He said that in those first years he had worked 'in an environment that was rather hermetically sealed against any Nazi party infiltration . . .' But in late 1937 when membership reopened, von Braun did join the National Socialist Party. He said he had been commanded to. Both his father Magnus and older brother Sigismund had warned him against joining the party. When Hitler had declared himself Chancellor in 1933, the Baron had resigned his public office and retired to a small estate in Silesia. He never joined the party. As a young diplomat at the German Embassy, Sigismund had had to swear allegiance to Hitler. He wrote in his diary that afterwards he felt anxious for his humanity. He spoke out against Hitler and the future world conflagration he said Hitler risked. Both Wernher von Braun and Dornberger were horrified by his recklessness. Sigismund would be sent to a backwater posting in Addis Ababa, then the capital of Italian-occupied Ethiopia. Wernher von Braun said that he only wore his uniform and swastika pin when there were visiting dignitaries. He was never openly critical of Nazism but he was careless sometimes, or casual in what he said. He had a habit of getting carried away and talking about his dreams of manned space flight. His exuberance would get him into trouble, but that was later.

The first rockets to be fired at Peenemünde were von Braun's A-3 missiles. They advanced the A-2 by adding an inertial guidance system, basically a way of making the missile self-steering. It was effectively a simple analogue computer, a step on the road that would lead to IBM's Apollo Guidance Computer.

Standing 22 feet 1 inch tall, the A-3 was designed to fly to an altitude of 15 miles carrying a payload of 100 lbs. The engine produced a thrust of 3,300 lbs. Four A-3s were built. One reached an altitude of 11 miles but all of them developed problems of one kind or another after launch. And yet despite the problems, von Braun and his team were by now far ahead of the rest of the world. An altitude of 11 miles was also 1 mile higher than Picard's record ascent of 1931, though Picard's record had been broken several times, most recently on 11 November 1935. As part of an expedition sponsored by the National Geographic Society, Captain A. W. Stephens and Captain O. A. Andrews from the US Army Air Corps ascended to an altitude of 13.7 miles in a gondola attached to a helium-filled balloon. It was one of the first attempts to explore the stratosphere scientifically. A photograph taken from the balloon as it sailed high above the Black Hills of South Dakota captured the curvature of the Earth for the first time: visual – rather than merely intellectual – proof that we live on a sphere. The photographs showed a panorama stretching 330 miles. There had been earlier photographs from lower altitudes that purported to show the Earth's curvature, but in those cases the curvature was only apparent, an artificial consequence of the use of wide-angled lenses.

Von Braun's team soon abandoned the A-3 in favour of the next iteration. The A-4 was to be a huge leap forward from the A-3. Years of testing lay ahead before one of the most famous rockets ever built became flightworthy. Late in the coming war the A-4 would be renamed *Vergeltungswaffe* 2, Vengeance Weapon 2, or more familiarly, the V-2.

On a cold, rainy day in the spring of 1939, Hitler made his first visit to Peenemünde, along with Field Marshal von Brauchitsch and General Karl Becker from Army Ordnance, Deputy Führer Rudolf Hess, Martin Bormann, and several others. They witnessed the firing of two types of rocket engine – not an actual launch. Hitler was not impressed. Dornberger had had to warn von Braun not to talk about space. Their job was to build weapons. Von Braun admitted later that he had been so keen to talk to the Führer about space exploration that he had had to bite his tongue. Instead, von Braun tried to describe to Hitler the technical problems the engines presented. Hitler walked away shaking his head, saying nothing. Later, Hitler became more animated and said: 'Even now I still don't know how a liquid-propellant rocket can fly. Why do you need *two* tanks and *two* different engines?' Von Braun was in despair. He had just explained why. Hitler had clearly been paying no attention at all. He explained it all over again, and this time, since Hitler still did not seem to have got the point, he emphasized the potential the rockets offered as carriers of weapons. Hitler seemed bored and fell silent. The party left. No further funding was offered but neither was their budget cut.

Later that year Britain and France declared war on Germany. With steel quotas cut and redirected to the manufacture of munitions, something drastic had to be done to protect the work at Peenemünde. The workforce at Werk Ost alone now numbered 1,200. It was clear that the operation would have to maximize its contribution to the war effort if it was to continue to justify its level of funding. Dornberger wrote a memo to his superiors declaring that Germany was in a rocket arms race with other nations. Dornberger may have been encouraged to pursue the deception because of the work of an inept intelligence operative working in America named Gustav Guellich. For years Guellich sent reports on Robert Goddard to a German military attaché in the US named General Friedrich von Boetticher. It had started early in 1936, when Boetticher sent a copy of the 4 January issue of *Science News Letter* to the army's General Staff in Berlin. The front cover of the journal was a photograph of a Goddard rocket. Inside was the transcript of a presentation Goddard had made at the American Association for the Advancement of Science (AAAS) the week before. The attaché noted that Goddard had the support of both the Guggenheim Foundation and the Carnegie Institute. General Staff forwarded the communiqué to Guellich, and told him to investigate further. Over the next years Guellich submitted regular reports, mostly fabricated.

On the last day of October 1939 an A-5 rocket (somewhat confusingly, a small version of the future A-4) was successfully launched. 'It was an unforgettable sight,' von Braun said. 'The slim missile rose slowly from its platform, climbing vertically with ever-increasing speed and without the slightest oscillation, until it vanished in the overcast.' He could hear the rocket thunder away into the distance. At 22 seconds after launch, he sent a radio signal to the rocket to cut the engine and deploy the parachutes. Then, five minutes later, the rocket came back into view as it landed, perfectly, 200 feet offshore. It was the first successful controlled flight of one of the army's larger missiles. Werk Ost was set a target: to make the A-4 operational within two years.

CHAPTER SIX

Charles and Anne Morrow Lindbergh were the most famous couple in the world, 'a dream couple, idealized, worshipped'. Their first child, Charles, was born in 1930, the year after their marriage. 'It is difficult to believe, or even to remember how little privacy we had,' Anne wrote of those first years of marriage, and yet it was also an idyllic time. They flew together, crisscrossing America, laying out passenger and mail routes, Anne sitting in the back navigating. The planes were often ill-equipped to carry instruments. Overladen, they hedge-hopped their way across the country. In 1931 they went on the first of two long expeditions for Pan American, on this first trip surveying the Pacific for future flights, via Alaska, to Japan and China. If they got a bad weather report, Charles liked to get close to the storm and take a look: 'we can always turn back,' he'd say. Anne said that 'he wanted to get as close as possible to the danger to assess it accurately,' a technique he applied in life generally.

On 1 March 1932 Charles Jr, then aged 20 months, was kidnapped. Charles probably died on the scene, perhaps accidentally dropped by the kidnapper as he descended the ladder he had used to gain entry to the nursery where Charles was sleeping. The case was so over-reported, and so much 'evidence' emerged, that the police operation was swamped. All kinds of oddballs stepped forward with advice. Al Capone offered a $10,000 reward. Charles

Charles August Lindbergh in 1931 with dogs Bogey and Skean.
He was kidnapped and murdered the following year.

Lindbergh's low opinion of the press sank further. The nadir came when a journalist broke into the morgue and photographed the baby's mutilated body. It took two and a half years to catch the child's murderer.

Richard Hauptmann, a German-born illegal immigrant, was electrocuted on 3 April 1936, proclaiming his innocence to the end. Several historians have wondered if such an elaborate crime could have been the work of a single individual. The journalist H. L. Mencken called the kidnapping and subsequent trial the biggest story since the Resurrection. And the public outpouring of grief was said to have been the greatest since the assassination of Lincoln. Anne said that her husband never cried again. She cried silently alone every night, hiding her emotions from her stoical husband and her equally stoical mother.

The Lindbergh's second child, Jon, was born less than six months after the kidnapping. Threats of violence against Jon were made soon after his birth. In a daze, Charles and Anne took off on a five-and-a-half month expedition, a second survey on behalf of Pan American. Jon – not yet one year old – was left behind with a nanny and under heavy guard. His parents flew up the North American coast to Labrador, and on to Greenland, Iceland, the Faroes and Shetlands, and from there to continental Europe and Russia, to Africa, and finally a 16-hour leg that took them back to the United States by way of Brazil: 29,000 miles in all. They studied the terrain and weather conditions and looked for possible landing sites for a future air service between America and Europe. 'I did not realize,' Anne wrote, 'that we were part of the revolution in twentieth-century transportation.' She was kept so busy decoding messages that there was hardly time to look out at 'the beauty of sea, sky, and mountains'. Anne established a long-distance record for decoding a message between an airplane and a ground station. Only later did she come to understand that in fleeing their grief she had left behind 'the most healing and nourishing element' in her life: her son Jon. Anne wondered if their fame had made it hard for her to grow up. Looking back on her younger self, she said she had become 'thoroughly sick of the ego of the adolescent and young married Anne Morrow Lindbergh'.

Though she had been cut off from her writing, in 1935 her account of the 1931 Pacific survey, *North to the Orient*, came out and was an immediate bestseller. It won her the very first National Book Award for non-fiction.

The intrusions made into the Lindberghs' private life by the press and by the public alike had hardly abated. In 1935 the Lindberghs decided they had no choice but to leave America for Europe. 'It is extremely distressing and discouraging,' wrote one Hearst-owned newspaper, seemingly oblivious to its own part in the Lindberghs' planned exile, 'that this grand country of ours is so overrun with cranks, criminals, and Communists that a splendid citizen like Colonel Lindbergh must take his family abroad to protect them against violence.'

The Lindberghs first stayed with friends in Wales, and then settled at Long Barn, Sevenoaks, Kent, at the invitation of Harold Nicolson, whose former home it had been. Nicolson and his wife, Vita Sackville-West, had since acquired Sissinghurst Castle. In early June 1936, Lindbergh received a letter from the American military attaché in Berlin, Truman Smith, inviting him to visit Germany's air factories. Truman Smith had been appointed to his position the year before, his chief responsibility being to report any evidence of militarization in Germany. Smith soon realized that he did not have the support, through lack of interest, of his superior William Dodd, the American Ambassador in Berlin. In a meeting with Hitler, Dodd had expressed the view that Jews 'held a great many more of the key positions in Germany than their numbers or talents entitled them to'. Lindbergh later wrote something remarkably similar in his diary: 'A few Jews add strength and character to a country, but too many creates chaos.' Dodd told Hitler that in America they had dealt with the problem 'of over-activity of Jews in university or official life' by introducing a quota system, 'in such a way as to not give great offense'.

Smith had met Hitler in 1922. After the meeting, he described him as a fanatical man whose 'power over the mob must be immense'. With the political situation in Germany now in ferment, with no funds forthcoming to finance any spying operations, and at odds with his superior, Smith realized that he would need to be creative. When his wife showed him an article in the *Herald Tribune* describing a visit Lindbergh had made to an aircraft factory in Paris, Smith saw an opportunity. He approached the German Air Ministry, floating the idea that Lindbergh be invited to tour German factories. Permission came back from the Air Minister himself, Hermann Göring. Smith rightly guessed that the famous aviator would be welcomed, that a visit from Lindbergh would be a chance for the Luftwaffe to show off; exactly what they had to show off was what Smith hoped Lindbergh would find out. Göring had announced that Germany was rebuilding its air force, not in itself a contravention of the Treaty of Versailles, but if the country was re-arming then that would be a different matter. In his letter to Lindbergh, Smith wrote that the tour would be 'of high patriotic value'. Lindbergh agreed to the proposal so long as he might also be invited to the opening of the Olympic Games.

The Lindberghs flew themselves to Berlin from a small airfield at Penshurst in Kent. At a state lunch, Göring made his great admiration for Lindbergh clear. Part of Lindbergh's appeal to Göring was his Swedish ancestry. Göring never recovered from the death of his first wife, a Swede. He loved Sweden and all things Swedish. It has even been suggested that this was the main reason Sweden was not invaded by Germany during the war. Göring spoke Swedish fluently and was disappointed to discover that Lindbergh did not. Throughout the meeting Lindbergh was polite but distant.

During the subsequent tour Lindbergh got to visit air bases that had never been seen before by an American. He was even invited to fly their newest planes. Lindbergh not only saw more than Göring suspected, but even more than Smith had hoped he would. It took Lindbergh a very short time to realize that Germany was not only rearming, but preparing for war.

Lindbergh was impressed by 'the organized vitality of Germany', by the relentless activity he saw around him. He admired the country's ability to build new factories, airfields and research laboratories. On the other hand he disliked Germany's regimentation and the crudeness of its political regime. For him, there was no comparison between Germany and England and France: 'I was stirred by the spirit in Germany as I had been deadened by the lack of it in England and disturbed by its volatile individuality in France.'

In May 1935 France had signed a treaty of alliance with the USSR. Hitler saw this as an act of aggression and cited it as an excuse to reoccupy the Rhineland, against the terms of the Versailles agreement. President Roosevelt failed to condemn the invasion. Lindbergh argued that France should attack Germany with all its power, saying that soon it would be too late. Another world war was coming, Lindbergh said, an inevitable consequence of the failure of the Treaty of Versailles; Germany should have been crushed at the end of the First World War. Germany's new-found energy, he wrote, had been built on the strongest of foundations, that of defeat. Lindbergh wondered, too, why Britain had not taken the opportunity of crushing Germany in 1934; though on what pretext, he did not say.

Lindbergh was to visit German air factories and airfields and fly German planes during a number of tours made over the next couple of years. He helped Smith prepare his reports. They estimated that within three years German air power would be ahead of all other European competition, and that the gap between Germany's and Britain's air power was already almost non-existent. The US began to modernize and expand its air fleet, in part because of Lindbergh's intelligence, but his role was underplayed

Truman Smith and Lindbergh in Germany

by the Roosevelt administration, almost certainly because by this time Roosevelt had come to dislike Lindbergh.

In 1933 Roosevelt had ordered a Senate investigation into air-mail contracts issued under the Hoover administration by the then Postmaster General, Walter F. Brown. The investigation delivered its report early in 1934, concluding that the contracts had not been awarded on the basis of competitive bidding but only to the largest carriers. There were charges of cronyism. Records had been destroyed. President Roosevelt insisted on cancelling mail contracts across the board. Lindbergh had been associated with most of these companies – Transcontinental & Western Airways was known affectionately as the Lindbergh Line. His name was now in danger of being tarnished. On 9 February 1934, Roosevelt called in the Air Corps to fly the mail. An army air pilot died almost immediately, and a further four within the first week. Six others were injured and eight planes written off. Ten army pilots died before three weeks had passed. The losses were a great embarrassment to Roosevelt. Eddie Rickenbacker, the First World War ace, called it 'legalized murder'. The losses were also an embarrassment to the army. Colonel Billy Mitchell, a popular figure, said that it showed the army pilots in a bad light if they could not fly as well as their civilian counterparts. 'The Army has lost the art of flying,' he wrote.

Statistically, the number of deaths had not changed significantly. Night flying was halted as being too risky, but it had always been highly dangerous. Even before the army pilots took over, United Air Lines had had four crashes in four months, resulting in the deaths of 11 crew and 11 passengers. But the public was not interested in whether or not the statistics were comparable, only the current reality: the death of a significant number of army pilots, and, what was incontrovertible, a much-reduced service. By the early 1930s the mail service had become fast and efficient, and businesses had grown to rely on it. As a result of the army's inability to manage the mail service, a number of businesses took the law into their own hands and began to send mail, illegally, on commercial passenger planes.

When Lindbergh personally intervened, the President was forced to back down and return the mail to private companies. After a speech Lindbergh gave in defence of the old contracts, the *New York Times* reported that Lindbergh 'seemed still to be one of the world's most fascinating figures'. As a way of saving face, the old companies were not allowed to bid. They got round this by simply renaming themselves. During the period of the kidnapping, Lindbergh's word carried more force than the President's. Roosevelt vowed to get his revenge. 'Don't worry,' he said to one of his advisers, 'we will get that fair-haired boy.'

In 1938, in order to be closer to their friends, the surgeon Alexis Carrel and his wife Anne-Marie, the Lindberghs left England and moved to France. Both Charles and Anne later wrote that the time they had spent living in England had been the happiest period of their life together, but Charles could not understand why Britain seemed to be voluntarily letting go of its Empire: 'There was a sense of heaviness in England that pressed like a London fog ... I felt that England ... saw not the future but the past.' He had no desire to return except for visits. 'England was aged and declining,' and only had itself to blame for being weak. His views, not surprisingly, disgusted his recent host Harold Nicolson. Lindbergh felt no more generously towards his new home country. The French were too much in love with their land and their government. The ideology of Germany might be intolerable but the weakness of Britain and France was more so. In any case, the greatest threat, Lindbergh believed, came not from Germany but from Russia and the Far East. He thought that Europe was doomed to repeat the mistakes it had made over thousands of years. He believed that America could learn from those mistakes and make a new beginning. Thinking about the world dispassionately, if he had total freedom to choose, he came to the conclusion that he would still prefer to be what he was, a citizen of the United States.

'Hitler is apparently more popular than ever in Germany,' Lindbergh wrote in a letter in 1937, 'and, much as I disagree with some of the things which have been done, I can understand his popularity. He has done much for Germany.' In a letter she wrote to her mother, Anne Morrow Lindbergh was more ambivalent about the Nazis: 'There are great big blurred uncomfortable patches of dislike in my mind about them; their treatment of Jews, their brute-force manner, their stupidity, their rudeness, their regimentation. Things which I hate so much that I hardly know whether the efficiency, unity, spirit that comes out of it can be worth it.'

In October 1938 Lindbergh flew back into Berlin: 'a healthy, busy, modern city'. At a dinner reception given by Göring he was presented with the Order of the Golden Eagle. Casually, Lindbergh passed the box containing the medal to his wife. Anne opened the box, looked at the medal: 'The Albatross,' she said. Truman Smith's wife Katherine said to her husband later that evening: 'This medal will surely do Lindbergh much harm.'

During the evening of 9 November, over a hundred synagogues were burned across Germany, Jewish-owned shops were destroyed, thousands of Jews were arrested, and dozens killed. Kristallnacht. 'My admiration for Germans is constantly being dashed against some rock such as this,' Lindbergh wrote in his diary. 'They have undoubtedly had a difficult Jewish problem,' he wrote a few days later, 'but why is it necessary to handle it so unreasonably?'

Lindbergh had planned to stay on in Germany as an observer, encouraged to do so by the US Ambassador to the UK, Joseph Kennedy, father of the future President, and vocal in his belief that America should not get involved should war spread across Europe. But Lindbergh's motives for staying in Europe were being questioned back home, and so he cancelled his plans, stating in a letter to a friend: 'I do not wish to make a move which would seem to support German action in regard to the Jews.' He made two last, secret missions to Berlin in December 1938 and January 1939. In a diary entry dated 2 April 1938 he wrote: 'Much as I disapprove of many things Germany has done, I believe she has pursued the only consistent policy in Europe in recent years. I cannot support the broken promises, but she has only moved a little faster than other nations have in breaking promises. The question of right and wrong is one thing by law and another thing by history.'

Lindbergh returned to America in April 1939. He was immediately invited by the army to survey American air power. He took the opportunity to make what would be his last visit to see Robert Goddard in Roswell. He told Goddard that though he had been given a free pass to review German air power, there was intense secrecy around their rocket capability. Lindbergh was convinced the Nazis were working on rockets but not willing to talk about it. 'Yes, they must have plans for the rocket,' said Goddard. 'When will our people in Washington listen to reason?'

Franklin D. Roosevelt won an unprecedented third term in office campaigning as a non-interventionist. Now, in 1940, he had the difficult task of persuading a nation overwhelmingly in favour of staying out of a world war to change its collective mind. Once again Lindbergh was in his way.

Lindbergh argued that America should stay out of the coming cataclysm and protect its hemisphere of the globe: 'My mind tells me that we better face our problems and let Europe face hers.' America would be safe from German bombers as they were not designed to cross the Atlantic. He acknowledged that such planes could be built, but he thought the cost would be prohibitive. If America intervened in a world war now, it would, he said, be condemned to intervene in world affairs forever after. If America became the world's guardian it would be at the expense of its own power. America needed to be strong in order to see off the greater threats that came from Russia and the Far East. The war correspondent H. R. Knickerbocker wrote that Lindbergh's idea that Britain should be allowed to fall was 'as cynical as any Nazi could invent'.

Lindbergh's first isolationist talk attracted probably the largest radio audience ever at that time, exceeding those of the President's 'Fireside Chats', his popular weekly broadcasts. 'I do not intend to stand by and see this country pushed into war,' Lindbergh told the nation in 1939, 'if it is not absolutely essential to the future welfare of the nation.' His quiet voice and gentle manner drew people in.

When, after the evacuation of Dunkirk and the Battle of Britain in 1940, Roosevelt supported Britain by 'lending' them armaments, Lindbergh spoke out against lend lease: 'First they said, "Sell us the arms and we will win." Then it was, "Lend us the arms and we will win." Now it is, "Bring us the arms and we will win." Tomorrow it will be, "Fight our war for us and we will win."'

The Interior Secretary, Harold Ickes, had also come to loathe Lindbergh. Ickes had begun to orchestrate a campaign with the intention of damaging Lindbergh. Reports began to surface that Lindbergh had been duped, that Germany wasn't as strong as he was being led to believe: what he had seen were the same planes ferried from airfield to airfield. A story was leaked to the press that Lindbergh was being investigated for tax irregularities. Few people come out of a tax investigation unscathed, but scrupulous Lindbergh did.

On 11 September 1941 Lindbergh gave a speech on behalf of America First, the main non-interventionist movement in America. Among the 800,000 members were prominent businessmen and politicians, as well as celebrities like Walt Disney, E. E. Cummings, Lilian Gish and Sinclair Lewis. Gore Vidal and Gerald Ford were student members. A young J. F. Kennedy sent a check for $100 and a note: 'What you are all doing is vital.' Friends claimed that Lindbergh was no anti-semite, but his speech that night made it clear that he was. Even if his anti-semitism was unconscious, even if it was the kind of casual anti-semitism that characterized the era – Roosevelt once told the Catholic economist Leo T. Crowley that the United States was 'a Protestant country and the Catholics and Jews are here on sufferance' – even if no one could know at that time what was going to happen to the Jews across Europe, Lindbergh's speech at Des Moines was by any test anti-semitic. 'Their greatest danger to this country,' he said that night, 'lies in their large ownership and influence in our motion pictures, our press, our radio, and our Government.' Even if his claims had been true, which they were not, he had separated out Jews as if they were other than American, as if they were one race, and as if their motives for war came out of self-interest. His attempts to justify his views only sunk him deeper into the mire: 'It is not difficult to understand why Jewish people desire the overthrow of Nazi Germany,' he said. 'The persecution they suffered in Germany would be sufficient to make bitter enemies of any race. No person with a sense of dignity of mankind can condone the persecution the Jewish race suffered in Germany. But no person of honesty and vision can look on their pro-war policy here today without seeing the dangers involved in such a policy, both for us and for them.' He said that war would make Jews vulnerable to attack in America: 'Instead of agitating for war the Jewish groups in this country should be opposing it in every possible way, for they will be the first to feel its consequences. Tolerance is a virtue that depends upon peace and strength ... Their greatest danger to this country lies ...' And on he went. That night he wrote in his diary: 'When I mentioned the three major groups agitating for war – the British, the Jewish, and the Roosevelt Administration – the entire audience seemed to stand and cheer.' Four days later, another diary entry ran: 'I felt I had worded my Des Moines address carefully and moderately. It seems to me that almost anything can be discussed today in America except the Jewish problem.'

Lindbergh and America First were much damaged by the speech. From the moment he delivered it, Lindbergh was reviled. Anne had foreseen the outcome. She had managed to temper some of the language of the original draft, but he wouldn't remove the sentences she found most offensive. She told him that he would be accused of being anti-semitic. His response was simply, 'But I'm not.' He said he was logically advancing an argument 'in an orderly fashion'. Anne told him that she would rather America went to war than that it was 'shaken by violent anti-semitism'. Charles told her that their opinions differed. In later years Lindbergh would come to trust and seek out Anne's judgement, but at this stage in their marriage he trusted only his own. Anne wrote in her diary that she agreed with those who had criticized him. She said that what had made him a great pilot was the unmaking of him as a politician. In the end, the chief architect of Lindbergh's downfall would prove to be Lindbergh himself.

He was never to apologize. In later years he tried to justify his stance: 'To me, the most important element in this situation is the future welfare of my country, my family, and my fellow-citizens.' And to hell with the rest of the world, he might have added. So much for being first citizen of the world.

Anne too would not emerge unscathed. In 1940 she had published her own defence of isolationism, *The Wave of the Future*. Her second book, *Listen! The Wind* (1938), an account of the second survey for Pan American, had won her another National Book Award. Her third book sold 50,000 copies in the first two months, and the reviews were positive. But then something changed, seemingly overnight. The press and readers turned against the book, and her. *The Wave of the Future* is barely 100 pages but it is hard to work out what argument is being advanced, or, rather, hard to believe that the argument can really be what it seems to be: that Fascism must run its course and while it does we must learn to live with it, seems to be the gist of it: 'The wave of the future is coming and there is no fighting it.' In one of the most obscure and confused sections of the book she tells her American readers that their '*first* duty is to ... family and nation', and yet she also asks that they take 'a planetary point of view of the world's troubles'. America should stay up above the world and not descend 'into the maelstrom of battle'. The Roosevelt administration called it 'the Bible of every American Nazi, Fascist, Bundist and Appeaser'.

Here was relativism at its most extreme and most dangerous. Charles's influence is clear. In a diary entry dated 10 October 1940 he writes: 'As I read through my diaries, I realize that the element I foresaw least clearly was the vacillation of the "democracies" and their complete inability to follow a consistent policy.' Consistency was a quality Lindbergh valued highly. He would often criticize the 'democracies' for their lack of it. In a diary entry dated 11 December 1941 he wrote, 'We talk about spreading democracy and freedom all over the world, but they are to us words rather than conditions. We haven't even got them here in America.' He said, 'let us make sure that the roots of freedom and democracy are firmly planted in our own country, starting with the Negro in our Southern States'. In this he may have been decades ahead of his time, but perhaps what was most dangerous about his thinking was his willingness to sacrifice one ideal for another, and through the agency of logic.

Only after her father had died did Lindbergh's daughter Reeve listen to a tape recording of his Des Moines speech of 1941. '"Not you!" I cried out silently to myself, and to him – No! ... You raised your children never to say, never to *think*, such things . . .' Did he reckon, she wondered, that it was enough to say, tight-lipped, that he was not anti-semitic? Did he know how hard it had been for her that he had remained silent forever? What did it mean, and how could it have happened, that he never repudiated or amended his words? Had he never come to understand the corroding power silence has? 'Was there,' she asked, 'in fact, such a thing as innocent, unconscious anti-semitism? Was it prevalent before the war, and did the Holocaust forever criminalize an attitude that was previously acceptable and widespread among the non-Jewish population of this country and others?' These were questions that she could not answer; questions she could only set alongside what she also knew to be true of her father, that he had been a good boy who had grown into a good man, and that he had 'continued to grow along with his own century'.

Lindbergh had undoubtedly been influenced by the ideas of his mentor Alexis Carrel, whom he had met fortuitously a year after his chance meeting with Goddard. In 1930, Anne's sister Elizabeth was found to have developed a heart condition, damage caused after a bout of rheumatic fever. Lindbergh asked Elizabeth's doctor why her heart could not be repaired. He was told that surgery on a beating heart was not possible. When he asked why the heart could not be stopped and the blood circulated using a pump the doctor could not say. He suggested Lindbergh talk to her anaesthesiologist, but he had no answer either. The anaesthesiologist in turn suggested Lindbergh talk to a surgeon he knew, Alexis Carrel.

Carrel explained to Lindbergh that no one had yet been able to build a heart pump because blood clots readily when it comes into contact with glass or metal surfaces, and blood cells get broken up as they pass through mechanical valves. A first step would be to work out how to study human organs *in vitro* (literally, in glass), which was what Carrel was working on at the time at the Rockefeller Institute. What was needed was some kind of pump that could keep organs free from infection and supplied with blood, as if otherwise alive, while they were experimented on. Carrel suggested that he and Lindbergh collaborated on the project together. The work absorbed Lindbergh for the next five years.

The press inaccurately called their invention 'the Lindbergh heart', properly known as a perfusion pump. The machine was a significant advance in the history of surgery and medical research. Transplant organs could be kept viable for longer periods of time. Organs could be repaired outside the body. Carrel planned to harvest insulin and growth hormones from perfused organs. Carrel and Lindbergh appeared together on the cover of *Time* magazine. Their invention was hailed as a step towards human immortality. It was a subject that interested Lindberg, who believed it would be a necessary development if humans were ever to explore the far reaches of the universe. If human beings had learned how to fly then why should they not discover how to live forever? The perfusion pump was widely used for decades. In 1966 a researcher invited Lindbergh to design an improved version. Lindbergh obliged.

When Lindbergh watched Carrel operate, he felt that he was standing on the threshold between scientific and mystical worlds. After years of working on the perfusion pump he had come to value 'the mechanistic qualities of life' more, not less. He could understand now why, for his great-grandfather Edwin, there had been no conflict between his work as a doctor and his religious convictions: 'His studies of flesh didn't convince him that all existence ends with the flesh. He had faith in some quality that is independent of body.' Lindbergh came to the conclusion that life does 'not lie in the material'. 'I found that any branch of science pursued to its peripheries ended in mystery. Man could neither explain the miracle of creation or the fact of his awareness, nor conceive the end of space and time. The miracles of science and technology become trivial in the face of the unknowable.'

'Carrel's mind flashed with the speed of light,' Lindbergh wrote, 'in the space between the logical world of science and the mystical world of God.' Under Carrel's guidance Lindbergh developed a philosophy similar to Carrel's own Lucretian notion that the material and the spiritual could and would ultimately be explained through the medium of some kind of subtle particle, and the configurations of those particles.

Carrel was strongly Darwinist in his thinking, but in its perverted form of Social Darwinism. The philosopher Herbert Spencer (1820–1903), inventor of the phrase, 'survival of the fittest', had applied the principles of Darwinism to society generally. His ideas had helped Victorians feel comfortable with the Biblical notion that the poor will always be with us. Like many Social Darwinists, Carrel was a firm believer in eugenics. The word was coined, and the doctrine first espoused by Darwin's cousin Francis Galton. The vogue for eugenics in America, in the form of mass-sterilization programmes, was currently inspiring the Nazis as they began their rise to power in Germany. Laws allowing the compulsory sterilization of mentally and physically disabled people were adopted in over 30 states in America in the 1930s. In 1934 similar laws were introduced in Germany. Hitler once said how impressed he was by America, a country that had managed to all but wipe out an entire race. 'An uninteresting, and one may almost say, a justly exterminated race,' said the reviewer of Hiawatha in the New York Times in 1855. By the end of the Second World War the National Socialist compulsory sterilization programme had overseen the sterilization of around 350,000 people. Some American states continued their sterilization programmes into the 1970s.

In 1935 Carrel published a controversial and best-selling book,

L'Homme, cet inconnu (Man, the Unknown). In it he argued that humankind would be best guided by a parliament of elite intellectuals, and by strong principles of eugenics. He believed in what he called 'a hereditary biological aristocracy'. He advocated the building of small euthanasia units where a poisonous gas was to be used to kill off all those who committed certain categories of crime: those who had betrayed public confidence, murderers, armed robbers, those who had robbed the poor, and those who had kidnapped children. In a preface to the German edition published in 1936, he wrote of the nascent practice of eugenics in Germany: 'The German government has taken energetic measures against the propagation of the defective, the mentally diseased, the criminal. The ideal solution would be the suppression of each of these individuals as soon as he has proven himself to be dangerous.' Carrel wrote that a nation might best be made strong not by protecting the weak but by developing the strong: 'only the elite makes the progress of the masses possible'. A friend wrote of Carrel that: 'He had no love for Nazism, Fascism, or Communism, but he knew that their ideologies gave those nations an ever-flowing source of energy.' It was because democracies had discarded faith and religion that they had become weak and inefficient, he believed. And yet Carrel also believed that every human being is 'a unique event in space and time'. If his philosophy was not directly anti-semitic, it was certainly, politically, dangerously naive. Carrel was himself strong evidence against his belief in a governing intellectual elite.

Lindbergh was an elitist too. He wrote in his diary that America had a future only if the best could be brought forward: 'If we must depend upon a complete mixture, if we must depend upon averaging together all the elements of which we are composed ... then I am doubtful of the future.'

🌍

Lindbergh's friend Henry Ford was both an anti-interventionist and profoundly anti-semitic. Ford had used his great wealth to publish a newspaper, *The Dearborn Independent*, that relentlessly attacked Jews as being at the root of all America's woes. Bad things did not just happen, they happened for a reason. Bad things happened because events were being manipulated and controlled by Jews, because Jews wanted power and control. Published between 1919 and 1927, *The Dearborn Independent* had a circulation second only to the *New York Daily News*, mainly because it was forced on all Ford workers, 75,000 of them at his 2,000-acre plant at River Rouge near Detroit. After a number of lawsuits following the publication of various anti-semitic articles, the paper closed. Ford had also republished the *Protocols of the Elders of Zion*, a book that claimed to be the minutes of meetings that took place in the late nineteenth century between a group of Jewish leaders intent on controlling the world's press and economies. It had first been published in Russia in 1903 and translated into many languages. Ford printed 500,000 copies and distributed them across America. *The Times* of London had exposed the document as fraudulent in 1921. After the Nazis came to power in 1933 the text was taught in German classrooms as if it were a historical document.

In 1938 Ford was awarded the Grand Cross of the German Eagle (Lindbergh had been given the lesser Order of the German Eagle). From 1937 the medal, at Adolf Hitler's instigation, was given to prominent diplomats and foreigners considered to be sympathetic to Nazism. Hitler was aware of Ford's writings, and admired Ford's role in the technological modernization of America.

Of course Lindbergh cannot be judged solely by the friends he kept. His views were usually more nuanced than Carrel's or Ford's, but he seems to have judged Carrel and Ford uncritically. In a diary entry dated 27 July 1940 he wrote of Henry Ford: 'I have the greatest admiration for this man; he has genius, understanding, fearlessness, optimism, humor, and a simple, open character that seldom survives success, especially great success. He combines firmness with consideration and kindness.' Ford 'will always remain one of the greatest men this country has produced'. He wrote that Ford's 'shadows serve to accentuate the heights'. In the fall of 1940 Ford tried to persuade Lindbergh to accompany him to Europe to negotiate a peace deal, presumably under the aegis of the American Ambassador to Britain, Joseph Kennedy, an isolationist and noted anti-semite.

In a broadcast made on 16 June 1941, clearly aimed at rallying support from America, Churchill said: 'The destiny of mankind is not decided by material computation. When great causes are on the move in the world, stirring all men's souls, drawing them from their firesides, casting aside comfort, wealth and the pursuit of happiness in response to impulses at once awe-striking and irresistible, we learn that we are spirits, not animals, and that something is going on in space and time, and beyond space and time, which, whether we like it or not, spells duty.'

On 7 December 1941 the Japanese bombed Pearl Harbor. Four battleships were sunk, other ships were sunk or damaged, 188 aircraft were destroyed; 2,402 people were killed and 1,282 wounded. Lindbergh had planned to give a speech on 12 December in which he would accuse Roosevelt of behaving like a dictator. He immediately cancelled the plan. He said in a statement that America must now retaliate with force. Lindbergh made no further public statements for several years.

The day after Pearl Harbor, Congress voted on whether or not to declare war on Japan. Roosevelt's Secretary of War urged the President to include Germany and Italy in the motion but canny Roosevelt decided to bide his time. There was just a single dissenter, Jeanette Rankin of Montana, who had also dissented in

1917. Everyone had been taken by surprise by the Japanese move, Hitler included. He had never intended to declare war on America. He had hoped that Japan would attack the Soviet Union. Even now he might have decided to side with America against Japan, but on 11 December Germany declared war on America, perhaps Hitler's single greatest blunder. History played into Roosevelt's hand. We were fortunate, as the historian Walter Ross observed, 'that Hitler wasn't always logical'. Congress voted again. America was now at war with Germany, Italy and Japan.

CHAPTER SEVEN

By 1941, after two further years spent developing the A-4, Hitler had now become convinced that the work of von Braun's team was of prime importance. Von Braun was summoned to the Wolf's Lair, Hitler's hideaway in East Prussia, to make a progress report. We do not know what von Braun said but after that August meeting Hitler pronounced that the war might be won decisively if tens of thousands of A-4s were to be produced. At this point the A-4 was far from ready to be mass-produced. It hadn't even been fully tested. Static tests of the rocket's engine that had taken place the previous November had been disastrous.

The first test flight of an A-4 didn't take place until March 1942, and it, too, was a decided failure. A second launch attempt took place in June. Albert Speer, who had recently been appointed Minister of Armaments, was in attendance: 'Wernher von Braun was beaming,' he wrote later. 'For my part, I was thunderstruck at this technical miracle, at its precision and at the way it seemed to abolish the laws of gravity...' And yet even this launch was not entirely successful. Control of the rocket was soon lost and it crashed into the sea 700 yards off the coast. Nor was the next test, in August, problem-free, though for the first time in history a machine broke the sound barrier. A number of scientists had predicted that pressure waves around the rocket would become so strong that as the rocket tried to push its way through the sound barrier it would break apart. In fact there is no physical barrier as such. The rocket did break apart but not for that reason.

With all these delays, it was feared that Speer's support might be lost, and that meant Hitler's support would be lost too. Worse, the Luftwaffe had a rival project at Werk West, the V-1, a jet-powered flying bomb, what the Allies would call the doodlebug or cherrystone. Dornberger's Werk Ost was in danger of losing out. Without the guarantee of a mass attack of thousands of rockets, Hitler was doubtful that the A-4 had a role to play in the war effort. He wasn't that excited about the V-1 either, and for the same reason. It was left to Speer to convince him that both projects were worthwhile and complemented each other.

An entirely successful launch of an A-4 finally took place on 3 October 1942. Carrying a one-ton (presumably dummy) warhead, the rocket reached a speed of 3,100 mph, arced 56 miles into the sky and came down under control, landing exactly where it was meant to land, confirmed by marker dye released into the water where the rocket hit the sea over 100 miles from the coast. That night in the officers' club, Dornberger told the crew that the spaceship had been born.

At that time, where the atmosphere ended and space began had not been universally agreed upon. The American definition put space at anywhere beyond 50 miles above the Earth's surface. By this definition Dornberger was correct. A later, international threshold, the Kármán line, defines outer space as anywhere 100 kms (61 miles) above the Earth's surface. It would not be long before an A-4 would break that barrier too.

On 11 December Heinrich Himmler, Chief of German Police and the main architect of the Holocaust, made his first visit to Peenemünde. Himmler had taken an early interest in von Braun, and would continue to do so. In the spring of 1940 von Braun had been invited/commanded by Himmler to join the SS. Von Braun asked the advice of his boss, Dornberger, who told him that the

SS had been trying to get a finger in the pie of the rocket business and that for the sake of the group's work he would have to join. He was given the rank of Untersturmführer, the equivalent of Lieutenant. He was promoted to Obersturmführer in November 1941 and to Hauptsturmführer (Captain) the month before Himmler's visit. Von Braun said that he was notified of the promotions by letter and never did anything deliberately to advance his career in the SS. In an attempt to impress Himmler during his visit, an A-4 was launched in his honour. The rocket crashed four seconds after it left the launch pad.

Even after a successful launch in April 1943, von Braun was still of the opinion that the A-4 was not ready for mass production. Dornberger thought von Braun was too slow and meticulous, and evidently so did Speer, who sent in Gerhard Degenkolb, an engineering manager, to take control of production. Degenkolb was a fanatical Nazi, dictatorial and bull-necked. He immediately saw that if Hitler's target was to be achieved the production process needed overhauling. Von Braun named Degenkolb and his cohorts Speer's muscle men. Dornberger resisted as best he could Degenkolb's attempts to wrest control of the project from Army Ordnance.

Recent victories by the Allied forces, at Stalingrad and in North Africa, put the A-4 back in the spotlight. Field Marshal Walther von Brauchitsch was put in overall command at Peenemünde and assigned 3,500 officers and enlisted men to boost the workforce and rocket production there. In June 1943 Hitler personally demanded that 1,000 A-4s be produced a month, ready for an attack on London planned to begin that October. In order to increase production, an assembly line was set up and 1,400 prisoners from concentration camps were brought in. They were mainly Russian POWs, employed against the terms of the Geneva Convention. They were imprisoned at Peenemünde in a small camp constructed in the basement of building F-1.

On 28 June Himmler made his second visit to Peenemünde. The following day two A-4s were launched. The first crashed into part of the site and destroyed several aircraft. The second launch was a success. Himmler promoted von Braun to Sturmbann-führer (Major). Dornberger had recently been made Generalmajor (Brigadier General).

On 7 July Dornberger and von Braun were commanded by Hitler to make a presentation that very same day at the Wolf's Lair. Von Braun brought along a film of a rocket launch he had commissioned from a professional film-maker. Hitler had never seen an A-4 in flight. The cameraman had employed a number of cinematic techniques. The film was in colour and reshot from various angles for dramatic effect. Von Braun spoke of the rocket with great enthusiasm. Dornberger told Hitler not to think of the ballistic missile as a wonder weapon. Hitler either wasn't listening or ignored Dornberger. 'A strange, fanatical light flared up in Hitler's eyes,' Dornberger later wrote. 'But what I want is annihilation – annihilating effect!' Hitler apologized twice for not previously believing in the work. He said that he had dreamed that their rockets would not work, but that he now revoked that dream. He ordered the production rate be increased to 1,800 rockets a month, with the threat of arrest if the target was criticized. He said that only German workers were to be used for fear of sabotage.

At the end of the meeting Hitler shook von Braun's hand and said, 'Professor, I would like to congratulate you on your success,' which was how von Braun learned that he been made a professor. Speer said that Hitler would often talk of von Braun after that, bringing him up during his increasingly maniacal monologues, comparing von Braun's invention with the achievements in their young years of Alexander the Great and Napoleon.

After Hitler had left, Speer introduced von Braun to the Nazi economist Hans Kehrl saying, 'Congratulate the youngest pro-fessor of the Third Reich.' Von Braun was so euphoric he forgot about the warning about talking only of the war and not of space, and began to speak animatedly to Kehrl about the future possibility of flying to the moon. When he got home Kehrl told his wife how at headquarters he had just met a madman.

To meet Hitler's production target, more labour was brought in from the camps. Degenkolb and Dornberger ignored Hitler's order that only German workers be used, and soon there would be 15 concentration camp workers to every German worker.

Increased activity at Peenemünde attracted the attention of the Allied forces. At 11pm on 17 August, after an evening of drinking, and soon after he had fallen asleep, von Braun was woken by the sound of sirens. Waves of British bombers had arrived. The attack was massive: 600 bombers dropped 1,800 tons of bombs, and 735 people were killed, half of them POWS and civilian forced labour from the camps. Walter Thiel, one of the designers of the A-4 rocket engine, was killed along with his wife and their children. Just days before, Thiel had dared to suggest that the A-4 could not be mass-produced and had recommended that its manufacture be abandoned. The Allied operation, named Hydra, had been pushed for by a young intelligence officer named Duncan Sandys, Winston Churchill's son-in-law and chairman of a committee for defence against German flying bombs and rockets. The damage to the plants, however, turned out not to be as bad as it first looked. The test stands survived intact. Only two months later, von Braun's operation was back in business.

The day after the raid, Hitler put Himmler in charge of overall A-4/V-2 production. An order went out that the manufacturing of V-2s was to be moved literally underground, to a series of storage tunnels that had been carved out of the Harz mountains, 250 miles to the south-west of Peenemünde. A new company was established named Mittelwerk, contracted to build 12,000 V-2s. The plan was to construct a truly vast underground factory, but by the end of the war the only operations up and running were the

V-2 plant, a small operation that mass-produced V-1s, and another that made Junker aircraft engines. Himmler placed SS Brigadier General Kammler in overall charge of the factory. Von Braun was told that he no longer reported to Dornberger – his protector and something of a father figure – but to Kammler directly.

Dornberger and von Braun both loathed Kammler on sight. Probably neither of them would have known at that point about Kammler's role supervising the construction of the Auschwitz gas chambers, but they had no trouble identifying a ruthless fanatic when they saw one. In turn, Kammler thought von Braun was 'too young and too childish ... for the job'.

At the end of August, labourers began to arrive from Buchenwald. For fear of sabotage, Jews were excluded, though some Jews slipped through the identification process. For the next three months the prisoners were set to work extending the tunnels using picks, drills and dynamite in an operation that continued 24 hours a day. Cathedral-like spaces were cut out of the granite mountains. Teetering wooden towers of scaffolding were constructed. The tunnels were thick with dust. Exhausted workers would regularly fall to their deaths: 'A cry, a thud, that is all, another man takes the victim's place.' When von Braun visited the site in October – he was largely based in Peenemünde – there were 4,000 prisoners working in the tunnels. When he came again in November there were 10,000.

For the first months the prisoners worked, ate and slept – on straw or bare rock – in the tunnels themselves. As the numbers grew, a sub-camp of Buchenwald named Mittelbau-Dora was built nearby. Three other local sub-camps would be constructed before the end of the war, at Nordhausen, Ellrich and Harzungen. There were no latrines in the tunnels, and no drinking water. Epidemics of dysentery, typhus and tuberculosis swept through the workforce from the start. Around 20 prisoners died every day, mainly from disease but also starvation, blastings that went wrong or the extreme cold. The workers were beaten with clubs and rubber-tipped copper cable. Some prisoners went mad from the noise. There were mass hangings, with workers often chosen at random, 12 at a time. The wire noose ensured death came slowly. The bodies might be left hanging for days or weeks. One doctor told Speer that he had seen Dante's *Inferno*.

At the end of 1943 the V-2 was still unreliable, often breaking up in the air. Of the 39 test firings of the missile that had taken place in the last two years, only 14 had been successful. During one launch, yet another test that attempted to fix the problem of the rocket exploding mid-flight, Dornberger and von Braun were closely following the trajectory of the rocket only to see it lose control and head straight in their direction. Dornberger threw himself to the ground, but von Braun, ever the scientist, held his ground in order to see what would happen. Dornberger said that afterwards von Braun was still standing there, but with a window frame around his neck. The account doesn't sound very plausible, but it was claimed that von Braun had been similarly stalwart during the Allied bombing raid of Peenemünde.

The production target for December had been 200 V-2s (the targets were forever changing), but only a handful made it through the entire production line, and all were deemed to be of too poor quality to be used and were sent back. A new goal was set: to mass-produce problem-free V-2s by the following April.

A portable crematorium was brought in to Dora in January 1944. Before that, the dead had been sent to be cremated at Buchenwald. Prisoners at Buchenwald were so horrified by the condition of the bodies that some committed suicide rather than be sent to Dora. Others went there to sabotage the factory. Some Dora workers managed to send out intelligence reports useful to the Allies.

On 21 February Himmler summoned von Braun into his presence. 'I trust you realize that your A-4 rocket has ceased to be an engineer's toy,' he told him, 'and that the German people are eagerly waiting for it.' He invited von Braun to call on his aid. Von Braun said he was confident in Dornberger and that it was technical problems not red tape that were holding things up. Apparently Himmler laughed and the conversation ended pleasantly.

A month later three SS officers knocked on von Braun's door at around two or three in the morning. He was instructed to accompany them. He was not being arrested, they said, but being taken into protective custody. Wernher von Braun's younger brother Magnus and Kurt Riedel (one of the founding members of the Berlin rocket group VfR) were also arrested.

During his period of captivity in a Gestapo cell von Braun turned 32. A car sounded its horn in the street. The driver got out and was allowed to bring in flowers and birthday gifts. One of von Braun's engineers was in the car waving.

At the end of March charges against him were read out and von Braun was formally interrogated. There was a fat file on him. One document referred to a story he had written as a child. Another document claimed that the reason he held a current pilot's license was because he planned to fly to England, taking with him plans of the A-4. Remarks he had made at parties that had taken place years earlier were read back to him. He had, apparently, and probably it was true, talked openly about how in the distant future a rocket would be developed and used to send mail between the United States and Europe. He was told that at a party just that month, he had talked about how the war was going badly, and had said that his main ambition was to create a spaceship. Von Braun said the interrogation was like a hallucination.

Dornberger came to the rescue, persuading Speer to intervene on von Braun's behalf. Dornberger arrived at the prison with a bottle of brandy and a signed document from the Führer's office granting von Braun conditional release for three months. Two months later Hitler was still grumbling to Speer about the trouble he had been put to, but the Führer told Speer that von Braun would be protected if he was that important to him. It seems likely that the arrest had been instigated by Himmler as a test of von Braun's value to the Führer. There are no surviving records of the arrest. The events exist as anecdotal accounts written later by Speer, Dornberger and von Braun.

On 13 June 1944 the first V-1 fell on London, damaging a railway bridge and a number of houses on Grove Road in Mile End. Six people were killed. The V-1 was technically a flying bomb – effectively an unmanned airplane, the world's first cruise missile – not a rocket.

On 22 June a V-2 was the first rocket to enter space proper, reaching an altitude of 109 miles.

On 29 August Hitler issued an order to begin the use of V-2s as soon as possible. Himmler told Kammler to accelerate the timetable for V-2 deployment. Just 10 days later, Kammler's rocket troops successfully launched V-2 rockets from The Hague towards first Paris and then London.

The V-2 was the world's first ballistic missile, meaning that it was initially guided into an arcing trajectory but then came to earth under gravity. It stood 46 feet tall and weighed 13 tons, of which 1 ton was explosive packed into a warhead. The rocket was powered by alcohol and liquid oxygen fuels delivered by pumps – driven in turn by a steam turbine – to a chamber where the fuels spontaneously ignited. The power of the controlled explosion delivered a thrust of 55,000 lbs, and accelerated the rocket to

3,500 mph. It traced a path up into space and back to Earth again which Thomas Pynchon, in his 1973 novel partly set in London during the period of the V-2 raids, called Gravity's Rainbow. The rocket was guided by an advanced gyroscope that relayed signals to the fins and to vanes in the exhaust. It had a range of 170 miles. A scientific reconstruction carried out in 2010 showed that a 1-ton V-2 warhead was capable of sending up 3,000 tons of earth, creating a crater over 60 feet wide and nearly 20 feet deep

The first V-2 fired that day exploded in Paris in the Porte d'Italie region of the recently liberated city, causing some damage. There were no casualties. Two more V-2s were launched in the direction of London.

6.43pm on a rainy, overcast autumn evening in London. Sapper Bernard Browning, on leave from the Royal Engineers, was walking down Staveley Road in Chiswick when, without warning, nearly 30,000 lbs of metal and explosive struck the ground nearby at three times the speed of sound. Browning would not have known what killed him. First came the simple presence of the rocket, only then the sound of the rocket arriving. From out of the upper atmosphere, catching up last of all, a sonic boom was heard all over London. There were two other fatalities: 63-year-old Mrs Ada Harrison, and three-year-old Rosemary Clarke (killed by the shock waves; there wasn't a mark on her body), and 17 people were seriously injured. Rows of houses were reduced to rubble. A few minutes later a second V-2 landed in Epping. There were no casualties. In central London, Duncan Sandys looked up and said, 'That's a rocket.' Only the day before he had assured Londoners that the city was safe: 'Except possibly for a last few shots the battle of London is over.'

Of the 2,500 rockets that were launched before the offensive came to an end on 17 March 1945, 500 fell on London. In England

2,742 people were killed and 6,467 seriously injured. The worst single attack had taken place at 12.26pm on 25 November 1944 at Woolworth's in New Cross in South London: 160 people were killed and another 108 seriously injured. Across Europe it is estimated that a further 5,000 people were killed. Around twice as many people died making the V-2s as were killed by them. Hitler had hoped to rain tens of thousands of V-2 missiles on England. In the event no more than 5,000 V-2s were manufactured. In the numbers game that is war, the attacks had done little more than briefly demoralize a country waiting to hear at any moment that Germany had capitulated. The V-2's role in the war was at an end, but its role in history was about to begin a new chapter.

An unexploded V-2 in Trafalgar Square, 1945

CHAPTER EIGHT

In protest against what he considered to be Roosevelt's warmongering, Charles Lindbergh had resigned his commission. He had briefly considered spending a year or two in contemplation, but now that America had entered the war, he regretted his hasty decision and realized that it was his duty to fight. He wrote to his friend General Hap Arnold, Chief of the Army Forces, offering his services. Arnold made the offer public. The *New York Times* was supportive, the rest of the press less so. Roosevelt personally intervened to prevent Lindbergh from re-enlisting. When, in 1942, the War Department allowed Lindbergh to work with Pan Am on their war projects, the White House was furious and put pressure on companies that had government contracts not to employ him. When Henry Ford offered Lindbergh a job in Detroit – where his factory, then the largest ever built, was making B-24C bombers at the rate of one an hour – even the Administration couldn't stand in Ford's way.

Lindbergh thought the B-24C was mediocre. After a number of test flights, he suggested some improvements. The Russian-born American plane designer Igor Sikorsky once told a woman at a dinner party that Lindbergh could not only fly any plane but could pinpoint and fix any design flaw on the drawing board. 'You mean to say that all of your test pilots can't do that,' the woman responded. 'None of them can,' Sikorsky told her. 'They can fly anything, and when they bring it down they can tell me how it handles, but they don't know *why* it behaves a certain way. Charles will know where the mistake is and have a suggestion about correcting it.' After some months working with the Ford Company, Lindbergh renegotiated the deal, deciding that he was unsuited to the work, and reduced his connection to that of part-time adviser.

For two weeks in September 1942 Lindbergh offered himself up as a guinea pig to Dr Walter M. Boothby, who was studying the effects of high-altitude flying at the Aeromedical Laboratory in Minnesota. New planes were being designed that could fly to 40,000 feet, but no one knew how pilots and crew would cope quickly descending from those altitudes, or what happened to reaction times at low air pressures and low oxygen levels. Several times he was tested to unconsciousness. Lindbergh used the results of the research to test and redesign existing procedures used during emergency parachute jumps from high altitudes. Lindbergh also taught young fighter pilots how they could reach higher altitudes, and engaged with them in mock aerial battles, but what he really wanted was to see active service. In January 1944 he was asked by United Aircraft to carry out research in the South Pacific on two planes that he had been testing for the company: the F-4U Corsair and Lockheed P-38 fighters, both of which had been adapted to carry bombs. By April he was in New Guinea, officially as a civilian technician and observer. When he wrote to Colonel Charles MacDonald, Commander of the 475th Fighter Group of the Fifth Air Force based in the Far East, for permission to go on a bombing raid, MacDonald's deputy said: 'My God! He shouldn't go on a combat mission. When did he fly the Atlantic?' MacDonald said, 'I'd like to see how the old boy does.' The old boy was 42. Lindbergh was not given formal permission, but an unspoken arrangement was reached that the military would turn a blind eye. Coincidentally, somewhere over the Mediterranean, Saint-Exupéry was also flying a P-38, and at

43 was by far the oldest pilot in the service.

It soon became apparent that when Lindbergh flew the P-38 his fuel consumption was much lower than that of his colleagues, pilots around half his age. News of his achievement reached the ear of General MacArthur himself, who asked to see him. Mac-Arthur said it would be a gift from heaven if the P-38 could be flown more efficiently. Lower fuel consumption meant longer flying time. Combat missions could be extended. Lindbergh was told to teach his fellow pilots how to fly as he flew. Because of Lindbergh, six- to eight-hour missions became ten-hour missions. The enemy was surprised deeper into its territory. 'Lindbergh was indefatigable,' MacDonald wrote. 'He flew more missions than was normally expected of a regular combat pilot.'

On 8 September 1944, Lindbergh dropped a 2,000lb bomb over Wotje island, one of the Marshall islands. In his diary he wrote, 'so far as we know, this is the first time a 2,000lb bomb has been dropped by a fighter'. Four days later he broke his own record and from an F-4U dropped a 2,000lb and two 1,000lb bombs. From the air he saw one of his 2,000lb bombs wipe out a Japanese emplacement on the ground: 'One moment the earth below me lay motionless; the next, a column of earth and dust appeared like magic in the air. On the razor-edge of time, an unknown number of human lives and bodies vanished by my pressing the red button on a control stick.' Such senseless death challenged his adherence to the law of natural selection. Evolution had used death as a way of advancing the fittest, yet highly trained airmen were among the first to die in war: 'No selection resulted from man's atomizing of his cities.' On another mission, flying over the coastline of Japan, he saw a naked man wandering on the beach. Lindbergh had been ordered to kill anyone he saw on the ground. As he drew closer the man did not speed up: 'I should never have forgiven myself if I had shot him – naked, courageous, defenseless, yet so unmistakably a man.'

On 12 April 1945, after 4,422 days in office, and just a few weeks shy of the formal end of the world war, FDR died. He remains the only American president to have served three full terms, and the only one to have been elected to serve a fourth term. With no objections now put in his way from the White House, on 11 May, three days after VE Day, Lindbergh was on a navy transport plane to Germany, as consultant to the United Aircraft Corporation and as part of a navy technical mission to learn as much as possible about German developments in high-speed aircraft.

Lindbergh was invited to fly the plane but mostly he slept: 'One might as well sleep,' he wrote in his journal, 'for the modern military plane is usually uninteresting from the passenger's standpoint – high above the earth – often above the clouds, so that no details can be seen (even if bucket seats and badly placed windows didn't make it so difficult to see anyway). Every year, transport planes seem to get more like subway trains.' Among the places Lindbergh visited as he made his way across Germany was Hitler's 'fabled mountain headquarter' near Berchtesgaden. He thought the view one of the most beautiful he had ever seen. During the afternoon of Sunday 10 June, Lindbergh arrived at Mittelwerk to inspect the underground factory where von Braun's V-2 rockets had been manufactured.

In the chaos of the last weeks and days of the war, thousands of concentration camp prisoners had suddenly been dumped at the Mittelbau-Dora camp. On the nights of 3 and 4 April an already hellish situation was made worse when Allied planes bombed the camp, which from the air had been mistaken for a munitions plant. A massive evacuation of the camps around Mittelwerk – Dora, Nordhausen and the two smaller camps, Ellrich and

Harzungen – began the following day. Several thousand prisoners were sent on death marches. Many thousands more were packed onto trains and sent to Bergen-Belsen. There were mass hangings, and mass shootings. Paul Tregman, a survivor of Ellrich, described one such train journey to Bergen-Belsen that took place early that April:

> For six days our train, consisting of forty-some cars, dragged nearly 4,000 prisoners along a route which could normally be covered within seven hours. One hundred and thirty persons were crammed into each car, pressed together like sardines in a can and reeking like rotten garbage. One SS man and one Kapo were assigned to each car to keep these pigsties in order. A tiny window heavily barred, let in a miserly bit of fresh air. No one was allowed to get off the train at any of the stops, and there was no place for the prisoners to relieve themselves. If a passenger finally lost control of his bladder or bowels, the Nazi overseers would beat him till he bled and toss him from the moving train, and the SS man would fire a bullet after him to make sure he died.

By the end of the war it is estimated that out of some 60,000 prisoners who were held at the camps around Mittelwerk, 25,000 perished. There may have been many more whose lives and deaths were not recorded. More than half of the 25,000 were killed en masse in the last days of the war by SS officers aware that the American military was close by and that this was their last opportunity.

By 10 April most of the camp guards had left the region. At 2.30pm on that day some of the remaining prisoners at Camp Dora saw a solitary figure, an American soldier, exchanging fire with one of the last fleeing guards. At 4.00pm the American soldier, Private John Galione, arrived at the camp gates.

Five days earlier John Galione had complained to his sergeant, Leonard Puryear, that there was a terrible smell, and that he and some of his fellow soldiers wanted to put together a search party to go and investigate. There had been rumours that there was a labour camp in the area, and perhaps that was what they were smelling. Sergeant Puryear refused permission; the risk of ambush was too great. German soldiers were scattered everywhere across the region.

That night, at around 9pm, Galione decided to leave the encampment on his own. On a hunch, he decided to start walking along nearby railroad tracks. He thought he had smelled that same smell before sometimes when they were passing German trains. He left word that he would return soon, and that if their sergeant noticed he was missing to tell him that he wasn't deserting but going on a mission. He thought he would be back by 12.30am at the latest, but he just carried on walking; first for hours, and then for days. He had to drag himself forward; one leg had been shot at some weeks earlier and had not yet fully healed. He said he never slept, for fear of ambush; he rested every so often, leaning against a tree for an hour, and then returned to the tracks. He said that energy came to him mysteriously. When he felt most like giving up it was as if hands came out of nowhere and pressed him on his lower back, literally propelling him forward. After more than four days Galione arrived at the mouth of a cave, where he found a railway truck filled with dead bodies. He was spotted by a German officer and they began to exchange fire, but the officer was clearly on his way somewhere and soon disappeared. Galione had walked 110 miles due west from his company's base in Lippstadt and had arrived at Mittelwerk.

An hour or so later he made a horrifying discovery. Pressed against the wire fence of what was clearly some kind of prison were dozens of emaciated bodies, scarcely breathing, and behind them piles of dead bodies. The gates to the encampment were padlocked shut. The sun was setting. Again fearing ambush, he left the camp intending to return in the morning with help.

By good fortune Galione happened upon two soldiers from his own 104th Infantry Division 'Timberwolves' (motto 'Nothing in hell can stop the Timberwolves'). They were working on a broken-down jeep, which Galione offered to fix. The three returned to the camp and forced their way in. On the other side of the gates they saw what Galione later estimated were a thousand dead bodies, and hundreds of people clearly barely alive, some taking no more than two breaths a minute. Galione said that some prisoners were so emaciated he could see their spinal columns through their stomach muscles. The three GIs decided that their best course would be to drive back to Galione's detachment at Lippstadt and get help. It took three hours to cover the distance Galione had taken four days to walk.

Sergeant Puryear radioed the Third Armored Division who were stationed closer to the camp and told them that something unusual was going on nearby and that they should investigate. Out of guilt that he had not acted sooner, Puryear asked Galione to keep the details of the discovery quiet. Galione agreed and kept his word for decades. In the 1960s his daughter – sensing that he had a story to tell – pressed her father to tell it. He said he would one day, but not yet. He finally told her in the 1990s. Most accounts of the relief of the camps around Mittelwerk begin with the mysterious message relayed to the Third Armored Division. John Galione's account has not been officially sanctioned by the US Army but has been confirmed by a number of Dora survivors, and in an affidavit signed by Leonard Puryear.

On their way to camp Dora, Combat Command B of the Third Armored Division stumbled by accident on the camp at Nordhausen. Some of the most shocking images and footage of the

Corpses at the Nordhausen concentration camp

Local civilians were forced to dig a long trench grave to
bury the many dead from the camps around Mittelwerk

conditions inside Nazi concentration camps were taken by US
soldiers soon after they first entered Nordhausen on 11 April 1945.
One soldier, Sergeant Ragene Ferris, described how he had come
across a crater made after the Allied bombing of 5 April filled with
bodies. At the bottom of the pile were three people who had been
struggling for five or six days to get out, 'but the weight of other

bodies on them had been too much for their starved emaciated bodies'.

Back at Camp Dora, Galione set out in search of food, forcing a local German woman at gunpoint to slaughter two pigs. Half a century later he told his daughter how, again, some superhuman effort had been required of him to drag each carcass back to the camp.

It was not long before American soldiers found their way into the network of tunnels at Mittelwerk that branched deep into the Harz mountains. One officer reported that being inside the tunnels was 'like being in a magician's cave'. Work had stopped there only the day before; eerily, the electric power was still on, the ventilation system still humming. When Galione had first entered camp Dora the day before, the crematorium there was still smoking.

The sheer scale of the V-2 operation came as a shock. Special Mission V-2 was quickly orchestrated from Paris by Colonel Holger Nelson Toftoy, Chief of the Army Ordnance Technical Intelligence, to recover as many V-2s as possible. A hundred complete V-2s were retrieved from the tunnels and rapidly broken down into their constituent parts. In the next few months, dozens of trains took the components from Mittelwerk to Antwerp, where they were boarded onto 16 Liberty ships destined first for New Orleans.

By the time Lindbergh arrived at Mittelwerk in June, the operation was well underway. There were hundreds of V-2 parts scattered around the grounds of the vast encampment: 'nose cones, cylindrical bodies, and big Duralumin fuel tanks ... A number of tail sections, shining and finned, were standing on end like a village of Indian teepees.' Some of the discarded parts had been made into bizarre makeshift shelters. Passing through the still-stinking grounds Lindbergh entered the tunnels, even now brightly lit. In otherwise empty offices he saw discarded identity cards strewn on the ground, thousands of them. In the heart of the mountain he came upon what looked 'like a giant grub', an entire gleaming V-2 rocket, in the process of being dismantled by Allied experts. The evidence of past efficiency and present desolation was jolting: 'The Nordhausen establishment seemed far from the earth I knew, as though I had ridden one of the missiles it produced and stepped out on a strange and terrible planet.' In his head rang the words of a man who had directed them to the tunnels. He had told them that for the workers the only way out had been as smoke.

A 17-year-old Polish survivor, still wearing a striped concentration-camp uniform, showed Lindbergh around the site. The boy looked down and Lindbergh followed his gaze. They were standing at the edge of a pit 8 feet long and 6 feet wide filled to overflowing with ashes and chips of bone. There were two more pits nearby. Lindbergh guessed that the pits might be 6 feet deep. 'Of course, I knew these things were going on,' he wrote in his journal the same day, 'but it is one thing to have the intellectual knowledge, even to look at photographs someone else has taken, and quite another to stand on the scene yourself, seeing, hearing, feeling with your own senses.' Tellingly, he now writes about his former belief in American exceptionalism in the historic past tense – 'I had considered my civilization as everlasting. It was too scientific, too intelligent, to break down as earlier civilizations had' – before moving into the present tense to describe what he witnessed at Nordhausen – 'This, I realize, is not a thing confined to any nation or to any people ... What is barbaric on one side of the earth is still barbaric on the other ... It is ... men of all nations to whom this war has brought shame and degradation.'

At the end of 1944 Wernher von Braun told a colleague that he had already packed his bags and that he intended to offer his services to America: 'And then I will build my space rocket.' On the last day of January 1945 General Kammler gave orders to begin evacuating Peenemünde. The Soviet army was advancing west across Prussia and Kammler realized that it might be expedient to have a bargaining chip. He rounded up 500 scientists and engineers, von Braun prize among them, and put them on a well-appointed train to Nordhausen, 250 miles to the south. How different were the trains crisscrossing Germany in those last weeks of the war! Kammler placed SS Major Kummer in command with orders that he was to shoot his charges rather than let them fall into enemy hands. The party arrived in Nordhausen in March.

On 17 March von Braun was on his way by car in a last attempt to try and raise more money for V-2 production. En route his driver fell asleep at the wheel while doing 60 mph. They could both have been killed. As it was, von Braun suffered multiple fractures to an arm and shoulder. The fracture knitted badly. In great pain von Braun returned to Nordhausen.

On 19 March Hitler issued his so-called Nero Decree, which ordered the destruction of everything that might be of value to the Allied forces. Von Braun was ordered to destroy all the records and blueprints of the Aggregat rockets. Instead he hid 14 tons of documentation in a mine and had the entrance dynamited shut. Von Braun, too, had realized that a bargaining chip might come in useful later.

On 1 April the 500 Peenemünders, along with 100 SS officers, were put on another train and sent to Oberammergau in the Bavarian Alps, where, supposedly, the Nazis were going to make a last stand. Oberammergau is most famous as the site on which an anti-semitic passion play has been performed since 1634. Henry Ford saw the play performed in 1930.

Once they had arrived it was clear that something needed to be done about von Braun's arm. He was sent to a hospital – the nearest one was 50 miles away – where his arm was re-broken, without anaesthetic. After two weeks' recuperation he returned to the group's hideaway with his arm in a cast. Von Braun was billeted with his younger brother Magnus and Dornberger at a ski resort in Oberjoch, not far from Oberammergau. Major Kummer had been persuaded that they would be less of a sitting target if the group was split up and distributed about the region. In fact von Braun was less worried about the enemy than he was about the unpredictable Kammler, who had disappeared. His return was feared each day. If Kammler thought the engineers might be about to fall into enemy hands there was no knowing what he'd do.

By 24 April Soviet forces had surrounded Berlin. In the next few weeks 325,000 Berliners would be killed. It looked as if the German engineers should prepare themselves – if they were lucky – for a Russian future.

Von Braun was listening to Bruckner's Seventh Symphony on the wireless when the performance was interrupted to announce the death of the Führer. 'Hitler was dead,' von Braun later wrote, 'and the hotel service was excellent.' Dornberger persuaded their SS captors that their best bet now was to burn their uniforms and ID cards and offer themselves to the Allied forces as POWs. News came through that there were American soldiers close by. Magnus, who spoke the best English of any of them, was sent off on a bicycle to search them out.

The 44th Infantry Division of the American Seventh Army had arrived in the region just the day before. They were as keen to find von Braun's party as von Braun's party were to find them. When Magnus came upon them he called out: 'My name is Magnus von Braun. My brother invented the V-2. We want to surrender.' He had to spend half an hour trying to persuade disbelieving army personnel that the V-2 engineers were up in the mountains nearby. He was given a safe conduct pass and told to bring back evidence.

Dornberger and von Braun put together an advance party of around ten. Von Braun said later that they hadn't known what to

First Lieutenant Charles Stewart, who had granted Magnus safe
conduct passes; Herbert Axsted, Dornberger's chief of staff;
Dieter Huzel, Wernher von Braun's assistant; Wernher von
Braun; Magnus von Braun; Hans Lindenberg, a V-2 engineer

expect when they arrived. In the event 'they immediately fried us
some eggs'. One member of the army division later said of von
Braun that he treated the US soldiers 'with the affable condescen-
sion of a visiting congressman'. He posed for endless photographs
with GIs: 'He beamed, shook hands, pointed at medals and otherwise conducted himself as a celebrity rather than a prisoner.'
He apparently boasted that if they had had two more years the
V-2 would have won the war for Hitler. One GI remarked: 'If we
hadn't caught the biggest scientist in the Third Reich, we had
certainly caught the biggest liar.' During that night a member of
the kitchen staff, a Pole, was intercepted as he attempted to shoot
the sleeping party of German rocket engineers.

Just a week after Germany had formally surrendered, von
Braun wrote an eight-page report for his interrogators translated
as 'Survey of Development of Liquid Rockets in Germany and
their Future Prospects'. He set out his dream of space exploration.
He described rockets that could travel between Europe and America in 40 minutes. He wrote of a future in which the 'whole of

the Earth's surface could be continuously observed'. He imagined how we might be able to control the weather (a presumed ability often associated with the deranged). Sunlight reflected from a giant mirror orbiting in space could be beamed down to where it was needed on Earth. He envisaged travel to the moon. He also showed himself to be politically astute (or was it opportunistic?): 'We are convinced that a complete mastery of the art of rockets will change conditions in the world in much the same way as did the mastery of aeronautics.' Von Braun was convinced that war with Russia was inevitable, and that whoever first mastered rocketry would gain the upper hand. He seized this first, early opportunity to propagandize on behalf of space travel: 'When the art of rockets is developed further, it will be possible to go to other planets, first of all to the moon. The scientific importance of such trips is obvious.' And yet it would not be obvious to everyone; in the immediate years that followed the war von Braun would have to spell it out repeatedly.

Colonel Toftoy was the man in charge of deciding who, out of the thousands of German scientists and engineers interned in Germany, should be cleared to start a new life in America. Von Braun was top of the list and was held under armed guard to keep him out of Soviet hands. The British asked that von Braun and Dornberger be handed into their custody, they said for a few days. The pair were driven across London and shown the devastation their missiles had caused. It took some pressure from the US War Department before the British military was willing to return von Braun into American custody. The British held on to General Dornberger for two more years, not because they wanted his technical knowledge – which was extensive – but because they wished to see him tried at Nuremberg as a war criminal. They had really wanted SS General Kammler, but he was nowhere to be found. Dornberger said he had heard that he died in the last days of the war in Prague, killed at his own request by a fellow SS officer, but neither where nor when he died is known for certain. In 1947, as it became clear that the case against Dornberger was weak, he was released. Dornberger then made his way to the States.

Von Braun had attempted to argue that all of the group of 500 Peenemünders should be given clearance – he always tried to push his luck. He was told he could choose 100, and in the end 117 Peenemünders made the list as part of Operation Overcast, later renamed Operation Paperclip because a paperclip was attached to the file of those selected.

Von Braun and six others prominent rocket engineers were the first of the 117 to arrive in America, landing at Fort Strong in Boston Harbour on 20 September 1945 from where they were moved to Fort Bliss, an army base near El Paso, Texas. By February 1946 most of their colleagues had arrived. The men were designated DASE, employees of the Department of the Army Special Employees. Only towards the end of that year was their presence in America made public. Einstein wrote to President Truman: 'We hold these individuals to be potentially dangerous carriers of racial and religious hatred.'

By the time von Braun arrived in America, Goddard had been dead a month. When peace came, Harry Guggenheim agreed to start funding Goddard again but Goddard died in August of throat cancer. He lived just long enough to inspect one of the first V-2s to arrive in America from Mittelwerk; and just long enough to complain, one last time, that his ideas had been stolen. The V-2, he said, was clearly based on his own designs. The National Geographic Society in Washington DC issued a statement in support of Goddard's allegation. Lindbergh told the press that he agreed and that the V-2 had plainly infringed American patents. He said that America ignored its visionary geniuses. The Germans had indeed been in touch with Goddard before the war, and had used whatever they could find from published papers, but you only had to look at the thing to see that the V-2 was technically way ahead of anything America could then achieve. V-2s had soared to 60 miles or more above the Earth's surface and broken the sound barrier: Goddard's rockets barely, and rarely, reached altitudes of 2 miles.

Goddard's last rocket launch before America entered the war had used a turbo-pump and possessed a more powerful engine than any he had used before, but the rocket nevertheless crashed soon after takeoff. The pumps worked well, but it was too little too late. During the war years, with Lindbergh away supporting the war effort, Goddard's work was curtailed. Both Guggenheim and Lindbergh attempted to interest the military in Goddard's rockets, but with little success. One meeting happened to coincide with the second day of the evacuation from Dunkirk; military minds were elsewhere, and anyway few in the military thought it likely that missiles would be used in this war. The navy offered Goddard a crumb: he was invited to work on a solid-fuelled jet engine they were developing, the JATO (jet-assisted take off) engine. It would enable a heavily loaded aircraft to take off from a short runway, like 'hitching an eagle to a plow' as one commentator put it. It was basically an adaptation of Goddard's pump rocket from a vertical to a horizontal plane. Goddard worked on refining the engine until the end of the war. It was almost perfected when the navy switched to a different technology.

Goddard was still defending his territory less than a year before his death. On 19 June 1944 Edmund Wilson gave Willy Ley's book *Rockets* a favourable review. Goddard wrote to the paper claiming that the Germans, as far as rocket science was concerned, had not been first in anything. Three days later the first V-1s struck London, and soon after the first V-2s.

In the report Lindbergh wrote for the navy on his return from Mittelwerk, he implied that but for some blunders made by Nazi High Command, the V-2 would surely have won Germany the war. This was nonsense. The Germans had banked on the wrong weapon, as Churchill perceptively pointed out. If they had encouraged Heisenberg to develop the atom bomb, history might have been rewritten. Even if von Braun had been able to perfect his A-9 winged version of the V-2 and his A-10, a souped-up version of the A-9, it is doubtful if Germany could ever have got the upper hand. The A-9 would have put New York within striking distance when fired from France, but the force of the atom bomb spoke for itself. So devastating was its effect when used by America at the end of the Second World War that it has never been deployed in war since.

The British had been promised 50 of the 100 recovered V-2 rockets but in the event all of them were shipped to America. Nevertheless, on 22 June 1945 General Eisenhower agreed that a British operation should attempt, right away, to build and fire some V-2s. The charmingly named Operation Backfire came into being. With most of the salvageable parts en route to America, the group had to scrabble about to find what they needed, but by September German engineers had eight V-2s in near-readiness for firing from the launch site at Cuxhaven in north Germany. Early in October, and after a couple of stalled attempts the day before, the first launch went without a hitch. The Germans and their British guards were jubilant. Speeches were made on both sides. A second launch two days later was also a success. A third launch – named Operation Clitterhouse, and set for 15 October – was to be for the benefit of observers from America, Russia, France, the UK, and the press. By this time von Braun had arrived in America, but among those gathered to witness what turned out to be yet another flawless flight were Russian rocket designers Sergei Korolev and Valentin Glushko.

CHAPTER NINE

The enthusiasm for space that swept through Russia in the 1920s was unlike that in other parts of the world in that it had a curious quasi-religious aspect. Cosmism and futurism, two Russian ideologies of the early twentieth century, underpinned Russian thinking about space and space travel. One of the leading proponents of both philosophies was Nikolai Fyodorov (1829–1903), a Russian Orthodox librarian and writer. Cosmism was a peculiarly Russian swirl of religion and philosophy mixed with up with ethics, theories of evolution and early rocket science; futurists were obsessed with machines and the idea of speed. In his *Outline of the Image of a Universal Task of Resurrection* (1902) Fyodorov described how humankind, because it is perfectible, will eventually overcome death in a final and universal resurrection. He wrote that it is the duty of humanity to strive to achieve immortality for all generations past, present and future. The first of the founding fathers of rocketry, Konstantin Tsiolkovsky (1857–1935) was heavily influenced by Fyodorov. Tsiolkovsky believed that the resurrection would occur in outer space. In one of his more esoteric articles, he imagined colonies of humans living in space free of all worry and forever. There would be no more wars, everyone would be happy in a society 'guided by geniuses'. He believed that life – because, according to the futurists, life itself is a form of motion – must be widespread across the universe, indeed that

Konstantin Tsiolkovsky

the whole universe was alive: 'I am not only a materialist, but also a pan-psychist, recognizing the sensitivity of the entire universe.' And because all physical and chemical processes are universal, he considered human beings to be more intimately connect-ed to the cosmos than to the Earth. He said the ultimate intangible universal law was happiness, and the ultimate tangible universal law was gravity.

Tsiolkovsky lived for most of his life as a recluse in Kaluga, a town some 120 miles south-west of Moscow. He was the fifth of 13 children, caught scarlet fever at the age of 10 or 11 and for the rest of his life was almost totally deaf. Because no school would admit him, he stayed at home and, as Goddard had, taught himself. Like all true rocketeers, it was reading Jules Verne that first fired in him an early obsession with the possibility of space travel. He was sent to Moscow when he was 16. He spent much of his time at the library at which Fyodorov was librarian. Fyodorov gave away much of the money he earned. He tried to give some to Tsiolkovsky, but without success. Tsiolkovsky lived on bread and water, his hair uncut, his clothes eaten away. He spent what money he had on equipment or chemicals. Whenever he wasn't in the library reading or writing, he was back in his room conducting experiments. As a young man, he had made a centrifuge and subjected a cockroach to 300 times the force of gravity, and a hen to 10 times. Both species apparently endured the experiment without harm. Tsiolkovsky was an avid reader of Tolstoy, who used the same library and was a friend of Fyodorov's. Like Tolstoy, Tsiolkovsky was a vegetarian. Like Tolstoy, he learned to ride a bicycle late in life. He studied the gospels and decided that 'of all the moral teachers Christ was the greatest of all'. He admitted that he had married without love, though the marriage lasted 55 years and there were seven children. Two of his four sons committed suicide; the other two died before he did. He said that he had placed the welfare of his family and associates on the lowest plain, with his own on the highest: 'I was often vexed and perhaps as a result this made life difficult and nerve-wracking for those around me.' Today he would likely be diagnosed, as might Lindbergh too, with Asperger's.

During his lifetime Tsiolkovsky wrote over 400 scientific papers. He remains best known for the paper written in 1903 on the mathematics of space flight. Ninety or so papers were about space travel. He wrote six science-fiction novels. He came up with designs, in words and drawings, for the multistage rocket, a space station, and spacesuits. He conceived the air-lock as the means of moving from a spaceship to outer space. He designed a spaceship and for the rest of his life could see the craft in his dreams: 'I imagined myself inside it rising high.' He built models of over a hundred aerostats (a class of airship) but for lack of funding never a full-scale prototype.

He never held an academic position nor received any funding (except once, briefly, early on from the Russian Academy of Science). He earned his living as a schoolteacher. After the 1917 Revolution, as he was entering his sixties, Tsiolkovsky was arrested by the new secret police on charges of treason and imprisoned for three weeks in the Lubyanka in Moscow.

In 1924, the Society for the Study of Interplanetary Communication – made up of 25 members of the N. E. Zhukovsky Air Force Union Academy – announced that a meeting was to be held to debate the future of space flight. It would be the first meeting of its kind anywhere in the world. Tickets sold out immediately. Tsiolkovsky declined to attend, but a paper he had written on rocket design was read out. The meeting was dominated by discussion of a curious rumour that had reached the city: that Robert Goddard was going to launch a moon rocket on American Independence Day, in less than three months' time. Goddard would never know that, during the summer of 1924, Moscow had been gripped by Goddard fever. The debate became so overheated that the militia had to be called out. When months passed with no news of the launch, a special meeting of the society – advertised as 'The Truth about the Dispatching of Professor Goddard's Projectile to the Moon' – was announced for October.

In 1926 Tsiolkovsky designed a launch pad for a two-stage rocket. In 1927 the Moscow Association of Inventors and Designers celebrated Tsiolkovsky's seventieth birthday, putting on an exhibit that celebrated his work, as well as that of Goddard and Oberth. The show attracted thousands of visitors.

A military facility dedicated to rocket research had been set up in Russia as early as 1921. By 1929 it was known as the Gas Dynamics Laboratory (GDL) and was carrying out pioneering work in rocket science under the leadership of Valentin Glushko, who was to become one of Russia's chief designers of rocket engines. Glushko had, inevitably, been inspired by the novels of Jules Verne. He had also been inspired by Konstantin Tsiolkovsky. Aged 15, he wrote to Tsiolkovsky, telling the reclusive seer that he wanted to devote his life to the 'great cause' of rocket science.

The Soviet's first rocket club, GIRD (Gruppa Izucheniya Reaktivnogo Dvizheniya; Group for the Study of Reactive Motion), was founded in 1931 with the aim of building a jet engine and a rocket engine. The members greeted each other with the rallying cry: 'Onwards to Mars'. Club motto: 'We were the first to penetrate space.' Aged 74, Tsiolkovsky corresponded with the club about their jet engine plans. England and Germany had developed functional jet planes by the end of the 1920s; Russia was lagging behind.

GIRD was folded into the GDL in 1933 and renamed the RNII (Reactive Scientific Research Institute). On 17 August 1933, the group launched the first Soviet liquid-fuelled rocket. It reached an altitude of 1,300 feet. The Russian rocket programme lost its momentum, however, during Stalin's Great Purge of 1937–8. Millions of Russians were labelled Enemies of the People and either executed or sent to the Gulag, the collective name of hundreds, perhaps thousands, of prison camps spread across an archipelago that stretched over 13 time zones. No one has been able accurately to calculate how many died in the camps during their 20-odd years of existence. Estimates vary from less than 2 million to much more than 10 million.

There were around a hundred scientists working at the well-funded RNII in 1937. Eight rocket scientists, including the director and deputy director, were shot during the next two years, many more were sent to the Gulags. Valentin Glushko was arrested on 23 March 1938 and charged with sabotage and being a counter-revolutionary, the usual vague formulation. Under torture he denounced his friend and colleague Sergei Korolev, who had been one of the founding members of GIRD. Glushko was sentenced to eight years in the Gulag. Korolev was arrested on 27 June. When his wife, Ksenia, returned to their sixth-floor apartment at around 9pm that night, she saw two members of the secret police, the NKVD, standing in the entrance. 'Well it's obvious they have come for me,' her husband said. To pass the time they sat and listened to a record of folk songs that he had bought that day, on one side 'The Blizzard' and on the other, 'In the Field Stood a Birch Tree'. They listened to the songs repeatedly. At 11.30pm there was a loud knock on the door.

Two days later Korolev signed a confession. His subsequent trial lasted one minute. He was sentenced to 10 years' imprisonment in the Gulag, with a further five years' loss of civil rights, and confiscation of his property. He was sent to the notorious

Kolyma mines in Siberia, infamous even among the many Gulag camps. A third of the inmates died each year. His mother wrote to Stalin, without success, pleading for his release. Ksenia remained silent since – according to the perverted logic of the time – a wife was more likely to be arrested than a venerated mother. In prison Korolev was regularly beaten, lost teeth, had his jaw broken. His face was permanently scarred. His day started at 4am in temperatures 60 degrees below. Within months he 'could barely walk or talk'. He suffered from scurvy, malnutrition and frostbite. Towards the end of 1939, he was ordered to leave the Gulag and make his way to Moscow. He missed a boat that was to take him part of the way and had to wait all winter before another arrived. At Central Design Bureau 29 in Moscow he was set to work on bomber design even though he was officially still a prisoner.

Korolev was finally released in June 1944 (though the charges against him would not be formally dropped until 1957). It was only now that Korolev saw his wife again for the first time since he had been arrested. She received him coolly. Why had she never visited him? Glushko's wife had visited her husband. Maybe she had heard the stories about Korolev and Glushko's sister-in-law. It was rumoured that they had had an affair years earlier, before his imprisonment, but there again the rumours may have been unfounded. More likely, six years apart had simply taken its toll. Even more distressing to Korolev, Ksenia wouldn't let him see their daughter Natasha.

A little over a year later, Korolev was ordered to attend a meeting of the People's Commissariat of Armaments in Moscow. He was given the rank of Lieutenant-Colonel in the Red Army, and sent to East Berlin to research German missiles. The V-2 generated 25 tons of thrust; Soviet rockets of the time couldn't even generate 1 ton. It was Stalin's personal wish that Korolev should get to the bottom of the V-2's efficiency. To that end, the Institute Nordhausen was created in East Germany and Korolev was appointed deputy director and chief engineer. He was to be aided by Valentin Glushko, who by then had also been released.

In October 1945 Korolev and Glushko were in Cuxhaven, north Germany to watch the Allied forces' launch of a V-2. The two men had once been good friends, but now they were barely cordial, each imagining the other had betrayed him. They were men of very different temperaments. Korolev had a fierce temper. He was always casually dressed. He would rarely be seen out of his black leather jacket. Glushko was always immaculately dressed, more handsome. He never raised his voice, but bore grudges.

The two were joined at Nordhausen by Helmut Gröttrop, one of von Braun's engineers. Gröttrop had been earmarked by the Americans but he decided to stay behind and side with the Soviets.

Even though the Americans had had first pick of the engineers from Peenemünde, the Soviets still managed to round up some 5,000 German engineers of their own, a number of whom were moved to a small facility outside Moscow in October 1946. Korolev had by this time come to the conclusion that the V-2 was already obsolete, but Stalin wanted a copy of the V-2 and Korolev

Left: Sergei Korolev; *right*: Valentin Glushko

had no choice but to comply. Korolev was sent back to Russia in 1947 and promoted to Chief Designer, head of his own division, SKB-3, within NII-88. Although the Americans had disassembled and removed 100 V-2s from Mittelwerk, and even after the British had constructed a few more to fly from Cuxhaven, the Soviets managed to find enough parts to make at least 12 more V-2s. They had the parts and the engineers, but what the Soviets didn't have were the blueprints, which had been recovered from their underground hideaway by American army personnel before the Soviets, or indeed the British, could get to them. The first V-2s were launched in the Soviet Union in 1947; some blew up, and some flew in the wrong direction. On Stalin's orders Korolev oversaw the construction of a Russian version of the V-2, or R-1 as it was renamed. Korolev's somewhat modified version, with an engine designed by Glushko, was flying without a hitch, and in the right direction, by 1948. It was taken up by the army two years later. Despite the lack of blueprints, Korolev, Gröttrup and Glushko, along with many unsung German engineers, had solved a complex 'reverse-engineering puzzle'.

During 1947 Korolev fell in love with Nina Kotenkova, a translator brought in to keep him abreast of technical developments in the West. He wanted to marry her but his wife Ksenia, despite not wanting to see him, would not at first agree to a divorce. By 1949 she had changed her mind. Korolev and Nina got married in September that year, by which time Korolev was at work on the R-2. The engine was again designed by Glushko. At its launch on 20 October 1950 it flew 370 miles, twice the range of the R-1. As far as long-range missiles were concerned, with the development of the R-2 the Soviets were now ahead of America.

In 1949 the Soviets detonated their first atomic bomb. With longish-range missiles *and* the atom bomb the Soviets were becoming more of a threat. If the Soviets could beat the Americans to truly long-range missiles, that might really upset the world's balance of power. In 1950 Korolev and Glushko were at work on the R-3. It was to have three times the thrust of the R-2.

Korolev couldn't understand why the Americans had made such little use of von Braun. It seemed as if they had merely wished to keep him out of Soviet hands. At the end of the war, America had intended to develop guided missiles, but for now the atom bomb and long-range bombers were enough. America had little appetite for building long-range missiles. It was then the only nation in the world that had the secret of the atom bomb, and had a large fleet of long-range bombers to deploy them. In the years immediately following the end of the war there had been little that America required of von Braun. 'Once it had them,' an article in *Time* magazine later noted, 'the US hardly knew what to do with the German rocketeers.' To while away the time, von Braun kept his space ambitions alive by calculating the trajectories needed to get a craft to the moon, and writing a science-fiction novel about a manned mission to Mars. Von Braun and his German engineers began to wonder if perhaps they shouldn't have come to America. They were being used as technicians and little more. Von Braun said they were POPs, Prisoners of Peace. A year ago von Braun had had thousands of workers reporting to him; now he was reporting to Jim Hamill, a 26-year-old major who called him Wernher, not Herr Professor, and mostly ignored his requests.

The main task set to von Braun's group was to teach American military personnel how to assemble and launch V-2 rockets, and to make technical improvements. A decision was made early on that the V-2 in America would be used mainly for scientific not military research, no longer missiles but so-called sounding rockets, peaceful tools of scientific investigation. In 1946 the US Army formed the Upper Atmosphere Research Panel. The V-2

rocket was to be its research tool. On 16 April 1946 the first American V-2 was launched from the White Sands Proving Ground, an
area of desert in New Mexico. It had taken von Braun's team eight
months to reconstruct the V-2. It might have taken less time had
so many of the parts not rusted after months of lying around. The
first American V-2 reached an altitude of a mere 3.4 miles, but a
V-2 launched in September rose over 60 miles above the surface
of the Earth, as they had been designed to do. That same year a
V-2 reached an altitude of 107 miles. Not all the launches went
well. On one flight the gyroscope had been installed backwards,
sending the rocket across the Mexican border, causing a diplomatic incident. More than 60 V-2s would be launched in America
in the coming years, modified to carry various kinds of scientific
equipment including cameras. Clyde T. Holliday, photography
specialist at the Applied Physics Laboratory at Johns Hopkins
University, was responsible for the difficult task of putting cameras on V-2s. Dynamite was used to blow the rockets up in mid-air
to slow them down and preserve the instruments from the impact
of the landing. The cameras in turn had to be tough enough to
withstand the aerial explosion. Finding the film canisters proved
to be a time-consuming task, but amazingly none was ever lost.
Holliday got his best set of aerial photos from a V-2 flight of 1948
taken at an altitude of 65 miles. Another set of photos was taken
one hour later from a navy Aerobee rocket flying at an altitude
of 70 miles. Taking both sets of photographs, Holliday patched
together a 2,700-mile-wide panorama of the Earth, a tenth of
the Earth's circumference. The mosaic image was published by
National Geographic. A popular newspaper reproduced the image
under the headline: COLUMBUS WAS RIGHT. It showed the Earth
from further away than it had ever been seen before. There was
keen interest from the press in the photographs, but cameras had
been added primarily for the benefit of meteorologists, not for
aesthetic reasons. It was frustrating to the weathermen that the

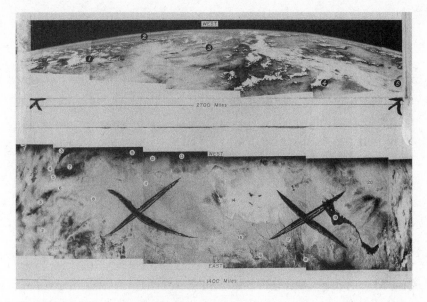

Clyde T. Holliday's 1948 photograph

rockets had to be launched during clear weather. They wanted to see clouds, not pretty pictures of the surface of the Earth.

Von Braun's V-2s took the first living creatures into space. In 1947, to test effects of radiation, fruit flies went into space and returned safely back to Earth again. On 11 June 1948 a Rhesus monkey named Albert became the first primate astronaut. A V-2 rocket took him to an altitude of 39 miles. Albert did not survive the flight, suffocating before the rocket returned to Earth. Albert II reached an altitude of 83 miles on 14 June, making him the first primate to reach space proper. He died on impact after the rocket's parachute failed to open. On 11 September the V-2 that was carrying Albert III exploded at 35,000 feet. Albert IV reached an altitude of 81 miles on 8 December. Again, the V-2's parachute failed on re-entry. Using a V-2, von Braun's team made the first television transmission from space – if only to an audience of a handful of technicians back in White Sands. A nascent space programme was underway.

In 1946 von Braun had made his first public speech in America. He told an audience at the El Paso Rotary Club that one day humans might fly to the moon and see its dark side. He got a standing ovation.

When his first cousin, Maria von Quistorp, turned 18, von Braun wrote to his father asking him to find out if Maria, 16 years his junior, was interested in marrying him. Maria happened to be visiting Wernher's parents when his letter arrived. The Baron came into the room waving the letter saying, 'I'm supposed to find out if you will marry Wernher.' Maria said, 'I've never thought of marrying anyone else.' They were married less than a year later on 1 March 1947. Military policemen escorted von Braun even on his honeymoon. 'Just pretend we're not here,' they told him. 'If you need us, we'll be in the kitchen.'

The Dora-Nordhausen War Crimes Trial was held during 1947. The camp leader, SS Obersturmführer Hans Möser, was sentenced to death, and 18 other defendants received sentences varying between five years and life. Colonel Toftoy, who had been responsible for bringing the Peenemünders to America, refused to allow any of them to be called as witnesses. At that time and for many more years to come the American public would not know that the V-2 had been mass-produced by a slave-labour force drawn from the camps.

In April 1949, von Braun – who was still reporting to Major Hamill – together with his team were transferred to a 40,000-acre site called the Redstone Arsenal at Huntsville, Alabama. There were 3,000 personnel at the site mostly working on projects related to industry. Around 700 of them reported to Hamill. Von Braun was given the title Project Director of the Ordnance Guided Missile Center. Large as the operation appeared to be, there were larger rocket operations elsewhere. The US Air Force

Arnold Engineering Development Center had been allocated 6,000 men and a wind tunnel for testing engines. As one army officer at Huntsville remarked, 'all we got were a hundred krauts'. Huntsville was at the time best known for its watercress.

Von Braun's group had just one military project. In 1946 von Braun had presented plans for a cruise missile, one of 21 different guided missile projects in the US at the end of the 1940s. Hermes II was a ramjet-powered two-stage cruise missile that used a V-2 as its first stage. When the combination was fired at White Sands, the second stage reached a record altitude of nearly 250 miles, and a speed of more than 5,000 mph. Though its significance was not clear at the time, Hermes II was the world's first multistage rocket.

In 1949 the army realized that they needed a larger testing ground than White Sands. The new site was established in a hostile and remote area of Florida. Von Braun said it reminded him of Peenemünde. He recalled how in those early pioneering days of rocketry, 'we resorted to all kinds of expedients because money was so tight. I recall one occasion when we rushed into Cocoa [Cocoa Beach, part of Florida's coastline at Cape Canaveral] and brought sewing tape, thread, and needles in order to get a missile off the pad.' The pioneering days of rocketry were, it seems, not so different from the pioneering days of aviation. The first rocket to be fired from the Cape, on 24 July 1950, was one of von Braun's two-stage Hermes II rockets.

Von Braun's fortunes changed soon after they moved to Huntsville. On 25 June 1950 the Korean War began. Here, it seemed, was another risk of world conflagration. The army now wanted new missiles fast, and one way to get them was to adapt the V-2. Von Braun once said that after the Second World War, America had built weapons 'at a tempo for peace'. Now that tempo was set to increase rapidly. Between 1950 and 1952 von Braun designed a family of rockets called the Redstone. They would be dubbed 'the Army's workhorse'.

By 1952 Von Braun was becoming well known to the public as America's chief advocate for space travel. There was much work to be done: a Gallup poll from three years earlier found that 88 per cent of the American population thought there would be a cure for cancer by the year 2000, 63 per cent that there would be atomic trains and airplanes, but only 15 per cent that 'men in rockets would reach the moon'. Von Braun's main outlet was *Collier's* magazine, with a circulation around 4 million. In the October 1952 issue von Braun wrote about a future three-ship, 50-man expedition to the moon. His imagined rocket was 160 feet long, 120 feet wide with 30 rocket engines, shorter and squatter than Apollo rockets would be and with far more engines. He wrote that 'the US must immediately embark on a long-range development program to secure for the West "space superiority". If we do not, somebody else will. That somebody else very probably would be the Soviet Union.' The article generated great interest. Though his space advocacy was not endorsed by the army, neither did they stand in his way. Not all of his colleagues, even those who had been with him at Peenemünde, approved of his writing for *Collier's*, but von Braun instinctively understood that if they wanted to push their agenda it would be the taxpayers who would be footing the bill, and that it was they who needed to be persuaded. 'We are living in a democracy,' he wrote, 'where the will and the mood of the people count. If you want to accomplish something as big as travel into space, you must win the people for your idea. Being diplomatic is necessary, but it is not enough. You have to be filled with a burning desire to bring your idea to life. You must have absolute faith in the righteousness of your cause, and in your final success. In short, you must be a kind of crusader.' Detractors said he was a propagandist, but he understood what distinguished a free country from an enslaved one: the voice of the people matters. His status as an idealist may be moot, but he was certainly an opportunist. He knew how things got done in his adopted homeland: 'You should know how advertising is everything in America,' he said. 'The way I'm talking will get people interested.' Among many predictions he made that got the attention of his audience was that 'within the next 10 or 15 years, the Earth will have a new companion in the skies, a man-made satellite that could either be the greatest force for peace ever devised, or one of the most terrible weapons of war'.

In 1953 von Braun wrote that America needed a coordinated space programme, and needed it now. That year also saw the publication of the book he had been writing – in German – during his spare time over the last four years. *Das Marsprojekt* had been published in Germany the year before and now – after 18 publishers had turned it down – in translation in America. The *Mars Project* is divided into two parts: a short science-fiction novel about a manned mission to Mars that supposedly takes place in 1965, followed by technical detail. Von Braun estimated that the expedition might cost $2 billion, about the same as the Manhattan Project. In the novel, three spacecraft – or boats as von Braun calls them – are launched from a large, rotating, wheel-like space station. The boats are named after the founding fathers of Rocket science: Oberth, Goddard and Ziolkowsky (the German spelling of Tsiolkovsky). The station – 150 feet in diameter, rotating once every 11 seconds in order to mimic the Earth's gravitational field – would later be the model for the iconic station imagined by Stanley Kubrick in *2001: A Space Odyssey*. Von Braun had, in turn, lifted the idea from Hermann Noordung, the pseudonym of Hermann Potočnik, a fellow member of the VfR. In his only work, *Das Problem der Befahrung des Weltraums – der Raketen-Motor* (*The Problem of Space Travel – The Rocket Motor*, 1928) Noordung describes a circular rotating space station named *Wohnrad* (living wheel). An editorial in *Collier's* put forward von Braun's rotating wheel as a possible future Manhattan-type project, estimating that it might cost $4 billion to build.

The novel and articles in *Collier's* were beginning to make von Braun famous. A year after *The Mars Project* came out in America, an official from the Disney corporation got in touch. Von Braun's advocacy for space travel was about to move to television. Disney made three TV films about space, each of them presented by von Braun, who also acted as technical adviser. *Man in Space* aired on the newly formed ABC network on 9 March 1955, attracting around 42 million viewers. Walt Disney introduced the programme himself. *Man and the Moon* followed at the end of the same year, and *Mars and Beyond* almost two years later. Disney invited von Braun and Willy Ley – a colleague from his days at the VfR, and with whom von Braun also collaborated for *Collier's* – to develop 'Tomorrow Land' at Disneyland: 'A vista into the world of wondrous ideas, signifying man's achievements ... The Atomic Age, the challenge of Outer Space and the hope for a peaceful, unified world.' It was to be an imagined representation of the year 1986, but when it opened on 17 July 1955 parts of the site were unfinished. The main draw was the TWA Moonliner, supposedly a commercial spacecraft from the year 1986. It had been developed out of the show *Man in Space*, and designed by von Braun and John Hench (one of Disney's so-called imagineers). At 76 feet, it was taller than Sleeping Beauty's Castle, and looked like a version of the V-2.

A month after *Man in Space* was broadcast, von Braun swore the oath of allegiance: 'I hereby declare, on oath, that I absolutely and entirely renounce and abjure all allegiance and fidelity to any foreign prince, potentate, state, or sovereignty of whom or which I have heretofore been a subject or citizen ... so help me God.'

In an interview he gave for an obscure religious magazine, von Braun revealed that he had had a conversion experience soon after arriving in America: 'A neighbor called and asked if I'd like to go to church with him. I accepted, because I was anxious to see if an American church was just a religious country club, as

I'd been led to expect.' He was taken to a 'small, white frame building ... in the hot Texas sun on a browned-grass lot'. The minister arrived in an 'old, battered bus' having driven 40 miles collecting parishioners who didn't have cars or the means to get to the church. 'Together these people made up a live, vibrant community ... This was the first time that I really understood that religion was not a cathedral inherited from the past, or a quick prayer at the last minute. To be effective, religion has to be backed up by discipline and effort.' Von Braun had been a perfunctory Lutheran in Germany, but here in America he embraced, as his biographer Michael Neufeld puts it, the 'welcoming inclusiveness of American evangelical Protestantism'. Von Braun said that he began to pray hourly: 'I took long rides out into the desert where I could be alone at prayer. I prayed with my wife in the evening. As I tried to understand my problems I tried to find God's will in acting on them.' Mostly he kept his religious beliefs to himself, but in another interview, which he gave soon after he had become an American citizen, he said: 'We should stop telling the world what we are against. We should tell the world what we are for. We must not fight Communist ideology with negative statements, but with the lofty ideals of the founders of this great Republic. The antidote to Communism is not anti-Communism, but the belief in God and the dignity of the individual.'

CHAPTER TEN

Aristid von Grosse, a German-born nuclear chemist, had been on Roosevelt's committee set up to study the feasibility of building an atom bomb. The Manhattan Project achieved success in a mere four years, and at a modest cost ($2 billion in US dollars of the time). In 1952, almost a decade later, President Truman commissioned Grosse to write a report investigating the threat Russia posed in space. Grosse's report floated around for a year or so and was not published until Eisenhower was in office. The report was largely the result of many hours' conversation with von Braun, who was convinced, as Lindbergh had been, that the greatest threat to the world's stability came from the East, a trope that was common in right-wing and fascist circles during the 1920s and 30s. Russia was not a threat to the West even in the years immediately after the Second World War. At that time only America had the atom bomb. During the Depression not only had Russia not been a threat, America had been keen to trade with both Russia and China.

Von Braun, however, wrote often about America's need to dominate in science so that it maintained the balance of power in armaments too. He imagined the day when the US might put a space station, or a series of satellites armed with atomic missiles into orbit – and so gain for America permanent military control of the globe. He foresaw the future nuclear armament build-up as a deterrent: 'Any attempt to maintain peace with a strong adversary who is in an expansionist mood can be successful only if the negotiations are conducted from a basis of strength. I am still convinced that World War II would have been avoided if, in 1939, Great Britain had sent a man of Churchill's bent, escorted by some British and Allied military leaders, to Munich, instead of a man with an umbrella willing to settle for what he called "peace in our time".' Once more von Braun shows that he was a skilful judge, and manipulator, of public opinion. Grosse, presumably through the mouthpiece of von Braun, asserted in his report that the launch of a Soviet satellite ahead of the Americans would be a considerable victory for the Soviets. Eisenhower was not convinced. He didn't think the public would care very much.

Von Braun of course thought otherwise. He got a chance to push for an American satellite programme when the American Second World War naval aviator and test pilot Frederick C. Durant III organized a private conference to explore the 'possibility of lofting an Earth-orbiting satellite in the near future'. Durant was also a director of the American Rocket Society and worked for a number of aerospace companies. Von Braun was to be the keynote speaker. He had been invited by Durant to present a paper at the Second Astronomical Congress in London, three years earlier. Von Braun had agreed to write the paper but said that he wouldn't be able to get the necessary clearance to attend in person: there were still fears in the military that he might be kidnapped. Durant agreed to present the paper on his behalf. They met up in New York later, and became good friends. Von Braun's presentation at Durant's conference in 1954 greatly impressed everyone in the room, including Commander George Hoover, chief of the Office of Naval Research in Washington. Von Braun 'went into the nuts and bolts of the thing', and said that, if given the go-ahead, his group could launch a 5lb satellite into orbit by the fall of 1956, certainly no later than November 1957. A multistage rocket would be needed, as Tsiolkovsky, Goddard and Oberth had envisaged. Von Braun said he would use a souped-up version of a Redstone rocket as the lower stage, with a cluster of solid-fuelled Loki rockets as the upper stages. The Loki was an anti-aircraft missile that could, like the Redstone, trace its ancestry back to Germany. It was an iteration of the Taifun, the German anti-aircraft missile first developed at Peenemünde in 1942, and after the war by German engineers in America. The modifications required to the Redstone and to the Loki would be relatively modest, von Braun said. The project was basically all set to go. The plan was subsequently written up by von Braun and other team members and sent to the highest echelons of his army employers. The project was duly taken up, and named Project Slug. It was later renamed Project Orbiter.

The competition from the Soviets was one thing, but von Braun faced greater competition from within. The air force had put forward a similar proposal but using the Atlas rocket as a launch vehicle. And the navy had its own Project Vanguard, which planned to launch a satellite using an upgraded Viking rocket.

The air force's plan had the distinct disadvantage that the Atlas rocket did not yet exist; the navy's Viking rocket had the advantage that it was a research rocket. Eisenhower wanted to present the project as purely scientific, untainted by military associations. The Viking had been designed to be used in weather research. It was more stable than the V-2 in the rarefied air of the upper atmosphere. It was only half the size of a V-2, but its design nevertheless owed almost everything to the rocket it replaced.

Von Braun's Redstone rocket was also a modified V-2, but its development history was more obviously martial. It had been commissioned to be a short-range ballistic missile (SRBM); that is, with a range of less than 1,000 kms (620 miles). The first Redstone – a single-stage rocket standing 69.3 feet tall – had been launched from Cape Canaveral on 20 August 1953. It flew for one minute and 20 seconds before crashing into the sea. Five months later, on 27 January 1954, an entirely successful launch of a Redstone took place. By 1954 von Braun had completed the design of a medium-range ballistic missile (MRBM; range between 1,000 and 3,500 kms). The PGM-19 Jupiter rocket would be the first rocket to carry a nuclear device. It should have been clear that von Braun's team was best placed to design and manufacture a rocket powerful enough to launch the first American satellite into orbit.

Despite von Braun's popularity as a result of his articles for *Collier's* and particularly after the first Disney TV show of 1955, Eisenhower had been told that von Braun would be a public relations disaster. On 29 July 1955, the President chose the Naval Research Laboratory's Project Vanguard. An upgraded Viking rocket, inevitably named Vanguard, was to be the launch vehicle of America's – hopefully the world's – first satellite. Among the reasons given for the decision was the bizarre one that Viking was more aesthetically pleasing than the Redstone. Von Braun was furious. 'This is not a design contest,' he said. 'It is a contest to get a satellite into orbit, and we are way ahead on this.' A Congress select committee even ruled that Project Orbiter was to be scrapped. As *Time* later reported, 'Wernher von Braun and his rocket team, the world's most experienced, were specifically ordered to forget about satellite work.'

The plan was to launch the satellite at some point during the upcoming International Geophysical Year (IGY), a long year that was to run from mid-1957 to the end of 1958. Some 67 countries were taking part, with the notable exception of China. The intention of the IGY was to show that scientists could work together in a spirit of international cooperation and peace. Shortly after the Americans disclosed their plan, the Soviet Union announced that they would launch their own satellite in the same timeframe. So much for cooperation. And yet despite what already looked like the beginnings of a space race, neither the Soviets nor the United

States at this stage seemed to have realized the propaganda potential of such an event. 'No one would get excited about a trifle like a little manmade object circling the Earth,' Presidential assistant Sherman Adams told Eisenhower. The Defense Secretary, Charles E. Wilson, told the *New York Times* that he wouldn't care if the Russians got a satellite into orbit first. Von Braun disagreed: 'The United States has no time to lose if it wants to be first in orbit,' he told reporters.

Vanguard and Orbiter were by no means the only rival satellite projects. Von Braun took an interest in another that aimed to put a satellite in orbit. His rivals in the navy seem not to have been aware of the existence of the plan by the Radio Corporation of America (RCA) to launch a weather satellite. When RCA put their TIROS project out for tender, only von Braun showed any interest. Von Braun's Ordnance Guided Missile Center at Huntsville, now renamed the Army Ballistic Missile Agency (ABMA), signed a contract with RCA in 1956. Among other satellite projects was a secret air force plan to get a military spy satellite into orbit using their Thor rocket. The Thor – not yet operational – was intended as an Intermediate Range Ballistic Missile (IRBM); that is with a range of between 3,000 and 5,000 kms (1,864 and 3,418 miles). The first launch of the Thor IRBM in 1956 was not a success. It rose 9 inches off the launch pad and fell back in a fireball.

Von Braun also had an IRBM project. On 20 September 1956 von Braun's team prepared to test the nosecone technology of their Jupiter IRBM – not the rocket itself, just its nosecone attached to a different type of rocket constructed specifically for the task, called Jupiter-C. Put together out of 'an elongated propellant tank and engine of a Redstone and a cluster of 11 scaled-down Sergeant solid-propellant motors for the upper staging', the Jupiter-C was a variant of a Redstone. (There had been a Jupiter-A – another variant of the Redstone – but seemingly no Jupiter-B. Rocket designations and variations are bewildering.) This first launch of a Jupiter-C was significant because in theory the rocket was capable of putting a satellite into orbit; indeed for this launch the rocket was carrying a dummy satellite weighing 30lbs in its nosecone. The Pentagon was suspicious that von Braun – despite the orders of the Congress committee – would replace the dummy with a real satellite and beat Vanguard. A simple command – 'Don't you dare!' – was sent to von Braun's new boss at Huntsville, General Medaris. Von Braun was told to inspect the nosecone personally to make sure the satellite was inactive. The rocket flew to an altitude of 683 miles and reached a top speed of 16,000 mph. Von Braun did a little dance. It hadn't quite achieved the velocity required to put a satellite into orbit (around 17,500 mph), but it was a significant achievement.

Rather than messages of congratulation, two months later an order came down from the Defense Secretary, who had become tired of the infighting, that army missile launches were to be restricted to altitudes not greater than 200 miles. This was just one of several attempts made by Wilson to try to crush Project Orbiter. The Huntsville team apparently burned the Defense Secretary in effigy.

On 8 August 1957 von Braun's team launched another Jupiter-C. Von Braun had teasingly renamed it Juno I in an attempt to disguise its origins as a missile, though everyone at Huntsville continued to refer to it by its original name. Jupiter-C missiles may have been banned from reaching altitudes of more than 200 miles but Juno I was, by fiat, a research rocket and so escaped Wilson's restriction. Juno I reached an altitude of 285 miles and covered 1,330 miles.

Three Jupiter-Cs had been built. The third of them – ignoring the Congress committee's order – von Braun put into storage in case Project Vanguard were to fail.

Project Vanguard was indeed failing. Since its inception it had suffered continual slippage and overspending. The first full-scale launch was put off until late 1957. In a letter dated 6 September 1957 von Braun was still sure of the significance of being first into orbit:

> I am convinced that, should the Russians beat us to the satellite punch, this would have all kinds of severe psychological repercussions not only among the American public, but also among our allies. It would be simply construed as visible proof the Reds are ahead of us in the rocket game. The fact that people would be able to *see* a Red satellite going around above their heads would impress most people far more than any assurances that the equipment our Western satellite, once successful, will carry, will be more sophisticated and refined.

In this he was, as he often was, extraordinarily prescient. When von Braun had been given an opportunity to persuade a congressional hearing of the significance of the threat, Senator Allen Joseph Ellender in response said that von Braun must be out of his mind. He had just come back from the Soviet Union and it was his opinion, judging by how few cars there were on the streets, and how ancient they were, that there was no way the Soviets could be about to launch a satellite. At the end of September one of von Braun's chief engineers approached General Medaris and pleaded with him to do what he could to let them launch a satellite. He was sure the Russians were about to make an attempt. Medaris told him not to get tense: 'You know how complicated it is to build and launch a satellite. Those people will never be able to do it. Go back to your laboratory and relax.' Medaris believed that the greater threat came from their own air force and navy.

Late on the afternoon of Friday 4 October, the telephone rang. It was an English journalist calling from New York. He said,

'What do you think of it?' 'Think of what?' said von Braun. 'The Russian satellite. The one they just orbited.' 'I'll be damned,' said von Braun. 'Damn bastards,' said Medaris. As the historian Matthew Brzezinski drily remarks in his book *Red Moon Rising*, it was 'unclear whether he was referring to the Soviets or to his own government'. Von Braun was in the middle of showing the new Defense Secretary Neil H. McElroy, Charles Wilson's replacement, around the Redstone plant. Von Braun said to him: 'If you go back to Washington tomorrow, Mr Secretary, and find all hell has broken loose, remember this: We can get a satellite up in sixty days.' General Medaris later remembered von Braun talking with 'driving urgency', his words tumbling over one another: 'We knew they were going to do it! Vanguard will never make it. We have the hardware on the shelf. For God's sake turn us loose and let us do something. We can get a satellite up in sixty days, Mr McElroy! Just give us the green light and sixty days!' Medaris, alarmed said: 'No, Wernher, ninety days.' Some accounts say that Medaris wanted to give the Jet Propulsion Laboratory a bit more time to come up with a satellite but in fact the JPL had already built one, and had put it on ice, just as von Braun had his spare rocket, ready for the events that were now beginning to unfold.

Over at the Soviet Embassy in Washington, 50 American scientists and their Russian counterparts, all of whom had been at the International Geophysical Year project, were at a reception when the news came through. One of the American scientists clapped his hands together to get the room's attention. He raised a glass and toasted the Soviet achievement.

'Listen now,' said the NBC radio network announcer that night, 'to the sound that forever separates the old from the new.' Beep, beep, beep. Each chirp an A flat lasting 3/10ths of a second with the same pause in between. Such dismay did the signal cast all around America, it might as well have been the broadcast of an arriving alien civilization intent on the destruction of the world. Three sounds, a message menacing in its simplicity, confounding but insistent proof of Soviet supremacy. The now former Defense Secretary Charles Wilson called it 'a neat technical trick'. Clarence Randall, Eisenhower's adviser, called it 'a silly bauble'. Admiral Rawson Bennett, Vanguard's commanding officer, clearly shooting himself in the foot, called it '[a useless] hunk of iron that almost anyone could launch'. The President downplayed the news. The Soviets had merely put 'one small ball in the air'. The press reaction was distinctly at odds with the official reaction. 'A grave defeat,' said the *New York Herald Tribune*. 'The Russians have just achieved,' wrote a journalist in that week's *Paris Match*, 'what the Americans have described so often and so prematurely.' Edward Teller, the so-called father of the hydrogen bomb, called it 'a technological Pearl Harbor'. The feat was likened to the splitting of the atom. 'Myth has become reality,' said *Le Figaro*, adding that America has had 'little experience of humiliation in the technical domain', and would now enter a period of disillusionment and bitter reflection. Chinese media said the event showed the superiority of Marxist–Leninist technology. An anonymous Huntsville scientist, perhaps it was von Braun himself, told the press how angry they were that Project Orbiter hadn't got the go-ahead in 1955. Washington's reaction was to impose a news blackout. In her diary, Anne Morrow Lindbergh wrote: 'I cannot understand why someone in the State Department didn't have the sense of form, manners, and statesmanship, or psychological insight, to *immediately* congratulate the Russians on their achievement. This would seem to me the line to take instead of this panic.' People across the world sat out at night hoping to spot the satellite pass overhead. But another shock was to come. The ball was far heavier than anything America was attempting to orbit. Sputnik weighed 184 lbs. The thrust of the Russian rocket must be far greater than any rocket America was building. In fact all their estimates of the size of the

Soviet booster that had put Sputnik into orbit were far too low, as they were soon to find out. Vanguard had already cost the taxpayer $110 million. What America could not have known at that time was that Sputnik had been put into orbit for an estimated $50 million.

At a meeting in Washington, von Braun made the most of this opportunity to propagandize on behalf of space, telling the assembled military brass that 'October 4, 1957 ... will be remembered on this planet as the day on which the Age of Space Flight was ushered in ... For the United States, the failure to be the first in orbit is a national tragedy that has damaged American prestige around the globe.' The Democrats of course were delighted at the opportunity to attack the opposition. Von Braun immediately had Lyndon Johnson on his side: 'Maybe it was all right with others in government,' Johnson said, 'but he for one didn't care to go to bed by the light of a Communist moon.' Somehow, if subconsciously, Johnson had understood that perhaps the moon itself was at stake. Johnson pushed for von Braun's Project Orbiter, mocking Vanguard's proclaimed elegance: 'It is not very reassuring to be told that next year we will put a "better" satellite into the air. Perhaps it will have chrome trim and automatic windshield wipers.' An article in the *New York Times* magazine later that month called von Braun a prophet. Eisenhower chose to stay silent, and his silence, as silence can, damaged him.

Eisenhower and his advisers had been taken by surprise by the response to Sputnik. Americans had become increasingly fearful of Russia once the Soviets had acquired the secret of the atom bomb; and then in 1956 the Soviet army had entered Hungary, which in turn was followed by the Suez crisis. Israel invaded Egypt in response to Egypt's nationalization of the Suez canal. A month later the French and British offered Israel their support, which elicited from the Soviets a thinly veiled threat that they would launch nuclear missiles against London and Paris.

With its overwhelming nuclear arsenal and vast fleet of long-range bombers it had seemed as if America could afford to be sanguine about Russia's development of the long-range missile. By 1955 the US had 2,280 atomic and thermonuclear bombs, a tenfold increase in four years. The B-52, which had been added to the fleet that year, could carry 70,000 lbs of nuclear warheads up to 8,800 miles at 500 mph. Of its fleet of 1,200 B-47 bombers, 400 were permanently loaded with nuclear weapons and ready to take off at a moment's notice. In 1954 the American air force had spent a measly $14 million on developing its Atlas ICBM (intercontinential ballistic missile). After the so-called missile gap had been identified that same year, Eisenhower raised the spending on missiles to $550 million, which in 1955 was still only 1 per cent of the military's overall budget.

Though Eisenhower had once assured the Soviet Union that America would never be the aggressors, from the early 1950s America had set itself on a course of building up 'massive retaliatory capabilities'. Early in 1956 Commander Curtis LeMay of Strategic Air Command – the model for Kubrick's cigar-chomping mad general Jack D. Ripper in *Dr Strangelove* – scrambled a large percentage of the bomber fleet in a simulation of all-out nuclear war. He 'even scared the CIA'. It was one of a number of operations undertaken around that time with the express intention of alarming the Soviets.

America overwhelmingly held the balance of power, yet the small ball in the air changed the public's perception: it made the Russian threat palpable. 'We failed,' von Braun said, 'to recognize the tremendous psychological impact of an omnipresent artificial moon.' If the Soviets could launch a heavy satellite into space they might soon be able to launch a nuclear device into space and back down again onto Earth, anywhere on Earth.

From America's perspective the Russians appeared to be near to producing the world's first ICBM, missiles with a range of 5,500 kms (3,400 miles). America had a vast fleet of long-range bombers and a huge arsenal of atomic weapons, but it did not yet have an ICBM. In the propaganda wars that followed the launch of Sputnik, both America and Russia would overemphasize Russia's missile capability and development of long-range bombers, while America would underplay its ability to deliver nuclear devices over long distances. Even at the height of its power the Soviet Union would have only a twentieth of the air power of America. And yet in 1957, because of Sputnik, America was decisively losing the propaganda war.

By 1951 Korolev and Glushko had freed themselves of the V-2 and were designing and building the first entirely Russian rocket. On 2 February that year, their R-5 rocket – as before, the overall design was by Korolev, engine designed by Glushko – carried an 80-kiloton nuclear warhead to its target. (A kiloton is a unit of explosive equivalent to 1,000 tons of TNT.) It was 'the world's first nuclear detonation delivered by a ballistic missile'. Soon afterwards, when he was informed that it would take only five nuclear warheads to destroy the UK, and seven or nine to wipe out France, Khrushchev became an avid supporter of a Russian rocket programme. 'In the high-stake arms race,' writes the political commentator Matthew Brzezinski, 'Khrushchev was a pauper playing at a rich man's table. Given the vast financial gulf between America and the Soviet Union, he had to marshal his resources more carefully than Eisenhower, and he saw that missiles were the cheapest, most cost-effective way to stay in the game.' Korolev and Glushko's R-7 was to be 10 times the size of the R-5. With 1 million lbs of lift it could fly four times faster than the V-2 and 40 times further. In theory it could reach the US in half an hour.

It appeared to be the ICBM Khrushchev craved. (What happened to the R-6 is to be something of a mystery. The R-6 may have been an early designation for a version of the R-7 that was subsequently abandoned.)

The early launches of the R-7 were problematic, as early rocket launches always are. On 15 May 1957 the R-7 didn't even rise above the launch pad. On 9 June an R-7 exploded less than a minute into its flight. The same thing happened two days later. On 21 August the launch was a success but the heat shield failed, which meant that the rocket would have been useless as transporter of a nuclear weapon or weapon of any kind. Without a working heat shield the R-7 could not function as an ICBM. On 7 September the nosecone failed again. At the time, America knew nothing about these setbacks, which made Russia's seemingly effortless successes all the more alarming. The R-7 had been meant to carry a payload of 2,700 lbs, Object OD-1 as it was called. Korolev suggested that in the meantime the R-7 be used to put a much smaller satellite into orbit in order to beat the Americans. Khrushchev was not particularly interested and his generals thought it was a distraction from the development of the R-7 as a long-range missile. Would Korolev's satellite PS-1, weighing a mere 184 lbs, impress or be an embarrassment? No one knew.

The PS stood for *prosteyshiy sputnik*, literally simple satellite. After the launch on 4 October Korolev decided to wait until Sputnik had completed a complete orbit before he telephoned Khrushchev. It was possible that the simple satellite hadn't quite made it into orbit. Finally, 93 minutes later, the satellite's signal – beep, beep, beep – was picked up again. Korolev and Glushko hugged each other. 'This is music no one has heard before,' said Korolev. When Khrushchev was told the news, he apparently muttered that it was just another Korolev rocket launch. Von Braun and Korolev had anticipated the satellite's effect, though Korolev wondered if Sputnik would stay aloft for long enough to make a real impact. There was a distinct risk that the orbit might degrade after a few days. In fact Sputnik remained in orbit for three months. Korolev had, however, been confident enough to insist that his simple satellite look beautiful, telling the designers that in years to come it would be displayed around the world. And so it is. An engineering copy of Sputnik can be seen today in the lobby of the United Nations building in New York, the first thing representatives of the world's nations see when they cross the threshold.

Almost no one seemed to have noticed that for the first time in history the gravitational pull of the Earth had been overpowered. Humans and their tools had been forever bound to the Earth and now one of its machines had entered into a new relationship with the Earth. In the lists of human firsts, it was one of the most significant moments in our history, both physically and metaphorically. The following day *Pravda* devoted just two short paragraphs to Sputnik. The main story was about the Union's preparations for the coming winter. Khrushchev was astonished by the scale of the rest of the world's reaction to the launch. Overnight – sometimes history really is that straightforward – the Soviet Union had become the most feared nation on the planet. And, as Matthew

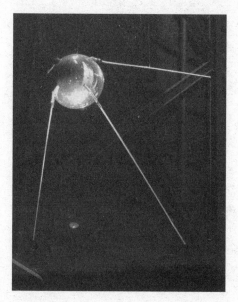

Engineering copy of Sputnik,
United Nations building,
New York

Brzezinski observes, they had done it by peaceful means, not, as had America, by vaporizing the populations of two cities.

Korolev and Glushko would remain anonymous all their lives. Even in his own country, Korolev was referred to in the press by his title, Chief Designer. The public face of the Soviet space programme was Leonid Sedov, whose comments made him something of a celebrity in America. He told Americans that they only cared about their cars, houses and electric refrigeration. The typical American, he said, 'has no sense of national purpose, nor is he receptive to great ideas that do not pay off immediately'. Like Korolev, Sedov assumed that von Braun was leading the American rocket programme. He wondered why von Braun had selected 'this other design' (he meant the navy Vanguard rocket). Why had he begun something new rather than follow through with his army rocket? The Soviets were shocked when they discovered that von Braun was not at the centre of everything, not running the space programme.

Within a month Korolev had another shock in store for the West. On 2 November, Sputnik 2 was launched into orbit. Not only was it six times as heavy as Sputnik 1 (American engineers wondered if the Soviets were using an even larger rocket than the one that had launched Sputnik 1: they were not), but, just as disturbingly, in its nosecone Laika the dog became the first living creature to orbit the Earth. If a dog, then why not a human being? It looked as if Russia could beat America at anything. The *New York Herald* ran the headline 'WHAT NEXT? A MAN ON THE MOON?' The *New York Times* took up Sedov's theme and wondered if Americans had become 'a little too self-satisfied, complacent, and luxury loving'.

In fact Sputnik 2 wasn't entirely the success it appeared to be. Once again the heat shield failed. Laika died a few hours after the launch. But America didn't know that. 'The first Sputnik had made Americans afraid for their lives,' writes Brzezinski, 'Sputnik

II made them question the American way of life.' Rather late in the day, Eisenhower ended his period of silence and made a series of national addresses in imitation of Roosevelt's 'Fireside Chats'. They were known as his 'Chin Up' speeches.

The first of them was broadcast on 7 November. On 8 November General Medaris was given clearance to get their satellite ready for launch in competition with the navy's Vanguard launch. Project Orbiter was back on. Here was a chance for Medaris and von Braun to dominate the space race, not just in the race against the USSR but more immediately within their own country. Within days of the announcement von Braun had booked a launch pad at Cape Canaveral. It began to look as if von Braun's gamble might just pay off. He said of the 'spare' Jupiter-C (aka Juno I) that he had kept in storage: 'All she needed was a good dusting.'

On 18 November von Braun was on the cover of *Life* magazine. Inside, an article titled 'The Seer of Space' described him as the 'thoroughly Americanized German', and quoted him as saying: 'I just wish someone had the authority to tell me, "All right, we'll

One of von Braun's Redstone rockets on display in Grand Central Station, New York

leave you alone for two years, but if you fail we're going to hang you.'" Russia wanted world domination, von Braun said. Here in a race to space was a contest that 'will demonstrate to all the world whether we are in fact, as we still have every reason to believe, the greatest nation on Earth and the shining hope of men everywhere'. Another article in the same issue, titled: 'Arguing the case for being panicky', made the same point but more histrionically: 'In short, unless we depart utterly from our present behavior, it is reasonable to expect that by no later than 1975 the United States will be a member of the Union of Soviet Socialist Republics.' Cold War propaganda was in full flight.

Disney's *Mars and Beyond*, the last of the three TV space programmes, each of them fronted by von Braun, was broadcast on 4 December. It was meant to have coincided with the launch of Vanguard, which because of weather problems was now scheduled for 6 December. Before the launch of Vanguard, General Medaris told his wife that he was praying for help. If Vanguard was a success, the army's role in the space programme would be over. If Vanguard was a failure and Orbiter a failure too, the army's role might still be over. What was needed was a navy failure and an army success. Medaris' negotiation with God was a tricky one.

Live on TV the navy's Vanguard TV-3 rocket rose a few inches off its pad, fell back and exploded. The headlines the following day were all of a theme: Flopnik, Kaputnik, Stayputnik, Splatnik, Dudnik, Puffnik, Oopsnik, Goofnik, Sputternik, Stallnik. America's dismay and fear deepened. Shares in Vanguard's manufacturer, Glenn L. Martin, plummeted and trading was suspended. Eisenhower's approval rating plummeted too.

Von Braun's team at Huntsville had a short window of opportunity in which to launch their Juno rocket and attempt to put a small satellite into orbit. The satellite was named Explorer and weighed 31 lbs, around a sixth of the weight of the first Sputnik. If they failed the navy would have another opportunity to launch their

Vanguard rocket. The weather was bad on Thursday 23 January 1958, the first launch slot, and conditions remained unfavourable all that week. On 29 January the jet stream was still too strong for a launch. On 31 January, the last day in the army's slot at Cape Canaveral, the speed of the upper air had reduced only a little, to 205 mph at 41,000 feet. Von Braun wasn't at the launch site, he was in Washington ready to brief the press should everything go to plan. No live television this time. Even the engineers' wives had been kept in the dark. The launch was being overseen by Kurt Debus, von Braun's 'unflappable crew chief'. In the file compiled on him during Operation Paperclip at the end of the war, his American interrogators had described him as 'an ardent Nazi'.

The prospect for a successful launch was generally judged to be marginal. The assessors conferred and advised that the launch be aborted. The Cape's chief meteorologist, 24-year-old John Meisenheimer, disagreed and predicted that the weather conditions would ease by late evening. Medaris said, 'To hell with it,' and decided to 'gamble the hopes of the nation, the future of his five thousand employees on the word of a twenty-four-year-old kid'. At 10.48pm Juno I was launched. Everything seemed to go smoothly, but as Korolev had, Medaris decided to make sure the satellite completed a full orbit before breaking the news. If the Explorer satellite really had been put into orbit the signal would be picked up again at 12.41am. There was nothing at 12.41am and nothing at 12.45am. It was highly unlikely that the calculations were wrong. It looked as if the Explorer's orbit had degraded, and the satellite fallen to Earth somewhere. And then at 12.49am a coded message was received from the tracking station: 'Goldstone has the bird.' Explorer had circled the Earth eight minutes late, in a slightly higher orbit than had been predicted. 'It makes us feel that we paid back part of the debt of gratitude we owed this country,' von Braun told a reporter. His father, the Baron, sent his son a telegram. It read: 'BEEP, BEEP, BEEP.' From that moment von

Braun was one of the most famous people in America, receiving thousands of letters a month.

That same month von Braun proposed putting a man, rather than a satellite, in a nosecone. He named the project Man Very High after the recent Project Manhigh. Three balloon expeditions in 1957 and 1958 had carried a pilot to around 100,000 feet in order to investigate the stratosphere. Yet again the air force put the kibosh on von Braun's proposal as it had its own project named – yet another unbelievably inappropriate acronym – MISS (Man in Space Soonest). In February Eisenhower created the Advanced Research Projects Agency (ARPA). It was given an initial budget of $520 million with the aim of trying to move ahead of the USSR in technology across the board. The idea was to expand the frontiers of science beyond the purely military. Perhaps after two world wars, scientific and military progress had become conflated.

On 17 March Project Vanguard finally got its own satellite, Vanguard 1, into orbit. It was 6 inches in diameter and weighed a mere 3 lbs. Nikita Khrushchev dubbed it the grapefruit satellite. As if to rub salt into the wound, on 15 May 1958 Sputnik 3, weighing in at 1,327 lbs, went into orbit. (Sputniks 1 and 2 and Explorer 1 are no longer in orbit, leaving Vanguard 1 as the oldest surviving orbiting satellite today. Vanguard was the first solar-powered satellite.)

The Soviet's success in lifting great payloads was a huge embarrassment to the US, but it played to von Braun's strength: his ability – for years now suppressed – to design and oversee the building of powerful rockets. The space race would essentially be about which nation could build the most powerful rocket. Reduced to individuals, it would be a competition between Sergei Korolev and Wernher von Braun.

CHAPTER ELEVEN

On 29 July 1958 NASA was created by federal statute specifically to develop aeronautics and astronautics. Its initial budget was a modest $100 million. Eisenhower never cared strongly about the space race but it was a stroke of genius to make NASA a civilian organization. Perhaps it took a military man to foresee the potential of a space race as an agency of peace. In any event it was as well that he was now Commander in Chief and not merely General Eisenhower: the generals were furious. Eisenhower had barely acknowledged the string of army successes in space, even though he had been a career army man. He admitted on a visit to Huntsville that he didn't even know how the engines on a rocket were arranged. Eisenhower thought that by taking the space industry away from the generals he could halt some of the excesses of the arms race; he was becoming increasingly worried that public policy was being controlled by a 'scientific-technological elite'. Later, he would say that specifically he meant Edward Teller and Wernher von Braun. His science adviser – the first Presidential science adviser – James R. Killan wrote that he repeatedly saw Eisenhower 'angered by the excesses, both in text and in advertising, of the aerospace-electronics press, which advocated bigger and better weapons to meet an ever bigger and better Soviet threat that they conjured up. I remember the shrill, hard campaigns by a few corporate lobbyists in support of their companies ... The Sputnik panic was being used to support an orgy of technological fantasies.'

On 7 October 1958 Project Mercury was announced. Its main aim was to put a human being in space before the Russians managed to, if indeed it was possible for a human being to survive in space. No one knew for sure. It was going to be called Project Astronaut, but Eisenhower thought that that put too much emphasis on the pilot. As it would turn out, the emphasis would be on the pilots regardless of what the project was called. Eisenhower had wanted the astronauts to live hidden in a secret village, but he was talked out of it. In the event, the American programme would be noted for its openness. 'We think that running an open show,' said von Braun, 'is within itself an important part of the message we are trying to get across about America and the Free World.' (In an address he made to the CIA in 1965, he added: 'So we would like to know more about what the Soviets are up to in space. This, of course, is where you people come in.') In the same month von Braun proposed a new series of more powerful rockets, to be called Saturn, the next planet further away from Earth after Jupiter.

The Russian equivalent of the Mercury programme was called Vostok. The Russians opted for what would have been Eisenhower's preferred way: secrecy. Russian cosmonauts lived with their families in the specially constructed Star City outside Moscow. The cosmonauts would not be allowed to tell even their wives about their missions.

At first, Project Mercury was a US air force initiative. The air force's MISS project was almost immediately folded into Project Mercury. And then not long after that the whole lot was folded into NASA, but not before the air force's Atlas rocket had been chosen over von Braun's Saturn rockets as Project Mercury's booster rocket. Lindbergh was on the committee that made the recommendation, along with John von Neumann. Von Braun, once again, had seemingly been cut out of the picture, but Saturn was in the earliest stages of its development, and it was thought that the rockets would be too costly. Atlas had been in development for years. The first test flight had taken place in June the previous year. The rocket blew up 24 seconds into the flight.

NASA had wanted to run an open competition to find the first astronauts, but again Eisenhower intervened and insisted that they should all be test pilots. Early in November 1958 the Space Task Group (STG) – a team of just 45 engineers – was set up under Bob Gilruth, 'a delightful, bald-headed teddy bear of a man, reedy-voiced, twinkly-eyed – certainly not impressive, but capable of putting together an impressive team'. On its very first day Gilruth wrote a one-sentence memo appointing himself director of Project Mercury.

NASA's first director, Dr T. Keith Glennan, made the decision that if he had the navy's Vanguard and the air force's Atlas rockets then he only needed part of von Braun's operation. Von Braun was livid and used his now considerable influence with the press to retaliate. If he pushed ahead with his plan Glennan would end up with nothing, von Braun said in public. He and his team would simply leave and take up jobs in industry. The threat was not an empty one: von Braun's workforce was incredibly loyal, and any of them could easily have taken much better-paid jobs elsewhere. Now it was Glennan's turn to be furious: 'I had not realized,' he said, 'how much of a pet of the Army's von Braun and his operation had become.' Glennan was forced to back down and agreed to take on von Braun's outfit wholesale. Rather than running the whole show as the Russians had supposed, von Braun had struggled, even given his recent success, to secure a prominent role in America's space programme at all. Now, as Head of Development Operations, he was in charge of the largest of NASA's operations by far, overseeing a workforce of several thousand. In 1958, 3,925 employees reported to von Braun; four years earlier it had been

950. Once it had been subsumed into NASA, the Army Ballistic Military Agency at Huntsville was renamed the Marshall Space Flight Center (MSFC).

Von Braun had had a good relationship with the army at Huntsville but was relieved to be out from under military command. For the first time in his life he could pursue space flight 'for space flight's sake, and not because of some temporary military objectives connected with outer space'. Von Braun wasn't the only one who was delighted by the new arrangement. Much of the team that had followed him from Peenemünde was still in place, and those who had joined him more recently were just as enthusiastic about the prospects that lay ahead of them. 'We knew if von Braun was leading it,' said engineer manager Bob Schwinghamer, 'things were going to get bigger and better.'

By this time von Braun had a number of celebrity friends, Arthur C. Clarke, Walt Disney and Walter Cronkite among them, but part of his great charm was that he gave everyone he came into contact with his closest attention. He would talk to anyone, was interested in everyone and everything. He remembered everyone's name. He was an excellent listener. One of the junior members of his team at Huntsville, Rutledge Parker Hazzard – later in life Brigadier General 'Hap' Hazzard – remembered the time von Braun had complimented him on coming up with a calculation that was 'basically correct' even though, in fact, it was an order of size too small. What Hazzard remembered all his life was the way von Braun treated him as if he were an equal. 'All of us were convinced', he said years later, that von Braun was 'not only human, but a human with empathy'. Von Braun had his human eccentricities too. He was a reckless driver, ran red lights while reading paperwork. He was a late riser, loathed early-morning meetings. He said that nothing of importance in world history had ever been accomplished 'before ten-thirty or eleven in the morning'. He was a hypochondriac, took handfuls of pills, and was continually phoning his doctor for additional medication for some imagined stomach ache, pain in the chest, sore throat or sinus trouble. He never lost his strong Teutonic accent. He made a lot of noise, was always humming or whistling. He had a photographic memory. He believed in the power of prayer. He gave brilliant as well as funny talks. He could talk fluently about Ancient Greek history or nuclear fission. If someone in the audience asked a tangential question, he might well launch into a detailed response that lasted 45 minutes. He was a hunter: shot all kinds of birds, deer, antelope, caribou, moose and bears; went on a panther hunt in the Yucatan. He was a scuba diver, on one occasion descending to 165 feet. He enjoyed swimming in rough seas. He piloted his own plane, and liked to fly straight into the nearest storm cloud for the thrill of it, particularly if he could also unsettle a nervous passenger. He said that he tried to live life at 'full bore'. But he couldn't operate a VCR and would tear out the tapes in frustration. When an engineer, Tom Shanner, visited von Braun's house and noticed that his TV was broadcasting in black and white, Shanner twiddled a few dials at the back of the set. Von Braun was amazed: 'Vhat did you do? Vhat did you do?' Shanner thought perhaps he had made a mistake and that von Braun preferred watching in black and white. 'No, no . . . I didn't know you could do that. It's been like that for five years.' He didn't know how to operate a household drill, but he did know how to build a powerful rocket.

The Soviets had been first to put a human-made object into orbit, and on 2 January 1959 they were the first to launch a human-made object – Luna 1, a craft weighing 800 lbs – beyond the influence of the Earth's gravitational field altogether and into outer space. Two earlier launches had failed, but they had been kept secret. After two days, Luna 1 flew by the moon. Korolev had intended that the craft hit the moon, but the guidance system turned out not to be accurate enough. Two months later America managed to get its first craft beyond the Earth's gravitational influence. Pioneer 4 was launched on one of von Braun's Juno II rockets (a version of the Juno I) and in the general direction of the moon. It managed to get no closer than 37,000 miles away before going into orbit around the sun. Russia was clearly still some way ahead. 'The space race may decide whether freedom has any future,' a commentator in *Time* magazine wrote.

In April NASA announced its first group of astronauts. The size of the Mercury craft meant that there were height and weight restrictions: no taller than 5 feet 11 inches, no heavier than 180lbs. Candidates had to be under 40 with a bachelor's degree or equivalent, at least 1,500 hours of flying time and qualified to fly jets. At first, 18 were chosen, all white men, and then out of those,

The Mercury
Seven
Front row:
Deke Slayton,
Gus Grissom,
Alan Shepard.
Middle row:
John Glenn,
Scott Carpenter.
Back row:
Wally Schirra,
Gordon Cooper.

seven. They were known from the start as the Mercury Seven: Alan Shepard, Gus Grissom, John Glenn, Scott Carpenter, Wally Schirra, Gordon Cooper and Deke Slayton: three air force, three navy and one marine.

Even though they were wearing civilian clothes, and NASA was a civilian organization, at their first press conference the Mercury Seven were introduced using their military rank. The press, however, questioned them not about their flying records but about their personal lives. John Glenn started talking about 'God, country and family', and the rest apart from Shepard followed suit. 'I don't mean to slight the religious angle,' Shepard said, 'but the Mercury Project is merely one step in the evolution of space travel.' At the end of the interview the reporters applauded. Glenn and Shepard had done most of the talking. It wouldn't be the last time Glenn and Shepard jostled for position of alpha male among this group of alpha males.

On 1 May 1959 it was announced that NASA's new centre of administration, then under construction in Maryland, would be named the Goddard Space Flight Center.

Lindbergh and Guggenheim, who had done so much for Goddard while he was alive, did not stop after he was dead. His legacy needed to be secured. Soon after the war, Lindbergh had agreed to become a civilian adviser to the Armed Forces on rocketry and space research. He was paid a salary of one dollar a year. He said all he really wanted was 'Justice for Bob Goddard'. Harry Guggenheim set about ensuring that all Goddard's innovations were protected, and 131 new patents were issued posthumously. All told, there were now 214 Goddard patents. In 1950, Harry Guggenheim brought actions against the American navy, air force and army. His lawyers argued that by using the V-2 on American soil, a number of Goddard's patents were being infringed. Needless to say, the military was resistant to this line of reasoning. In a bizarre twist (as if the whole thing wasn't bizarre enough already) the army secretly called on von Braun to assess the case. Von Braun, who had once claimed that he had never seen a Goddard patent, now argued that inevitably Goddard patents had been infringed, even if many of those infringements had been accidental. 'Goddard's experiments in liquid fuel saved us years of work,' he wrote, 'and enabled us to perfect the V-2 years before it would have been possible.' He was astonished how many of the questions they had had to address had already been addressed by Goddard.

Von Braun had become close to Goddard's widow Esther. He told her that Goddard had been a childhood hero of his, though Goddard's biographer, David Clary, points out that whenever he spoke to Germans, von Braun was always clear that his boyhood hero had been Oberth. Clary argues that it was all part of von Braun's 'lifelong effort to make himself an American hero, hoping his adopted country would forget his Nazi past'. One thing is clear: the silver-tongued von Braun knew what to say to whom. 'We can only wonder what might have been if America realized earlier the implications of his work,' von Braun said at the opening ceremony of the Goddard Roswell Museum in 1959. 'I have not the slightest doubt that the United States today would enjoy unchallenged leadership in space exploration had adequate support and recognition been provided to him.' Others were less generous. Goddard's rival Theodore von Kármán was harsh in his assessment: 'There is no direct line from Goddard to present-day rocketry. He is on a branch that died. He was an inventive man and had a good scientific foundation, but he was not a creator of science, and he took himself too seriously. If he had taken others into his confidence, I think he would have developed workable high-altitude rockets and his achievements would have been greater than they were. But not listening to, or communicating with, other qualified people hindered his accomplishments.' Kármán thought that

Goddard had become bitter in his later years 'because he had had no real success with rockets, while Aerojet-General Corporation and other organizations [presumably he meant ones like his own Jet Propulsion Laboratory] were making an industry out of them'. At the other extreme Lindbergh overestimated Goddard's achievements. 'Probably no figure in the history of science had a greater vision than that of Robert Goddard,' he wrote in a letter to Esther, 'or more courage and tenacity in translating his vision into fact. The determination of a youth to conquer the universe in spite of doctors' warnings that he had only a few more weeks to live; the vision of a man physically projecting himself off his Earth and into space, the design and construction of prototypes by which this project would shortly be accomplished, what more is needed to carry mortal man into the fields of the immortals?' Lindbergh apparently came to realize that the V-2 had largely been designed independently of Goddard's work, but he diplomatically skirted the question whenever he was asked.

When it was announced that the new unmanned spacecraft facility would be named the Goddard Space Flight Center, NASA decided it would look churlish to continue fighting Guggenheim's infringement suit, which had by then been trundling through the courts for eight years. All parts of the military capitulated, agreeing to pay $1 million between them, the bulk coming from the air force. The money went into further research in Goddard's name.

In the late 1950s the space race really gathered pace. Almost every month seemed to bring some significant new achievement.

At the end of May 1959 one of von Braun's Jupiter rockets sent the monkeys Able and Baker into space. They experienced a nine-minute period of weightlessness, and were afterwards recovered in good health although Able, a rhesus monkey, died four days later while undergoing surgery to remove an electrode that had become infected. His body was preserved and is on display at the Smithsonian Air and Space Museum in Washington. Baker, an 11-ounce squirrel monkey, lived on at the centre in Huntsville – where she was known affectionately as Miss Baker – until her death on 29 November 1984. Abel and Baker were the first primates to survive a journey into space.

The goal of Mercury was to get a human being into orbit, but if we were ever to assume ourselves into our wider family of primates the history books would be rewritten and Abel and Baker, along with Alberts I–IV, would be among our heroes. If we widened our circle of compassion yet further, we might even include flies among the first space pioneers.

In August 1959 the sixth in what had become a series of Explorer satellites was launched into a deliberately highly eccentric orbit, putting the satellite between 147 and 26,000 miles away from the surface of the Earth. The Explorer programme (what had started out as Project Orbiter) had by now been folded into NASA. The first stage of the booster rocket was no longer von Braun's Juno I but the air force's Thor IRBM in combination with a second-stage Able rocket from the Vanguard series; the combination – Thor-Able – was used frequently for satellite launches. A camera on

board took the first photographs of the surface of the Earth seen from orbit, though the main purpose of the mission had been to photograph not the ground but cloud cover. The resulting photographs, taken from thousands of miles away, needed to be enhanced and

Photograph taken by Explorer 6 on 14 August 1959 of the Earth seen from 17,000 miles above Mexico superimposed before anything could be discerned by the naked eye. It took weeks of work. The photograph that was released to the press had the outline of the Pacific Ocean and Hawaii physically drawn onto it in an attempt to identify what is otherwise cloud cover. NASA added a caption: 'Scientists can discern cloud banks in the large white areas.'

The following month the Soviets' Luna 2 intentionally ploughed into the moon, the first human-made object to make it to the surface of another celestial body. A few weeks later Luna 3 took the first photographs of the dark side of the moon. The images were reproduced all over the world.

On 1 April 1960 the TIROS weather satellite was put into orbit. TIROS, the satellite project in which von Braun had shown an early interest, had been taken over by NASA in 1959, and the launch vehicle reassigned from one of von Braun's Juno I rockets to the Thor-Able combination. TIROS weighed 270 lbs, still a fraction of the weight of the satellites being lofted by the Soviets. There were two TV cameras attached to the satellite, one with a wide-angle and the other a telephoto lens. As it neared completion of its first orbit, and at an altitude of 450 miles, TIROS sent pictures of the Gulf of St Lawrence back to Earth. NASA pasted together stills from the TV cameras to make a composite strip depicting the entire circumference of the Earth, a remarkable first even if the photographs were fairly low-grade. The image only ever appeared in a technical journal, but the photographs allowed meteorologists to see the Earth's cloud structures in detail for the first time.

Four months later, the American spy satellite Discoverer 14, orbiting at an altitude of 100 miles, took photographs of the Earth that were the first to be developed from film. All earlier images taken from orbit had been beamed back in electromagnetic form. In an elaborate operation, the capsule was scooped up by a C-119 recovery plane moments after the capsule had re-entered the Earth's atmosphere.

NASA announced Project Apollo in July 1960, a three-man space flight programme intended at that time to follow on directly from the one-manned flights of Project Mercury. Future goals included sending crews to an orbiting space station, manned orbital flights around the moon, and – at some unspecified time in the future – manned landings on the moon. Von Braun's Saturn rocket was chosen as the Apollo launch vehicle.

Mercury's first test launch at the end of that month was a disaster. Only 58 seconds into the flight the telemetry signals were suddenly lost. The air force's Atlas booster rocket suffered structural failure (some observers claim they heard an explosion) at 30,000 feet. The Atlas was clearly too unreliable to be used to send human beings into space any time soon. For the next test (four months later) one of von Braun's Redstone rockets was used. It was almost as much of a washout as the Atlas launch. The Redstone rose a few inches and then came down again, leaning to one side. It was known afterwards as the four-inch flight. The problem turned out to be a plug that had one prong shorter than the other because a worker had filed it down to make it fit the socket. A further test in December, also using a Redstone, was a qualified success.

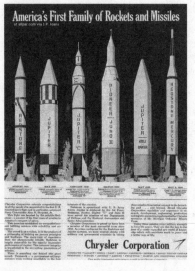

On 31 January 1961 a Redstone sent a chimpanzee – Chimp No. 65 – into a sub-orbital flight. He was returned safely but had a rough ride. Electric shocks were administered to the soles of his feet as a way of getting him to respond, but something went wrong with the mechanism and he received far more shocks than

Chrysler won contracts to build Redstone rockets and later Jupiter rockets, achievements celebrated in this advertisement had been planned. He was also shot further into space than was intended, and it took hours to find the capsule when it returned to Earth. Afterwards, when it was clear that he had survived, he was given the name Ham. According to NASA's official log, on his return 'Ham appeared to be in good physiological condition, but sometime later, when he was shown the spacecraft, it was visually apparent that he had no further interest in cooperation with the space flight programme.' Because of the problems with this flight, the first manned flight planned for the end of March was delayed.

Ham in his spacesuit

On 21 February 1961 an unmanned Mercury flight once more boosted by an Atlas rocket was only a qualified success.

On 9 March the Soviets launched a dog named Chernushka (Blackie), 80 mice, guinea pigs, various reptiles and a man-size dummy given the name Ivan Ivanovich into orbit. Inside Ivan a tape recorder played Russian choir music. Chernushka was recovered unharmed from the capsule. Ivan Ivanovich was automatically ejected and made a parachute landing.

On 23 March Russian cosmonaut Valentin Bondarenko was killed in a fire. He was undergoing tests when the oxygen-rich environment in which he was being held ignited. The next day Korolev heard that von Braun had successfully launched a Mercury capsule using a Redstone rocket. It had originally been scheduled to be the first manned Mercury flight but ended up being the last unmanned test flight.

The director of Project Mercury, Bob Gilruth, had told the Mercury astronauts to vote among themselves who should be the first American, perhaps even the first human in space. The press thought it was a foregone conclusion that John Glenn would be chosen. In fact Alan Shepard was the astronauts' choice. None of them voted for Glenn. When Shepard was told that his flight had been postponed, he said that von Braun was being too meticulous and that they should have risked it.

On 12 April 1961, Korolev's Vostok 1 was ready for liftoff. Korolev had said that his first cosmonaut should be less than 5 feet 9 inches tall, weigh less than 160 lbs, aged between 25 and 30, and 'above all, he should be a man with a smile'. Inside Vostok 1 sat a cosmonaut named Yuri Gagarin, 27 years old, 5 feet 2 inches tall; indeed chosen partly for his smile, but also because of his peasant background and because he was so popular among his peers. He had also proved to be one of the most proficient during his accelerated training.

> *Sergei Korolev at Mission Control:* 'Preliminary
> stage ... intermediate ... main ... lift off! We wish you a good
> flight. Everything is all right.'
> *Yuri Gagarin:* 'Поехали!' ('*Poyekhali!* Let's go!').

Afterwards, 'Let's go!' became a catch phrase in Russia that symbolized the beginning of the space age.

When the craft re-entered the Earth's atmosphere, the capsule was supposed to separate from the instrument module, but that didn't happen and both began to spin dangerously fast. Gagarin was also experiencing eight times his normal body weight in g-forces and came close to losing consciousness. Suddenly the bundle of wires that connected the capsule and module to each other snapped. The capsule stabilized. But then, as it continued to accelerate through the atmosphere, surrounded by flames, there were loud cracking sounds coming from the direction of the heat shield, chunks of which were breaking away. The sunlight was blinding. A parachute opened, the re-entry velocity began to

Yuri Gagarin

decrease. At around 7,000 feet, Gagarin ejected and parachuted to Earth. Gagarin landed in a potato field outside Engels, a city in the Saratov region of south-western Russia one hour and 48 minutes after liftoff. Local farmers were the first to find him. The following day was declared a national holiday.

A journalist who first heard the news as it arrived at Washington at 5.30am phoned NASA HQ. The hapless person who answered the phone said, 'Hey who is this? We're all asleep down here.' Headline the next day: 'Soviets put man in space. Spokesman says US asleep.' In fact at that period NASA personnel were working 18-hour days. The President had been asked if he wanted to be woken up if there was news. He said no. Shepard said, 'We had 'em. We had 'em by the short hairs and we gave it away.'

Days after the world's first orbital flight, Gagarin and Khrushchev appeared on the cover of *Life* magazine. A few months later Gagarin toured the world, visiting Czechoslovakia, Bulgaria, Finland, England, Iceland, Cuba, Brazil, Canada and Hungary, not the USA. He had never been abroad before, not at ground level anyway. In London he met the Queen and the Prime Minister, Harold Macmillan. In Manchester he was granted honorary membership of the Amalgamated Union of Foundry Workers, then one of the world's largest unions. 'Manchester's toiling masses,' an article in *Pravda* reported, 'accorded Major Gagarin a reception unsurpassed in its cordiality.'

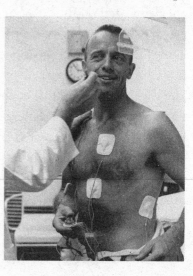

On 5 May 1961, after 20 unmanned Mercury launches, Alan Shepard was strapped into position. The Mercury capsule, named *Freedom 7* – under 11 feet long and just 6 feet wide (a launch escape system added another 15 feet to the overall length) – sat atop one of von Braun's Redstone boosters, now known affectionately as Old Reliable, partly in contrast to the Atlas rocket. Nine out of ten Atlas launches had been unsuccessful. Von Braun was known to make disparaging remarks about Atlas, as were the astronauts. Despite earlier attempts to keep von Braun out of Mercury, here he was at the heart of it.

Shepard was getting impatient. There had been delay after delay. 'Why don't you fix your little problem,' Shepard told the flight control centre, 'and light this candle.' He had to wait so long that he was desperate to urinate. Though he was told not to, he couldn't wait any longer and urinated into his suit. At 9.34am the candle was finally lit and Shepard was sent into a sub-orbital

Mercury capsule

flight that reached an apogee of 100 miles and brought him back to Earth a mere 15 minutes and 22 seconds after takeoff. The capsule landed in the Atlantic 300 miles away from where it was launched. It was hardly a match for Gagarin's orbital flight of one hour and 48 minutes, but it was a start.

Alan Shepard

CHAPTER TWELVE

On 1 May 1960 an American U-2 spy plane was shot down in Russian airspace. To save face, the Eisenhower administration encouraged NASA to release a press statement saying that the plane was a weather-research plane, and that it had accidentally flown into Russian airspace after the pilot had lost consciousness. One can only imagine how much Khrushchev must have relished his next move. He produced both the CIA pilot, Francis Gary Powers, who had been captured unharmed, and the plane's intact spying equipment. Eisenhower had been advised that the plane would have blown up mid-air and that the pilot was almost certainly dead; now he was caught out in a lie, just months before an election. Tensions between America and the USSR escalated. The lie also fed into what was becoming a growing rift between press and government in America, a rift that would only widen during the 1960s.

On 8 November 1960 John F. Kennedy beat Richard Nixon to the Presidency. America's oldest-ever President was succeeded by its youngest. During his campaign Kennedy had taunted Nixon about Eisenhower's caution on defence, and his lack of investment in space. In his farewell address to the nation Eisenhower 'highlighted the danger of allowing the political and economic interests of military contractors and bureaucrats to hijack the national security agenda for their own gain'. Kennedy had played to public fear, encapsulated in the notion of a missile gap, a term he had begun to use frequently from 1958 onwards. The truth, however, was somewhat different. Russia was still some way away from having an operational ICBM when Kennedy took office, whereas America now had 160 operational Atlas ICBMs and nearly 100 Thor and Jupiter IRBMs. The UK had taken the bulk of the Thors, from where they could reach Russian soil. Bizarrely, the agreement was that the weapons, once on British soil, would become British property and were to be deployed by the RAF, but the nuclear warhead each carried would continue to be American property and come under American control. No one wanted von Braun's Jupiter missiles, except with great reluctance Italy and Turkey, which took some in 1959.

Korolev's 'workhorse' R-7 rocket was a crucial and reliable booster rocket for the Soviets but there were practical considerations – even with a fully functioning heat shield – that ruled it out as an ICBM. The rocket was so large it could only be launched from one location, and so was easily identifiable by an American spy plane. It took so long to fuel that American bombers would have been on the scene well before the R-7 ever got off the launch pad. Once Khrushchev began to realize the R-7's limitations as a missile he looked to other rocket designers. In the end only seven R-7s were ever deployed and none for military purposes. The R-16 ICBM was the missile on which Khrushchev was pinning his hopes, the work of the rocket designer Mikhail Yangel.

The Soviets had attempted to test-launch an R-16 ICBM in October, the month before Kennedy won the Presidential election. The massive rocket blew up on the stand and vaporized the Chief Marshal of Artillery, Mitrofan Nedelin, and 71 officers and engineers (some accounts give a much higher number). Yangel, who was several hundred yards away smoking a cigarette, survived. Later, Khrushchev brusquely asked him why he was still alive. Yangel suffered a heart attack shortly afterwards. An Italian news agency reported the disaster a couple of months after the event but the source was unconfirmed. The accident remained a rumour in the West until it was confirmed by Russia in 1989.

NASA's first Director, T. Keith Glennan, left his post in January 1961, coincident with President Kennedy assuming office. In just a few years Glennan had turned NASA into an enormous umbrella operation that had subsumed the Naval Research Laboratory and Medaris and von Braun's Huntsville operation, the 8,000 employees of the National Advisory Committee for Aeronautics (NACA), the Langley Aeronautical Laboratory, the Ames Aeronautical Laboratory, the Lewis Flight Propulsion Laboratory, the Jet Propulsion Laboratory, ARPA and several satellite and lunar programmes, including Mercury. Even so, NASA was still fairly modest in size compared to the behemoth it would become.

The new President seemed, for the moment at least, no more interested in space than Eisenhower had been. Kennedy's science adviser, Jerome Wiesner, was even less sympathetic. At least 17 people turned down the job of NASA director when approached by the White House before James Webb accepted.

On 14 April, two days after Gagarin's orbital flight in Vostok I, Webb met with the President and Lyndon B. Johnson. Kennedy still wasn't ready to commit to a space race. Less than a week later, he sent a memo to Johnson asking what America could do decisively in space to beat the Russians. The reason for the change of heart and mind was political. In the middle of that week a brigade of over two thousand Cuban exiles had attempted to invade their homeland, landing at the Bay of Pigs. The operation had been originated by Eisenhower and the CIA, supposedly in secret, in an attempt to wrest Cuba from its increasingly Communist-leaning government, and had been approved by Kennedy in the first month of his administration. America's aggressive involvement had, however, become obvious. Kennedy rejected calls for direct air strikes against Cuba and disastrously failed to provide the air cover the invading forces needed and which had been promised. Over a thousand exiles were arrested, hundreds were injured, and over a hundred killed. The Cuban army suffered a greater number of casualties but the American invasion was quashed. Kennedy's reputation suffered a severe blow just a few months into his Presidency. It was America's first defeat in the Third World. Relations between Cuba and the Soviet Union became much stronger. Emboldened by his success, and angered both by the presence of von Braun's Jupiters in Italy and Turkey, and by the U-2 spy plane incident, Khrushchev made the decision to base R-12 and R-14 IRBM missiles in Cuba.

Johnson already knew what his answer was to be to Kennedy's question, but decided anyway to ask von Braun what he thought America could do in space to trump the Russians. If Johnson wanted something from NASA, he sometimes circumvented Webb and phoned von Braun directly. Even if Johnson had already made up his own mind, the fact that he was asking von Braun a question he hadn't asked of Webb or of the generals is significant. Von Braun said that America did not stand a chance of beating the Soviets to a manned laboratory in space. There was a 'sporting chance' of beating them to the soft landing of a probe on the moon and of sending a three-man crew in orbit around the moon, though the Soviets might beat the US by sending a single man with minimal consideration of his safety. 'We have an excellent chance,' von Braun wrote, 'of beating the Soviets to the first landing of a crew on the moon (including return capability, of course).' He predicted that the US could, 'going hell for leather', achieve this objective by 1967 or 1968. Johnson had already replied to Kennedy by the time he received von Braun's letter, but their thinking exactly coincided, and some of von Braun's language resonates with the language Kennedy would use when he made his plan known to the American people. A manned moon landing it was to be. Johnson was fired up and so was von Braun. Webb, however, took some persuading, and only agreed so long as there was a long-term commitment to space exploration generally. Early drafts of the President's speech set the target date as 1967, but Webb asked for the vaguer date of the end of the decade.

To lift three men out of the Earth's gravitational pull would require rocket power 10 times larger than any that existed at that time. If it could be pulled off, the achievement would be the result of a show of literal power, which played, of course, into von Braun's hands as pre-eminent builder of powerful rockets.

On 25 May 1961 Kennedy proposed, in a 'Special Message to the Congress on Urgent National Needs', that America shoot for the moon. His science adviser, Wiesner, who remained opposed to the idea, asked Kennedy never to refer to Project Apollo as a scientific enterprise. And he never did.

These are extraordinary times. And we face an extraordinary challenge. Our strength as well as our convictions have imposed upon this nation the role of leader in freedom's cause ... I believe that this nation should commit itself to achieving the goal, before this decade is out, of landing a man on the moon and returning him safely to the Earth. No single space project in this period will be more impressive to mankind, or more important in the long-range exploration of space; and none will be so difficult or expensive to accomplish ... Now it is time to take longer strides – time for a great new American enterprise – time for this nation to take a clearly leading role in space achievement, which in many ways may hold the key to our future on Earth.

'Everyone knows where the moon is,' von Braun said of Kennedy's clear objective, 'what this decade is, what it means to get some people there – and everyone knows a live astronaut from one who isn't.' Some scientists thought that Kennedy's plan was over-ambitious and would take at least 30 years. Eisenhower told reporters that it was 'a mad effort to win a stunt race'. He said it was 'just nuts!' In response von Braun said that Lindbergh's trip across the Atlantic had been a stunt too, but look what happened afterwards, the whole aviation industry took off. He said that Apollo would be the 'wisest investment America has ever made'. It wasn't about getting to the moon any more than for Lindbergh it had been about getting to Paris, it was about being first – 'No one remembers the second man to fly the Atlantic ocean' – and the technological spin-offs that would follow such a race. Eisenhower told the press that it was a waste of $40 billion. NASA calculated that Project Apollo would cost $10 billion. Webb argued that there were bound to be overruns and suggested $13 billion. When he went to Capitol Hill, on the spur of the moment, he came out with a figure of $20 billion, perhaps on the basis that everything always ends up costing twice as much, whether it's installing a new kitchen or flying to the moon. It would eventually cost around $25 billion.

Kennedy expected the proposal to be voted down, but it was carried after a debate that lasted for just one hour with only five senators choosing to speak. NASA's annual budget was raised from $1 billion to $5 billion. If von Braun had got his satellite into orbit first, as he might well have done if he had been given free rein, the space race would surely have taken a different course. It seems highly unlikely that the American public of the time would have so readily funded such a costly enterprise without the motivating force of fear.

In the 1960s, the pace of the space race gathered even greater momentum.

21 July 1961. Second Mercury flight. Capsule: *Liberty Bell 7*. The booster was a Redstone rocket. Pilot: Gus Grissom. Sub-orbital flight lasting 15 minutes and 37 seconds.

6 August 1961. Second Vostok flight. Pilot: Gherman Titov, at 26 years old even today the youngest astronaut to go into space.

(When I don't distinguish between astronauts and cosmonauts in the usual way, I use the word astronaut to mean spacemen of any nationality.) Flight time: 1 day, 1 hour, 18 minutes; 17½ orbits. After he parachuted out of his craft on re-entry, he almost landed on a train track. A gust of wind blew him into a field at the last moment.

29 November 1961. In a dry run for the upcoming first manned orbital flight of an American, Enos, a chimpanzee, was sent into orbit on an Atlas rocket. Like Ham, he was controlled by being given electric shocks. He

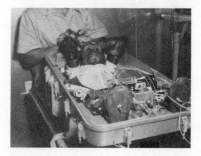

made two orbits of the Earth before the mission was aborted due to technical problems. After he was picked up Enos ran around the deck of the rescue ship ecstatic, shaking the hands of his rescuers, and masturbating. He died of dysentery, unrelated to his experience of being in space, on 2 November 1962.

20 February 1962. Third Mercury flight. Capsule: *Friendship 7*. The booster for this and all subsequent Mercury flights was an Atlas. Pilot: John Glenn. Flight time: 4 hours, 55 minutes, 23 seconds; 3 orbits. John Glenn was the first American to orbit the Earth.

24 May 1962. Fourth Mercury flight. Capsule: *Aurora 7*. Pilot: Scott Carpenter (pictured left with his wife, Rene). Flight time: 4 hours, 56 minutes, 5 seconds; 3 orbits.

11 August 1962.
Third Vostok flight. Pilot: Andriyan Nikolayev. Flight time: 3 days, 22 hours, 28 minutes; 64 orbits.

12 August 1962. Fourth Vostok flight. Pilot: Pavel Popovich. Flight time: 2 days, 22 hours, 56 minutes; 48 orbits. The first time that two spacecraft were in orbit at the same time. The two cosmonauts came within a mile of each other in space.

3 October 1962. Fifth Mercury flight. Capsule: *Sigma 7*. Pilot: Wally Schirra. Flight time: 9 hours, 13 minutes, 15 seconds; 6 orbits.

15 May 1963. Sixth and last Mercury flight. Capsule: *Faith 7*. Pilot: Gordo Cooper. Flight time: 1 day, 10 hours, 19 minutes; 22 orbits.

14 June 1963. Fifth Vostok flight. Pilot: Valery Bykovsky. Flight time: 4 days, 23 hours, 7 minutes; 82 orbits.

16 June 1963. Sixth and last Vostok flight. Pilot: Valentina Tereshkova, the first woman in space. Flight time: 2 days, 22 hours, 50 minutes; 48 orbits.

It would be a long time before America got a woman into space, although 13 women made it through NASA's training programme. Jerrie Cobb, who had set records for speed, distance and altitude, petitioned Congress, but Lyndon Johnson said, 'Let's stop this, now.' John Glenn testified before a House Space Committee in 1962 against sending women into space. His main argument seemed to be that no man would want to see a woman urinate, or worse, in a confined space. Glenn took along his wife, Annie, who was in agreement with her husband's recommendation. Such a ruling is of course unenlightened from our perspective, but other decisions NASA made seem surprisingly liberal, particularly for the time. Some women were employed by NASA as so-called computers. In 1952 Katherine Johnson heard that the National Advisory Committee for Aeronautics (before it became part of

NASA) was looking not just for women mathematicians but black women mathematicians. She got a job working in their guidance and navigation department. Years later she calculated the trajectory and launch window of the first Mercury flight. Shortly after, though NASA began to use electronic computers Glenn insisted that she personally check the computer calculations for his upcoming flight.

In 1961, the first contract awarded after Kennedy's announcement of the moon goal went to the MIT Instrumentation Laboratory. A team headed by Margaret Hamilton, then aged 25, went on to create Apollo's on-board flight software.

In 2015 Katherine Johnson was given the Presidential Medal of Freedom, America's highest civilian award. The same award was given to Margaret Hamilton in 2016.

Each Mercury Seven pilot was paid a salary by NASA of $7,000 a year plus life insurance. No one else would insure them. The modest salary was boosted by a share of a $500,000 deal with *Life* magazine. Three full-time *Life* reporters were assigned to the project. One of them was Loudon Wainwright II, the father of the singer Loudon Wainwright III, and grandfather of Martha and Rufus Wainwright. NASA tried to direct what the astronauts' wives should wear for the first photo-shoot. Pink lipstick, they insisted. *Life* manipulated the photograph and changed the colour to red.

NASA wanted their astronauts squeaky-clean. They were told how to behave at social and media events: long drink, make it last; knee-length socks, no visible calves; answer the question, not too long, not too short. As one of them wryly observed, the world may have been sloppy but at NASA everything was very precise. If only they could all be like John Glenn: Bible-reader, Sunday-school teacher. Glenn had been disappointed not to be the first American astronaut into space; he was even more disappointed that Shepard was chosen. He was furious when it came out that Shepard had been having an affair with a young Mexican woman. Glenn had become the unspoken father of the Seven. He called a closed meeting and lectured them on how they had a moral duty not to philander. Shepard's affair was hushed up. There would be more meetings in the future. The astronauts called them 'séances'.

Shepard may have been the first American in space but Glenn, as first American to orbit the Earth, got the greater attention. He came back to the largest tickertape parade since Lindbergh's. He insisted that all seven of them should parade together, along with their wives.

NASA didn't always get its own way, not even with John Glenn. Vice President Lyndon Johnson wanted to be seen congratulating Glenn's wife and was trying to get into their home along with three major network news crews. Loudon Wainwright was with her at the time, and Johnson sent in word that he was to leave. Annie Glenn wasn't having it and wouldn't let them in. Webb told Glenn he needed him to get his wife to cooperate. Glenn told his wife she should do whatever she thought was best. She told Wainwright to stay where he was. Johnson was furious and later let it be known that he would be happy to see the *Life* contract cancelled. It wasn't, but Glenn had to get to the President himself to save the deal.

Other perks followed as the astronauts quickly became household names: low-rate mortgages, new-ish Chevrolets offered at a huge discount. For whatever reason – perhaps it played better with the public – the car had to be seen to be a used car. By today's standard's these perks might seem modest, but in some quarters there was dissent. NASA may have been a civilian operation but the Mercury Seven astronauts were all from the military, and some of their colleagues were fighting or would soon be fighting in Vietnam.

There were seven of them but only six flights. Deke Slayton

was meant to be the second American to orbit the Earth, but he was grounded when it was discovered he had an irregular heart rhythm. He was, instead, given what would turn out to be a powerful position, the first Chief of the Astronaut Office. It was he who got to decide which astronauts went on which flights. He would not make his own first flight into space until 1975. After Mercury came to an end, the Seven wrote a collective memoir published as *We Seven*, the title seemingly in homage to Lindbergh's first autobiography, *We*.

Almost anything anyone did in space during this time was a first of some kind. Gagarin was the first man in space, the first to orbit the Earth, the first to see the curvature of the Earth, the first to experience a substantial period of weightlessness. Previously he had only experienced weightlessness for a second or two in a specially adapted freefalling elevator.

Gagarin had been weightless for less than two hours; Titov was weightless for more than a day. The experience made him unwell for much of his flight. Titov has the dubious honour of being the first human being to vomit in space, a detail the West would only learn much later. None of the Mercury Seven astronauts suffered from space sickness, but roughly half of the future Apollo astronauts would. Something about how snugly the Mercury astronauts fitted inside their capsules seems to have made a difference. Vostok capsules were between two to three times the size of Mercury, still not exactly roomy, but that extra degree of freedom of movement seems to have been what instilled nausea and vertigo in Titov and other cosmonauts. Titov eventually managed to go to sleep during his flight, but after he returned to Earth continued to suffer balance problems for a few days. Such tests of endurance were vital if there were ever to be manned expeditions to the moon, and beyond.

Shepard had to make do with being the first American in space, Glenn with being the first American to orbit the Earth. Nikolayev, the third Russian in space, and the seventh man in space, was the first astronaut to unstrap himself and experience weightlessness unfettered by straps. And so on. The recording of firsts was one way the momentousness of what was being achieved could be acknowledged, but soon even firsts lost their ability to instill wonder or awe.

Another way was through the astronauts' own words; their descriptions of what they had experienced. Gagarin said that weightlessness had made him feel euphoric. Something about the experience was almost childlike. Glenn said it was amazing how quickly he had got used to it: 'You feel completely free. The state is so pleasant ... that we joked that a person could probably become addicted to it without any trouble. I know that I could.' Gagarin had attempted to describe the extraordinary colours he had seen at the horizon. 'Such a pretty halo,' he said. 'The horizon became bright orange, gradually passing through all the colors of the rainbow: from light blue to dark blue, to violet and then to black. What an indescribable gamut of colors! Just like the paintings of the artist Nicholas Roerich.' Roerich was a mystic and follower of cosmism. His paintings were often described as being hypnotic, reflecting the artist's own interest in hypnosis. He is most famous these days for having designed the costumes and sets of the first production of Stravinsky's *Rite of Spring*. Gagarin's appeal to art is telling. It wouldn't be long before other astronauts also appealed to poetry.

Glenn's description of sunset in space was almost identical to Gagarin's: 'What can you say,' he said, 'about a day in which you get to see four sunsets?' As the sun descended, the light in the sky reduced to a band at the horizon, at first 'almost white in color. Then, as the sun sank deeper over the horizon, the bottom layer of light turned to bright orange, with layers piled on top of it of red, then purple, light blue, darker blue, and finally the darkness of space. It was a fabulous display.' After his flight, Carpenter said he took comfort from the fact that out there every sunset, spectacular as it was, looked exactly the same.

From the start there was a desire among the first astronauts to bring back photographic evidence of what they had seen (another attempt to share with the public what they had experienced out there). Titov was the first to take a camera into space. Glenn took a camera, too, but he had had to insist. In the end he bought his own: $19.95 from a local drugstore, and took it on board as personal equipment. NASA hadn't approved, not wanting him to get distracted during what was going to be a relatively short flight. Walter C. Williams, in overall charge of the launches, thought that cameras were a hazard, took up space and added unnecessary weight. It set the tone for what would be NASA's ongoing antagonism to what they sneeringly called 'tourist' photographs. One of the first designers of the Mercury capsule had even wondered if it might not be easier to build one without any windows at all, or perhaps with a periscope. Certainly taking photographs of the Earth had not been a high priority for Mercury. During Shepard's first Mercury flight, he and the instrument panels were recorded almost continuously, but little effort had been put into recording the flight as seen from the Shepard's perspective. A single camera with a wide-angle lens was suspended on a periscope and took whatever photographs it could of the world outside. The photographs were of poor quality and, because of the shape of the lens, distorted. They provided no corroborating evidence of the 'beautiful Earth' Shepard said he had witnessed. Glenn had found it hard to take photographs wearing thick gloves. Though he had managed to take 70, not surprisingly none of them was of great quality. In any case how could he possibly capture on film colours that he described as being more brilliant than any seen in a rainbow?

During the fourth Mercury flight Scott Carpenter became so overwhelmed by the view that he manoeuvred the craft so he could get a better view of the Earth, and ignored repeated warnings from the ground that he was using up a lot of fuel. When he landed, he was over 200 miles short of where he should have been, and couldn't at first be found. 'I'm afraid we may have lost an astronaut,' said anchorman Walter Cronkite. After a few hours a helicopter found him in his life raft, hands behind his head chewing space food. A photo of Carpenter in the raft appeared in the press the following day. Mercury flight director Chris Kraft, who was known for his 'forthright views on almost everything', was heard to say: 'That sonofabitch will never fly for me again.'

NASA was generally suspicious of intellectuals, and even more suspicious of a gathering world force: the 1960s. Carpenter and his wife represented both. They liked literature and philosophy, and skiing. In an interview with *Life*, Carpenter talked of his 'inner weather', words taken from Robert Frost. He worked out on a trampoline, which he described as being like a kind of earthbound flying. It was all a bit beatnik for NASA.

Beatnik: After Sputnik there was a short-lived craze for neologisms ending in –*nik*. Beatnik is the only survivor. It was coined in 1958 by Herb Caen, a journalist working out of San Francisco; the beat in beatnik is as in Beat generation, a term invented by Jack Kerouac just a few years earlier.

Astronauts weren't supposed to be part of the counter-culture. They weren't supposed to get overwhelmed by the view. It wasn't very right stuff. The right stuff was not to care about such things. Scott Carpenter never did go into space again.

🌍

Despite their reservations about 'tourist' photography NASA had set up a small advisory group responsible for photography led by Richard Underwood.

As a young man serving in the navy, Underwood had volunteered to observe the first Bikini Atoll atomic test of 1946. Cameras were banned but Underwood built, and surreptitiously used, a pinhole camera, using photographic film his mother smuggled to him inside a bar of chocolate. He was paid danger money of seven months' double pay, which he used to finance a university degree. He said he learned aerial photography from a light aircraft, hanging out the door with a rope holding him in place. He served in the Pacific in the Second World War; and then, after he had graduated, joined the US Army Corps of Engineers as part of an army project to study the Earth's geological features as seen from the air. After 4,000 hours in B-17s he became deaf in one ear but could identify any region of the Earth from an aerial photograph. 'Every little piece of ground has a signal that it gives off,' he said.

When he married a woman from Honduras, he lost his army security clearance and needed to look for new work. One day the phone rang. A man with a heavy German accent was on the other end of the line. It was Wernher von Braun:

'They tell me you know something about cameras that fly high,' he said.

'Well, up to 70,000 feet,' Underwood replied.

'Do you think one could work at half a million feet?'

'That would be in space.'

'Well, let's give it a try. Come to Alabama. I want to talk to you.'

Underwood didn't need the same security clearance to work for von Braun, and so for a couple of years in the late 1950s he was in Huntsville finding out how to attach cameras to some of the first Redstone rockets fired from Cape Canaveral. The images that came back were almost entirely of the ocean, but the experience got him thinking about photos of the whole Earth. From there Underwood found his way, via the Space Task Force, to Project Mercury, where, at first, because of the objections from senior NASA personnel, he was able to do very little.

In the run-up to the penultimate Mercury flight, Wally Schirra had at first been sceptical about the value of photographs of the Earth seen from space but then made a complete volte face – presumably Underwood's doing – and insisted on taking his own Hasselblad, not a cheap store-bought camera like the one Glenn had taken. He was told that he couldn't for a list of baroque reasons: the leather might emit a gas; without the leather casing, sunlight striking the stainless steel camera might blind him; if the glass broke it would float around the craft in zero gravity, and so on. Schirra was close to tears, reacted so badly that NASA management had a change of heart. He could take the camera so long as modifications were made. Victor Hasselblad himself agreed to make them. The camera was very precise and reliable but Underwood had been allowed to spend just three hours training Schirra in the art of space photography. He came back from the fifth Mercury flight with photographs that were either over-exposed or images of cloud cover.

Schirra, too, had been taken by the sunsets, but, presumably with Carpenter's fate in mind, he said later that he was aware that if he had got lost in wonder he might have wasted a flight and maybe his life.

Gordo Cooper was the youngest and most inexperienced of the pilots but the most experienced photographer of the Seven. He had exceptional eyesight. During the last Mercury flight, he said he could see, from 101 miles up over Tibet, the smoke rising from the chimneys of individual buildings. His reports were at first dismissed. Experts said he must have been hallucinating. We now know that the human eye can isolate linear features in a way a camera cannot, human sight being, of course, nothing like the way a camera sees. We also know that vision is less distorted in space than it is in the Earth's atmosphere. Afterwards his descriptions were independently verified.

Photographic emulsion couldn't capture what the first astronauts saw, nor mere words; perhaps then it is no surprise that, right from the start, God inserted Himself into the space race. Yuri Gagarin said that he had 'looked and looked and looked' but hadn't seen God. Except that Gagarin – a member of the Russian Orthodox Church, who baptized his daughter into the faith soon after his orbital flight – hadn't said anything of the sort. The words, or words like them, were Titov's. 'Some say God is living there [in space],' Titov had said after the Soviet Union's second space flight, 'I was looking around very attentively, but I did not see anyone. I did not detect either angels or gods ... I don't believe in God. I believe in man – his strength, his possibilities, his reason.' Khrushchev had put Titov's words into Gagarin's mouth during a speech the President made in support of the Soviet state's anti-religion campaign, presumably because they would seem more forceful coming from the more famous Gagarin. When John Glenn was asked for his reaction to Gagarin's supposed comment, Glenn pointed out that God, being everywhere, would be no closer in orbit than he was on Earth: 'God ... will be wherever we go,' he said. In the years that followed, almost every returning astronaut would be asked to respond to Titov's observation.

That the space race mirrored the Cold War's fight for technological supremacy was clear; that it also mirrored the Cold War's battle over belief was less obvious. Khrushchev had once ridiculed America's inability to launch anything other than 'grapefruit' satellites; now he was using the space race to ridicule America's God. That God might be more apparent in space than on Earth was crude theology, but Khrushchev had nevertheless thrown down a theological challenge. From this unpromising opening salvo a subtler debate would develop; not just between the USSR and America, but within America itself. Khrushchev's taunt would highlight just how little clarity there was about the place of religion in America.

In a landmark case of 1947 – *Everson v. Board of Education* – the desire of the Founding Fathers to keep church and state apart in America had been affirmed. In his ruling, the Supreme Court Justice Hugo Black stated that laws must not 'aid one religion ... or prefer one religion to another ... In the words of Jefferson, the [First Amendment] clause against establishment of religion by law was intended to erect "a wall of separation between church and State" ... That wall must be kept high and impregnable. We could not approve the slightest breach.' But during the 1950s, the separation had not always been clearly adhered to. In 1954 the words 'one nation' were changed to 'one Nation under God' in the Pledge of Allegiance. 'From this day forward,' President Eisenhower said, 'the millions of our school children will daily proclaim in every city and town, every village and rural school house, the dedication of our nation and our people to the Almighty.' The year before – the year he had been sworn in as President – Eisenhower chose to be baptized into the Presbyterian Church. He had been raised into a family of Jehovah's Witnesses, or perhaps they were Mennonites; opinion seems to be divided. He renounced the family faith before he joined the army, but had now re-embraced faith in a different form. Just as confusing was the change in 1956 of the nation's motto. It had been *E pluribus unum* (One out of many) since 1782 but was now changed to 'One Nation Under God'. How were these changes supposed to square with the idea that church and state were separate? Judgments around the country, too, seemed to show a lack of clarity in law. In New Jersey a school lunchtime blessing – God is great, God is good, and we thank Him for this food – had been ruled illegal because it asked for divine intervention. But for some reason the Lord's Prayer was ruled non-sectarian and was suggested as an alternative, until that too was opposed, and at last silence was put forward as an option. But then it all became decidedly Jesuitical: what if the children were to say grace silently to themselves, and then what would be the point of a silence during which you were to think of anything but God?

Whether or not the constitution mandated a separation of church and state had again become a hot issue during Kennedy's Presidential campaign. As a Roman Catholic he had needed to reassure the electorate that his faith would play no part in his role as President. He repeatedly asserted during his campaign that church and state were separate, and would remain so if he were elected. In a speech he gave in 1960, he said that because he was Catholic and no Catholic had ever been elected President, 'the real issues' had been obscured during his election campaign: 'So it is apparently necessary for me to state once again – not what kind of church I believe in, for that should be important only to me – but what kind of America I believe in.' He went on to reassert that he believed 'in an America where the separation of church and state is absolute; where no Catholic prelate would tell the President – should he be Catholic – how to act, and no Protestant minister would tell his parishioners for whom to vote; where no church or church school is granted any public funds or political preference, and where no man is denied public office merely because his religion differs from the President who might appoint him, or the people who might elect him.' In an address to the Greater Houston Ministerial Association in September 1960, he said that he believed 'in an America that is officially neither Catholic, Protestant nor Jewish ... For while this year it may be a Catholic against whom the finger of suspicion is pointed, in other years it has been – and may someday be again – a Jew, or a Quaker, or a Unitarian, or a Baptist.' Or an atheist, he did not add.

Some Protestant American churches were set against a Catholic President and were publishing anti-Kennedy literature. The IRS had made it clear that those churches risked losing their tax exemption status. The Revenue Service's action showed that it was indeed possible by law to keep church and state apart.

During the year of Kennedy's election, an American atheist by the name of Madalyn Murray filed an action against her son's school, which had expelled her atheist son for refusing to attend the school's daily prayer meeting. By 1963 the case – *Murray v. Curlett* – had made its way to the Supreme Court. On 17 June, just over a month after the last Mercury flight, the state make its strongest ruling yet in its affirmation of the separation of church and state. Eight of the nine Supreme Court justices ruled in Madalyn Murray's favour, bringing to an abrupt end the practice of religious observance in the country's public schools (called state schools in the UK). Madalyn Murray had set out her life's work.

INTERLUDE

Madalyn Murray began her life as Madalyn Mays. She was born on 13 April 1919, the younger of two children. Their father John was one of 13, and so was their mother Lena. John and Lena had married young and fled to Pittsburgh in an attempt to escape the desperate poverty of Appalachia. A few months into her second pregnancy, Lena tried to abort her child. The first birth had been so painful she did not want to go through that experience again. She took hot baths, ingested herbal abortifacients, jumped out of a second-floor window, threw herself down the stairs. The foetus stayed put and Madalyn emerged at full term. Madalyn was born 'in the caul', which in medieval times was taken as a sign of good fortune or of a great destiny. Caul births are rare – a one in 80,000 event. The caul usually covers just part of the skull and face (the word is a variant of cowl); rarer are cauls that cover the whole body as Madalyn's did. Rarer still, her caul was not translucent but dark. Lena, a psychic, must have wondered at the omen. It had once been the custom to present the caul to the mother as a keepsake, a tradition surely no longer common in Lena's day; but whether or not he knew that he was keeping a tradition alive, the doctor did offer the caul to Lena, and she kept it for many years.

John's construction business thrived for a time, well enough anyway that he could build his family a small brick house, but as the Depression deepened the business collapsed, and John turned to bootlegging. His drinking establishment also served as a brothel. When Franklin Roosevelt repealed the Prohibition laws, John lost his way of making a living yet again. In the years to come he would often say that Roosevelt had ruined his life. Madalyn remembered that as a girl she was encouraged to lie on the back seat of the truck, the hooch hidden underneath, pretending to be asleep.

In her early teens Madalyn had been religious. She claimed somewhat improbably that she had read the Bible voraciously over a weekend, from cover to cover. One day, as she was walking to her grandparents, she was blinded by the sun reflected off snow. She had been thinking irreligious thoughts at the time and wondered if God was punishing her. And then she wondered what kind of God would punish a young girl. This was the story she told in later life of her conversion, curiously paralleling that of Saul on the road to Damascus, but in reverse.

Madalyn was a student at the University of Toledo for a while after the family moved to Ohio, and then at the University of Pittsburgh when the family moved briefly back to Pittsburgh. She worked as a secretary part-time in order to pay the fees.

In 1941, at the age of 22, she met and fell in love with John Henry Roths, a steel worker. They ran off to Maryland together and got married. Very soon after they went their separate ways, both joining the military. Pearl Harbor had been attacked a few months earlier. Roths joined the marines and Madalyn the Women's Army Corps (WAC). Roths was posted to the South Pacific, and Madalyn to a number of placings in Europe and North Africa. She claimed she worked as a cryptographer on Eisenhower's staff in Rome and had high security clearance. Madalyn enjoyed the camaraderie of the army, and wanted society to be like that. She embraced socialism as the means. She was made a Lieutenant. She fell in love with William J. Murray Jr, a bomber pilot in the Eighth Army Air Corps, and became pregnant in September 1945. She had by this time divorced Roths. Murray came from a wealthy Long Island Catholic family and was already married. Murray ignored Madalyn's pleas that he should divorce his wife and marry her. He refused, even, to admit that he was the father of her child. And then he disappeared. Madalyn railed at and cursed God, challenging Him during a storm to strike her down with a lightning bolt. That He did not was further proof that He did not exist.

She gave birth to a boy on 26 May 1946. She named him William Joseph Murray III and changed her own surname to Murray, then sued the father for paternity. It was the first of many legal actions she would file during her lifetime. She had stumbled on her modus operandi. The jury laughed when Madalyn presented a pair of Murray's pants as evidence of their intimacy, but she won the case and Murray was ordered to pay $15 a week until Bill was 18.

After the war Madalyn used her veteran status to take out a loan and buy a house. She was back once more in Ohio. The whole family moved in, including her brother Irv and her mother and father. She went back to college, her choice limited to what was closest at hand, which was Ashland, an evangelical university. As a veteran her fees were paid for, but she was looking after the entire family and took a job at the Akron Rubber Company. Her brother Irv and her father John never had a full-time job the rest of their lives. In 1948 at the age of 29 she received a degree in history, ranked second in her class of 43. For several years she studied law, taking night classes at a law school run by the YMCA. She left in 1952 with another degree but failed her Bar exams. She got a job at Glenn L. Martin Company, the aircraft company that would later manufacture both the Vanguard and Titan rockets. Madalyn loved books and reading, but suburban life got her down. The satirical Malvina Reynolds song 'Little Boxes' didn't come out until 1962 but Madalyn already characterized suburban life as 'identical houses with identical people in them'. She wanted her life to mean something. She was intelligent and restless. She moved from job to job, falling out with her bosses. She said they could not cope with her wit and intelligence.

Madalyn was never very motherly. Bill said it wasn't until he was seven years old that he realized that Madalyn actually was his mother. He had taken to calling her by her first name. It was only when she told him to stop that the penny dropped. She may not have been maternal in a conventional way, but she was ambitious for her son. That was the year she gave him a microscope for his birthday. At the age of nine he was attending political meetings at home.

In 1953, Madalyn met a local Italian man and was pregnant again sometime in 1954. It led to a row at home that Bill remembered particularly, even though fights were a common occurrence. They were not a typical family, he said: 'We argued about the value of the American way, whether or not the workers should revolt, and why the Pope, Christians, and Jews – anybody who believed in God – were morons.' Everyone shouted and swore at each other all the time. There were physical fights too. Things were thrown across the room, blows were struck. But on the night Madalyn told her father about her pregnancy things escalated. Madalyn got hold of a kitchen knife, ran across the room with it, and managed to slash both her brother Irv and her father. She then ran out of the room cursing her father: 'I'll get you yet. I'll walk over your grave.' 'It was just another day,' Bill Murray wrote in his autobiography, 'in my life without God.' Bill said the only time there was peace at home was when his mother was depressed, which was, however, fairly often.

Jon Garth Murray was born in November 1954. Madalyn was as distant with Garth as she had been with Bill. She didn't touch him, and he was left for hours banging his head against the headboard of his crib for attention that never came. Around the time of his birth Madalyn began to talk about moving the family – from Baltimore, where they were they were now living – to Russia. After the launch of Sputnik in 1957 Madalyn wrote in her diary: 'I am aglow with joy...and have enough pride in it that one would think they were my own accomplishments.' Later, Madalyn encouraged Bill to clip out every mention of Gagarin. A signed framed photograph of the cosmonaut hung on their wall. Not that Bill needed much persuading: 'I had begun to indulge in fantasies about flying machines,' he wrote. 'The dream of flight, of mighty rockets propelling me into space toward far-off worlds, thrilled me. I dreamed of the day I would climb aboard one of those sleek ships and find my hands at the controls. I regularly read one or two science-fiction books each week'. Madalyn had an idea that she could use Bill's massive scrapbook – larger than anything, Bill said, Gagarin's own mother might have compiled – to impress officials at the Soviet Embassy. In the meantime, she encouraged Bill to use his school reports to write approvingly of the Soviet Union whenever the opportunity presented itself. In one report his mother persuaded him to write glowingly of holidays on the Caspian Sea, so much better than anything that America had to offer; in another, why the Soviet government and economic system were so effective. In the playground Bill was bullied for being a Commie traitor. At home Madalyn subscribed to *USSR* magazine, produced in the likeness of *Life* magazine. It was printed in Virginia to a quality far superior than anything that came out in the USSR at the time.

Madalyn managed to get a grant from the National Institute of Heath to study at Howard's University Graduate School of Social Work. That put her plan to defect on hold for a while. But her thesis was rejected, and she inevitably fell out with her professors. At home she continued to study and read, feeding her particular love of history.

She had begun to despair of Bill's poor performance at school: 'He was failing utterly,' she wrote in her diary. 'He cannot learn, he cannot read. He cannot translate thought into words ... I've engaged in an ugly brutal battle with the school because of it, but ... I grieve almost as if for a lost cause.' She would criticize his spelling, and then say to him: 'Men are stupid. No matter what the subject, I've always picked it up with ease. But not you, or your father, or your uncle, or your grandfather. There's not a quick-witted man in the world.' Bill said that what was really holding him back was that he was the only boy at school still wearing short trousers. Madalyn had transferred the ambitions she had held for Bill to Garth. Before Garth was three years old, she wrote in her diary that she had 'a mystic assurance ... Garth will have as much effect on the world as Jesus Christ, Freud or Marx have had on the total Western historical development'.

Even when she was at her most despairing, Madalyn retained her ability to self-aggrandize. If her life sucked, no life had ever sucked quite like hers. 'Here I am on the edge of 40,' she confided to her diary in 1959, 'with no references ... with all this social work background ... with two kids and no husband ... has anyone anywhere been such a glorious failure.' And yet her depressions were real enough and she contemplated suicide a number of times. She was restless and needed to find an outlet for her energy and intelligence. Perhaps in Russia she would be appreciated. She wrote to the Soviet Embassy in Washington expressing her desire to emigrate. When the staff at the embassy proved to be less than helpful, she took matters into her own hands. She had an idea that if she got the family to France it would offer an easier way into the USSR. She and the boys departed America on 24 August 1960 on the *Queen Elizabeth*. She left a letter officially renouncing her citizenship with her mother, telling her to mail it after they were safely out of the country. Lena consulted her tarot cards and saw that the family would be back. She decided to ignore her daughter's instructions. It was probably around this time that the FBI began to take an interest in Madalyn. Years later, when she got hold of her files through the Freedom of Information Act, she said that the first of several files was a bundle of some 650 pages.

Needless to say, getting the USSR to accept them as citizens was no more easily accomplished through the Soviet agencies in France than it had been in America. Bill said they had as much trouble getting into the USSR as most people had trying to get out. After about a year away Madalyn decided to call it a day. Back at the US border the immigration officer asked them if they had been to France on vacation. 'No, sir,' Bill said, 'trying to defect.' 'Very funny, young man,' the immigration officer said as he stamped their passports and waved them through.

The day after they returned home, Madalyn took Bill to be enrolled at school. They arrived just in time to hear morning prayers being recited. According to Madalyn's version of events, Bill was visibly disturbed and so she asked the enrolment officer if her son, an atheist, might be excused prayer. The officer said that would not be possible. An argument developed, at the end of which he said, 'Then why don't you sue us?' And so history was made. In Bill's version, his mother walked straight up to a young official at the school and started swearing at him.

Madalyn told Bill to keep a log of all religious observance at the school. Madalyn put her request that Bill be excused Bible reading, prayer and the reciting of the Pledge in writing. The school said no, and the battle began. Madalyn wrote to the *Baltimore Sun*. In her letter she touched on prayer in schools, the legend 'In God We Trust' on coins, and the words 'under God' in the pledge of allegiance, as three examples of unconstitutional intrusions of the church into the workings of the state. She cited the First and Fourth Amendments. She had done her homework, and she had a stroke of luck. Her letter caught the attention of a sympathetic journalist at the *Baltimore Sun*, and he decided to take up her cause. He did a bit of investigating and discovered that her challenge would be the first test since 1905 of the ruling that church and state were to be kept separate. (He seems to have overlooked the 1947 ruling in *Everson v. Board of Education*.)

At Bill's school, and in every school under the board's jurisdiction, the Bible or the Lord's Prayer was invoked before lessons every school day. Similar but varying rules applied across the nation. When the paper announced that Madalyn Murray was to challenge the ruling, calls poured into local radio stations and TV networks.

Madalyn had been keeping Bill away from school but the American Civil Liberties Union (ACLU) came on board and advised her to send him back. In order to test the judgment, the school would need to expel Bill for staying away from morning prayers and Bible study classes. Only then could a lawsuit follow. The next day Madalyn and a horde of pressmen accompanied Bill to school. Bill heard one reporter ask his mother what hospital he was to be sent to if he were injured. Schoolboy Bill played his part to perfection. He can be seen on film saying, in a sweet unbroken voice, 'I'm an Atheist. I wish to be an Atheist.' On the way someone shouted out, 'Why don't you move to Russia.' Bill said he had wanted to shout back, 'We tried, but they wouldn't have us.' On the way home, when there was no press around, he was attacked and could have been seriously injured.

For a while the school refused to play along. And then, worn down by the effort of trying to exclude him from morning prayers before he excluded himself, they played into Madalyn's hands and expelled him. A lawsuit against the school was filed, marking the beginning of a series of legal challenges. The Board of Education asked the Attorney General of Maryland to rule on the complaint. The case was dismissed on 27 April 1961. Round one to the Baltimore Board of Education. The next day Madalyn filed an appeal.

The Blue Marble.

On May 5, 1961, Alan Shepard becomes the first American in space. His Mercury capsule, named Freedom 7, sits atop one of von Braun's Redstone rockets (known affectionately as Old Reliable).

During the Gemini IV mission, Ed White takes the first American spacewalk.

Wernher von Braun's Saturn V rocket remains, to this day, the most powerful rocket ever launched.

The first color photographs of the whole Earth were taken in 1967 by the DODGE (above) and ATS-3 (below) satellites.

Earthrise in its original orientation as released by NASA, and as the Apollo 8 crew saw it. The mass of white at the bottom left-hand side of the Earth, as seen in this image, is Antarctica.

Earthrise as it is usually reproduced.

Jack Schmitt collects lunar rocks during the last manned mission to the moon.

She was energized. She found new things to protest: the 'Pray for Peace' Stamps which the Post Office had been issuing from 1956, tax exemption for religious institutions, the altered wording of the Pledge . . .

Her new-found notoriety attracted the attention of Paul Krassner, founder of *The Realist* magazine. On the masthead of the first issue, which had appeared in the spring of 1958, the magazine declared itself to be a forum for 'social-political-religious criticism and satire'. Over the years contributors included Ken Kesey, Richard Pryor, Woody Allen, Joseph Heller, Lenny Bruce and Terry Southern. Krassner invited Madalyn Murray to contribute. In her first article Madalyn set the tone of her future writings: 'I am against religion. I am against schools. I am against apple pies. I am against "Americanism". I am against mothers. I am against adulterated foods. I am against nuclear fission testing. I am against commercial television. I am against all newspapers . . . I am against Eisenhower, Nixon, Kennedy, Lodge. I'm even against giving the country back to the Indians. Why should the poor fools be stuck with this mess?' Over the next few years her combative style hardly changed.

Krassner allowed Madalyn to use her column to solicit funds to support her action, and she received thousands of dollars as a result. One of the largest donations was from Carl Brown, an 'atheist-nudist wheat farmer from Kansas'. His first cheque was for $5,000. He also sent bonds as a gift to her two boys, as well as deeds to a 160-acre site in Kansas on which to set up an atheist colony and atheist university. Madalyn visited the land with a news crew in tow. The locals couldn't have cared less and Madalyn soon lost interest. She only really thrived on adversity.

In January 1962, the Maryland Court of Appeals ruled four to three that the First and Fourth Amendments weren't meant 'to stifle all rapport between government and religion'. The appeal was thrown out. But the school board had made a major

Malice in Maryland

by Madalyn E. Murray

[Editor's note: Longtime Realist readers will recall the Kafkaesque misadventures of Madalyn Murray and her son, Bill, in their challenge of compulsory Bible-reading and Lord's Prayer-recitation in Maryland (and consequently all) public schools. Now that the case is going to be decided by the U.S. Supreme Court, Mrs. Murray has become a commodity of the mass media, a target of public hostility and an object of private pygmalionism; her family is under police protection.]

If I can't come through this case the same offensive, unlovable, bull-headed, defiant, aggressive slob that I was when I started it, then I'll give up now. My own identity is more important to me. They can keep their gawd-damn prayers in the public schools, in public outhouses, in public H-bomb shelters and in public whorehouses.

I am being tainted, corrupted, worked over, used, remolded, undone, reworked and tarnished. I am under pressure, suasion, used in power politics, maligned, abused, damned with faint praise and otherwise handled. I don't care what people think—and I am tired of the voices closing in from all sides:

"Madalyn, for Christ'sake, don't greet those TV men in your bare feet." "Madalyn, didn't you wear a bra for the Supreme Court, even?" "Madalyn, you can't say 'Fuck the religionists' on a national radio network." "Madalyn, you don't dare give this speech tonight when you offend the Jews' religion in it." "Madalyn, put down that daquiri, there are photographers here." "Madalyn, get the hell out of that picket line before you are recognized." "Madalyn, you just can't belt those boys merely because they belted Bill." "Madalyn, if you pull that again, he'll charge you with contempt of court." "Madalyn, you can't be seen walking arm in arm with a Negro." "Madalyn, you have got to say you are an agnostic, this atheist bit makes you into a Communist." "Madalyn, why in the hell don't you wear some make-up?" "Madalyn, will you please quit talking about Anarchy." "Madalyn, you have to quit lampooning Kennedy." "Madalyn . . ." "Madalyn . . ." "Madalyn . . ."

And then the mail rained down. These are authentic quotes from letters I have received:
• Two weeks ago I asked you to send me a picture of yourself in a G-string. I have not yet received. . . .
• I own only 4,000 shares of A.T. & T., and my problem is whether to convert to. . . . P.S. Here is my 50c a month pledge. Please acknowledge by airmail, special delivery, registered, certified mail.
• Dear Sister Murray: I am in my seventeenth reincarnation. I vividly remember that day they nailed me to the cross. . . .
• Your picture was in the Evening Bulletin. I am sitting here with a hard-on, writing this letter. . . ."
• I notice you own your own home. If you could sell this and send me the proceeds in a certified check, I would. . . .
• Christ is ever with us. . .
• You Communist whore. How dare you spread your vile filth in this nation, the greatest, the most wonderful, God given. . . .
• You have a lot of nerve expecting people to help you with money. If you are not in this for your principles. . . .
• Hurrah for you! We need more people who are willing to stand up and be counted. However, I must remain anonymous because. . . .
• I see you are divorced. If you can support me. . . .
• Three months ago I sent you a manuscript to edit, print, and have published. I felt that by now I would be receiving royalty checks, but you slimy. . . .

• I am instructing my attorney to sue you for three-quarters of a million dollars. . . .
• Don't you see the folly of trying to be open about this? You should join the Unitarian church and. . . .
• I know you are only in this for the publicity, so I'll be frank. If you could include me somehow in your next TV interview. . . .
• Your manuscript for our October issue has been received. I am deleting the first 15 paragraphs as being too militant. The next section dropped because of being entirely too bold is. . . .
• I see your son is 16 now. I have this wonderful 13 year old daughter he should meet. She is only two months pregnant. . . .
• I understand you need a job. Our Negro maid quit us because of our overworking her. This position is open to you because we are on your side. . . .
• You are a brave and wonderful woman. I know you get a lot of contributions for what you are doing, so I need $300 by the end of the week. . . .
• The issue of Atheism vs. Religion is one which divides the masses. In these perilous times, we Communists feel you must drop your case so that the masses can unite against the Capitalist swine who. . . .
• As a Birchite, I know you must understand the urgency of turning the tide against the Communist swine. . . .
• The real battle today is not with religion. We Trotskyites discern the basic issues. . . .
• Don't waste time on fighting religion. The most noble issue of Racism is before us now. . . .
• I understand that you paid your attorney $1,000 to appear in the United States Supreme Court for just one day. This is a ridiculous waste of the money we have contributed to you and I want my $2.00 returned. . . .
• Last month I sent you a newspaper clipping which you have not acknowledged to date. If this is how you treat those who are fighting with you. . . .
• You are handling your case all wrong. What you must do immediately. . . .
• Enclosed you will find 17 pages of suggestions on how to be more effective in your present case. . . .
• My husband receives your newsletters and I want you to know you are breaking up our home. What will my poor children do. . . .
• In June of 1961 I sent you $3.00 [but] after reading Mr. Arnoni's column. . . .

(Continued on Page 15)

Some Of My Best Friends Are Artlovers

From Bill Slocum's column in the New York Daily Mirror:

I tell another man's story here. He has just returned from his native Austria and when he told me this tale it startled me. I tell it without comment. I doubt that any comment is possible.

This man returned from Austria told me he had come across an old friend. He asked the friend how he had fared during the war and the friend replied, "They put me in a concentration camp in 1943. But it could have been worse. The German guards treated me well. Or better than they treated the others."

The man asked his friend how he had accomplished this miracle of decency on the part of the concentration camp guards. The answer was, "I carved statues for them. Dozens of statues. All of the same thing."

And what was it that all the German concentration camp guards wanted?

The fantastic answer was, "Statues of Jesus Christ."

A typically trenchant article by Madalyn Murray from the *Realist*, June 1963 misjudgement. In order to uphold their First Amendment defence they had had to insist that school prayers were not religious in nature. On this, the Court of Appeals said only the Supreme Court could make a final ruling. An appeal was lodged with the Supreme Court on 15 May 1962.

Madalyn was in her element. She had been given a small printing press by one of her supporters, and on 1 July 1962 she brought out the first issue of *American Atheists* magazine.

Madalyn had appointed Leonard Kerpelman as her attorney. He turned out to be an inspired choice: he had once taken on the case of a man who had been denied permission to hold a bullfight in Baltimore's Memorial Stadium. He was an effective lawyer but when, over a year later, the day came when he was standing outside the Supreme Court building about to go in and put his case, he got cold feet. 'I can't go in. I'm afraid,' he said to Madalyn. She had to grip his arm and push him into the building. In the event Kerpelman had nothing to fear. Chief Justice Earl Warren was warm and encouraging, and anyway, it soon hardly mattered what either party had to say because the nine Supreme Court justices – seven of whom were practising Christians – began to argue and discuss the case between themselves. When Justice William O. Douglas asked the attorney representing the City of Baltimore if the school board would allow readings from the Koran, it was clear which way the judgment would have to go. The ruling came down on 17 June 1963, eight to one in Madalyn's favour. The practice of public school prayer was brought to an abrupt end, and it had cost Madalyn less than $20,000. More than

Madalyn Murray

40 per cent of public schools were in violation of the ruling; in some states the proportion rose to 80 per cent. The judgment read:

> The place of religion in our society is an exalted one, achieved through a long tradition of reliance on the home, the church, and the inviolable citadel of the individual heart and mind. We have come to recognize through bitter experience that it is not within the power of government to invade that citadel, whether its purpose or effect be to aid or oppose, to advance or retard. In the relationship between man and religion, the State is firmly committed to a position of neutrality. The breach of neutrality that is today a trickling stream may all too soon become a raging torrent, and in the words of Madison, 'It is proper to take alarm at the first amendment on our Liberties.'

In his statement Justice Clark quoted Jefferson and his 'admonition against putting the Bible and Testament into the hands of the children at an age when their judgments are not sufficiently matured for religious inquiries'. In his concluding statement he said that the application of the rule that the state be neutral required 'interpretation of a delicate sort', but added that 'the rule itself is clearly and concisely stated in the words of the First Amendment'.

The *New York Times* ran more than two full pages of coverage with no ads. There was a photograph of Madalyn, her mother Lena and Garth, but bizarrely Bill was not in the photograph.

The judgment was widely criticized. The most reactionary critics accused the judges of being anti-American, and used the ruling to further inflame anti-Communist sentiment: only America and God could stand up to Russia and godlessness. Liberal elements, particularly those who were also religious, welcomed the judgment. Here, finally, was some clarity: the state need only be neutral in regard to religion, not hostile to it.

After the judgment the cheques began to arrive, tens of thousands of dollars from donors across the country. But Madalyn could hardly leave the house without being abused. Stones were thrown through the windows. She was shot at. It was to be the year of her greatest fame, and a year packed with incident.

Bill and Madalyn after the judgment

Not two weeks had passed since the judgment and her father Pup, as Madalyn called him, was dead. The morning of the day he died, Pup was sitting in front of the television as he did every day of the last decade of his life, the volume turned up full as he liked it. Madalyn got into an argument with him, and soon both were shouting to make themselves heard over the television. Madalyn stormed out of the house, her parting shot, 'You old bastard, I hope you drop dead.' Nothing particularly unusual about the day so far, but when Madalyn returned home later there was an empty chair in front of the TV. 'Where's the old man,' she asked? 'You got your wish,' her mother said tersely. 'Well, I'll be,' said Madalyn, matching her mother's sangfroid, 'Where's the stiff?' Madalyn at first insisted on the cheapest funeral arrangements possible, but then like many a sentimentalist she had a change of heart and bought the most expensive casket on offer. She kept vigil next to the body for two days and nights.

Without asking her son Bill's permission, Madalyn invited his girlfriend Susan to come and live with them. Madalyn thought they should get married. Bill had thought he was just having fun. They were both 17 years old. Anyway, Susan needed the permission of her father, Dr Abramovitz, to marry, and he wasn't about to give it; not because Bill was so young, but because he wasn't Jewish. Dr Abramovitz filed a petition for the couple's arrest. The petition was granted and the police turned up at the house. Madalyn told Bill to distract the police while she drove off with Susan out back. Bill, well-built and 6 feet 3 inches tall, shoved an officer and the officer fell over. Bill was beaten and cuffed. His grandmother Lena came out and joined in the fight, and then fell into a pretend swoon. Irv came out and dragged Lena back into the house, where she came to again. Rather than drive off, Madalyn had decided to enter the fray and tried to spray the officers with her tear-gas pen, but she couldn't get it to work. She scuffled with an officer or two. Bill remembered hearing police shouts of, 'Hit her again.' In court a high bond of $12,000 each was placed on Bill and his mother for assault. One of her supporters paid it. The two of them faced long jail sentences. For five counts of assault on five police officers, Bill was looking at 100 years in jail. Madalyn had eight counts of assault filed against her.

Madalyn announced that they had no choice but to flee the state; they were ready to leave for Hawaii in the next 24 hours. Madalyn had considered Cuba, but Hawaii was easier to get into in a hurry. 'Irv, you don't have to come,' she said to her brother, 'but if you don't, you'll have to learn to cook and make your own bed.' In his memoir Bill wrote: 'Irv left the room to go pack.'

Madalyn had notified the press of their plans. There were journalists at the gate to see them off. Madalyn told them that they had chosen Hawaii because 80 per cent of the population is Buddhist, and Buddhists, she said, are practically atheists. Her remarks made it to Hawaii before she did. She was front-page news before she had even stepped off the plane. Despite the fact that Buddhists are in a minority in Hawaii and that the Governor was a devout Roman Catholic, the party was warmly received. Madalyn told reporters that they were seeking religious asylum there. In their absence, the court in Maryland gave Madalyn a sentence of a year in jail for contempt of court, Bill six months, plus a $500 fine. Procedures for their extradition were also put into place. In August 1964 Hawaii's Governor Burns approved Maryland's call for the Murrays to be extradited. They were arrested, but allowed bail awaiting an appeal hearing.

Apart from the threat of extradition, their time in Hawaii was relatively uneventful. In 1964 *Life* magazine ran a feature on Madalyn, naming her 'most hated woman in America'. The article said she looked like a peasant. She didn't care. She had given up wearing make-up, went bare-legged, her weight ballooned. In photographs she is invariably caught laughing. People who knew her said she was never boring.

In 1965 Madalyn fell for the charms of the self-styled Reverend Keith Rhinehart. He was slim with fine features, not her ideal physically at all, but they soon became inseparable. Rhinehart had set up his own church as a non-profit organization. He was a medium and able to commune with the dead, but his particular superpower was his ability, when the conditions were right, to pull semi-precious stones from about his person; from his chest, his thighs, even from his eyes. The stones conferred healing powers on those who purchased them from him. Rhinehart's church was a successful business, and with all the tax advantages of a not-for-profit. To his followers he was known as Christos Logos Kumara. He was handsome, smart, charismatic, a flatterer and a fraud. Madalyn was in love. Despite her atheism, Madalyn was intrigued by spiritualism. She had been in the habit of visiting her father's grave and talking to him for hours.

Madalyn knew Rhinehart had to be a conman but he was a good one: she couldn't work out how he did it. At one meeting he boldly chose Madalyn herself as the person whom the spirit guide that he had conjured from the other side – Mr Kensington – wished to address. Mr Kensington said that he had three people with him who wished to say something to Madalyn: Laura, Suzanne and

Marie. They were the names of three of Madalyn's relatives, sisters who had remained close to each other all their lives and who had never married. Madalyn had just been thinking about them. As she listened to what they had to tell her, she realized that she was perspiring heavily. Madalyn tried to get Rhinehart to give up his secret but he insisted that he had a real ability to communicate with the departed. She tried, too, to persuade him to stay so that she could perform some controlled tests on him, but Rhinehart soon left Hawaii to try his luck elsewhere.

Next time Madalyn tried to get hold of him he was in prison in Seattle, having been turned in by a 16-year-old hustler. Years later he returned to the jail to put on a show, an event that was written up in the *Seattle Times*. As master of ceremonies, wearing a number of different women's dresses during the course of the evening, he doled out presents to the inmates to the value of $35,000, one a gift certificate of $2,000 to be put towards sex-change surgery. 'Nobody's straight and nobody is gay in our religion,' he told the prisoners at the close of the show. 'They're just plain sexual.'

Madalyn realized it was only a matter of time before the extradition order was upheld; better to leave Hawaii now – and break bail again – before she was forced out. This time she settled on Mexico. She had been told that forged identity papers would be enough to get her across the border. Her papers identified her as Mary Jane O'Connor, an Irish nun. She crossed the border wearing full religious habit.

Madalyn soon got into trouble in Mexico. She schemed to take control of a small college populated by American students and turn it into an atheist university, but the plan went terribly wrong. Anonymously, she reported the students for drug abuse. All the students were arrested, and they could have faced long jail sentences if strings had not been pulled. In turn, Madalyn was implicated, arrested and deported. The Mexican authorities seem not to have known about Maryland's judicial interest in her, and she was sent to Houston, Texas. Within a week she was back in Mexico. She needed to get married.

Richard O'Hair was no more Madalyn's type physically than the Reverend Rhinehart had been, but something between them clicked. He claimed to be an artist, though what kind of artist was never made clear. What O'Hair failed to tell his wife to be was that he was also a CIA informer. For years to come he would file reports on her. Madalyn got an inkling early on that something was amiss when a journalist reported the intended date of their wedding. Only she and Richard knew the date. She came to the conclusion that the apartment was being bugged, and wrote a column about it for *The Realist*. Richard had been brought up Catholic, and after the wedding he received a postcard from his mother formally disowning him. It wasn't long before Madalyn was arrested and deported again. The Mexican police told her that if she attempted to leave the plane she would be shot. Madalyn was inclined to take the warning seriously.

The O'Hairs settled in Austin, Texas. On 12 October 1965, the Maryland extradition procedure caught up with her. The Governor of Texas had signed the papers. Madalyn would have to face the music in Maryland. But then, as Bill later wrote, an angel intervened. At the same time as the Governor of Texas got round to reviewing and signing the order, in the Maryland Court of Appeals the charges of murder against a Buddhist were dropped on the grounds that the jury had been required to swear to a personal God. The defendant's lawyers argued that in Buddhism there is no such being as a personal God. His First Amendment Rights had been violated. The appeal court judges agreed. Between 2,000 and 3,000 grand jury indictments in Maryland were wiped off the books. Madalyn O'Hair's was one of them. Just like that, she was free. She was to spend the rest of her life in Austin, Texas, working hard to keep religion out of public life and strengthening her organization, American Atheists.

On 6 November 1967 Madalyn O'Hair, now aged 48, was one of the guests on the first ever *Phil Donahue Show*. The audience was entirely female. There were murmurs of disapproval from the start when Donahue announced her as the woman responsible for ending prayer in public schools. She ripped out a page from the Bible on air. During the commercial break Donahue walked among the audience. The audience was animated, asking question after question. Donahue realized that they were better questions than the ones he had prepared. After the break he decided, on the spur of the moment, to make a change: he would walk through the audience with a mike, bringing the audience into the show. And so Phil Donahue's distinctive style was born: provoke your audience, then walk among them with a microphone. It was a style of TV that was much imitated, the classic daytime chat-show format.

Madalyn O'Hair had become a fine orator in the style of evangelists like Jimmy Swaggart, Jerry Falwell, Pat Robertson and Oral Roberts. Her arguments were always clear. She said at one point that atheism is not just about challenging the ultimate authority of God, it is about understanding that if there is no such authority then all authority should be challenged, whether it is the state or an employer. By June 1968 O'Hair would have her own radio show, American Atheist Radio Series. In December 1968 Madalyn Murray O'Hair would use her radio show to challenge the authority of NASA.

PART TWO

CHAPTER ONE

For a couple of weeks from the middle of October 1962, the world came close to all-out nuclear war. The Soviets denied that they had a missile base in Cuba and were still denying it on 25 October. The public at the time was unaware that America had missiles based in Italy and Turkey directed at the Soviet Union. Two days later Soviet submarine captain Valentin Savitsky ordered a nuclear torpedo to be 'made combat-ready'. China aligned itself in support of Russia. The world stood on the brink. Kennedy told his aides that the chance of war was between one in three and even. Khrushchev blinked. He agreed to remove his missiles from Cuba, and America agreed to remove theirs from Italy and Turkey, though America's side of the deal was never made public. Khrushchev looked weak. His power at home never recovered.

And yet even by the time Project Mercury came to an end in May 1963, the perceived wisdom was that the Soviets were still way ahead in the space race. How much of this was propaganda from the likes of von Braun, who had an interest in promoting the idea that America was falling behind, or of Khrushchev, in whose interest it was that Russia should be seen to be ahead, will no doubt be debated by historians for decades to come. Even the Defense Secretary, Robert McNamara, wondered if there was something wrong with America as a whole that it could be so easily beaten by the USSR. He thought the country's industries should become more centralized, to the benefit of national strategic goals, as seemed to be the strategy in Russia. NASA's director, James Webb, shared some of the same reservations. They wondered, as Eisenhower had, if perhaps the country had 'over-encouraged the development of entrepreneurs and the development of new enterprises'. In a letter to his predecessor, Dr T. Keith Glennan, Webb wrote: 'My own feeling in this and many other matters facing the country at this time, is that our two major organizational concepts through which the power of the nation has been developed – the business corporation and the government agency – are going to have to be re-examined and perhaps some new invention made.'

When Webb offered George Mueller – then head of research and development at a large electrical engineering company – a top job at NASA, Mueller said he would accept it only if he had a free hand to restructure the organization, a vision that chimed with Webb's call for organizational reinvention. Mueller arrived on 1 September 1963 and soon concluded that a lunar landing would not be achieved by the end of the decade without wholesale reform of the way NASA worked. Mueller decided that Project Apollo's booster rocket Saturn V was taking too long to develop.

Project Apollo's first official mission had been the launching of the first Saturn I rocket live on TV on 27 October 1961. Standing 162 feet tall and weighing 460 tons, it produced 1.3 million lbs of thrust. It was the biggest and most powerful rocket ever launched at the time, 10 times as powerful as the rocket that had put the Explorer satellite into orbit. By the time of the launch, NASA had already requested a more powerful version, the Saturn IB, which would produce 1.6 million lbs of thrust. Shortly after the launch, Saturn V was announced as the launch vehicle of the manned missions to the moon. It would produce over 7.5 million lbs of thrust. (The key Saturn rockets are Saturn I, Saturn IB and Saturn V. There doesn't appear to have ever been, for example, a Saturn IV. There was an S-IV, the designation given to the second stage of the Saturn I, which I mention only to illustrate how baroque, or seemingly arbitrary, the art of rocket classification is.)

Instead of incremental testing, which was von Braun's preferred way, Mueller insisted on all-up testing of Saturn V; meaning a fully fuelled rocket should be flown as soon as possible rather than as a series of flights using dummy stages in varying configurations. Von Braun was by nature risk-averse, or more risk-averse than Mueller anyway, but he was persuaded. And yet von Braun was also perfectly capable of making his own grand gestures. At one point in the development of Saturn V he apparently casually told the engineers to increase the number of engines from four to five. His power may have been limited compared to Korolev's, but this kind of authority would be inconceivable today when, across all businesses, autonomy has been ceded to group decision-making.

For Webb the lunar mission hadn't even been a priority, but it was for Kennedy. Publicly Kennedy had declared that travelling to the moon was 'in some measure an act of faith and vision, for we do not know what benefits await us', but when Webb started talking about the benefits of Apollo to science, Kennedy got irritated. If it was about science, then why, Kennedy asked at one meeting, 'are we spending seven million dollars on getting fresh water from saltwater, when we're spending seven *billion* dollars to find out about space?'. Webb said that Apollo should be about pre-eminence in space, not just one short-term project. Kennedy said: 'We've been telling everyone we're pre-eminent in space for five years and nobody believes it!' From now on – Kennedy made it very clear – NASA's main goal was to get three men to the surface of the moon and safely back again. Everything else was peripheral. Webb, von Braun and others at NASA were disappointed. Von Braun had for years promoted the idea of a space station, and of manned missions to Mars. All these secondary long-term goals were now to be put aside for the short-term, if ambitious, goal that was Project Apollo. The largest casualty was the air force's rival to Apollo, the X-20 Dyna-Soar (another of those painful appellations), which from its inception in 1957 to the time it was cancelled at the end of 1963 had cost over $600 million. The Dyna-Soar was a rocket/plane hybrid. It was designed to be launched into space vertically, powered by one of the air force's Titan rockets, but when it returned to Earth it would glide and land on an airfield. That's how the Space Shuttle would work, but the Dyna-Soar (unlike the dinosaur) was ahead of its time.

How to land on the moon was a key problem that NASA needed to solve. But even before a manned mission to the moon had been planned, NASA had sent out a series of probes to test their ability simply to find the moon; that is, to test their ability to calculate the necessary trajectory. Rangers 1 and 2 failed to escape Earth orbit. Ranger 3 missed the moon by 22,000 miles. Ranger 4 missed the moon by 450 miles and suffered electrical failure. Ranger 5 also failed to find the moon. Ranger 6 was disabled at launch. Finally, five years into the project, in 1964, Ranger 7 crashed into the moon, sending back sharp TV images before it did so. How to land softly on the moon was a challenge yet to come. As to landing men softly onto the moon's surface, it was clear that however it was going to be achieved it would not be by gliding them down in a vehicle like the X-20. An alternative direct approach would be to launch a rocket whose upper stage could somehow back up and land on the moon stern first. If that approach had been taken, Kennedy's date would probably have been missed by a decade. Only now has that technology been developed. The preferred method of landing was called EOR (Earth Orbital Rendezvous), an approach not dissimilar in principle to the method von Braun described in *The Mars Project*. The idea was to launch several small rockets – much easier than launching one large rocket – and assemble the moon landing craft in Earth orbit. Only modest thrust would then be required to get it out of Earth orbit and on a trajectory to the moon.

But there was another solution. Tom Dolan, an engineer at the firm Vought Aeronautics, brought up the possibility of LOR (Lunar Orbital Rendezvous), an idea that was being touted by John C. Houbolt, an engineer at the Langley Center. Why not go all the way to the moon and then launch the lander from moon orbit? That way the moon lander could be much lighter than anything sent directly to the moon from out of Earth orbit.

Von Braun preferred the Earth orbital solution. He thought it was safer and provided a long-term future, as did Max Faget, the designer of the Mercury capsule. The danger with LOR was that if the rendezvous failed, there was no way to get the stranded astronauts back. Kennedy's science adviser, Jerry Wiesner, was against it too: 'They're risking those guys like mad,' he said. But, when he saw which way the wind was blowing, von Braun – always a team player – stepped into line and accepted the lunar orbital solution. Wiesner was furious. Kennedy went around saying, 'Jerry's going to lose. It's obvious. Webb's got all the money, and Jerry's only got me.' Though the rendezvous would take place in lunar orbit, the technology and techniques would need to be tested out first in Earth orbit. Project Gemini, a sequence of 12 two-manned flights, was inserted between Mercury and Apollo to ensure that two craft could meet up and dock in orbit. If two capsules could meet in Earth orbit then there was no reason to suppose that they could not do so in lunar orbit.

Project Gemini had 27 astronauts to draw from: the Mercury Seven and two later intakes. Late in 1962 NASA had introduced a further nine astronauts to the world. Unimaginatively, they were dubbed the New Nine: Neil Armstrong, Frank Borman, Pete Conrad, Jim Lovell, Jim McDivitt, Elliot M. See, Tom Stafford,

New Nine astronauts: *Back row*: See, McDivitt, Lovell, White and Stafford. *Front row*: Conrad, Borman, Armstrong and Young

Ed White and John Young. Lily Koppel in her book on the astronaut wives writes that 'NASA had a protocol officer conduct a New Nine orientation, where he prattled on about how astronauts needed a good breakfast before flying off to work ... Feed him well. Praise his efforts. Create a place of refuge.' All just as it had been before when they were officers' wives. Apparently some of the New Nine wives treated the Mercury Seven and their wives like royalty, called them Sir and Ma'am.

A year later NASA introduced a third intake of astronauts into the programme, NASA Astronaut Group 3. They don't seem to have been given a nickname. There were 14 of them: Buzz Aldrin, Bill Anders, Charlie Bassett, Al Bean, Gene Cernan, Roger Chaffee, Mike Collins, Walt Cunningham, Donn Eisele, Theodore Freeman, Dick Gordon, Rusty Schweickart, Dave Scott and Clifton Williams.

Kennedy had earmarked 29 January 1964 as the day America

Group 3 astronauts. *Back row*: Collins, Cunningham, Eisele, Freeman, Gordon, Schweickart, Scott and Williams. *Front row*: Aldrin, Anders, Bassett, Bean, Cernan and Chaffee

would conclusively show that it had moved ahead of the Soviets. That was the date on which von Braun's upgraded Saturn I rocket, the two-stage Saturn I SA-5 (an intermediary between Saturn I and Saturn IB), would be launched, putting into orbit the largest payload ever. In a speech made on 21 November 1963 at NASA's Manned Space Center, Kennedy referred to the day on which the US would fire 'the world's biggest rocket, lifting the heaviest pay-roll into . . .' He paused, realizing his error, '. . . that is, payload.' He paused again. 'It will be the heaviest payroll, too.' The following day Kennedy was assassinated. 'What a wonderful world it was for a few years . . . with men like you to help realize his dreams for this country,' Jackie Kennedy wrote to von Braun, in response to his letter of condolence. 'Please do me one favor – sometimes when you are making an announcement about some spectacular new success – say something about President Kennedy and how he helped to turn the tide – so people won't forget.'

Meanwhile the Soviets had their own rival to Gemini – Vokshod.

12 October 1964. Vokshod 1. Crew: Vladimir Komarov, Konstantin Feoktistov and Boris Yegorov. Flight time: 1 day, 17 minutes, 3 seconds; 16 orbits.

The flight has been described as the most dangerous of all missions undertaken during the moon race. After pressure from Khrushchev – in order to be seen to be beating the Americans yet again – three cosmonauts were crammed into a craft designed for two. NASA wouldn't send up a three-man crew for another four years. There was so little room that the crew had to fly without spacesuits, a risk NASA would never have taken. One of the crew, Boris Yegorov, was a doctor. It was another first for the Russians: first to send a civilian passenger into space. Vokshod was designed to bring the cosmonauts back to Earth without the need to eject from the capsule beforehand as the Vostok cosmonauts had had to do. The flight lasted just over a day. During that time, back on Earth, Khrushchev was ousted from office. He was replaced by the dour Leonid Brezhnev, who would prove to be less supportive of the space programme than Khrushchev had been.

Vokshod crew
seen after landing.

Pavel Belyayev and
Alexey Leonov

18 March 1965. Vokshod 2. Crew: Pavel Belyayev and Alexey Leonov. Flight time: 1 day, 2 hours, 2 minutes, 17 seconds; 17 orbits.

Alexey Leonov won another first for the USSR when he made the first spacewalk, or in NASA-speak the first EVA (Extra-Vehicular Activity) – the collective term for any activity that takes place outside a spacecraft, including a moonwalk. It lasted 12 minutes and nine seconds. He had kept his primitive spacesuit pressurized using elastic bands, but when he tried to get back into the capsule the spacesuit ballooned and he almost didn't make it. He had to partially deflate his suit, a very dangerous procedure. He lost 12 lbs in body weight trying to get back in his craft. New Nine astronaut Gene Cernan, who would soon have his own problems during an EVA, described Leonov as 'one of the gutsiest men alive'. And yet afterwards, Leonov said of his spacewalk that, 'Nothing will ever compare to the exhilaration I felt.'

23 March 1965. Gemini 3, the first manned Gemini mission. Capsule: *Molly Brown*. Crew: Gus Grissom and John Young. Flight time: 4 hours, 52 minutes, 31 seconds; 3 orbits.

Two unmanned Gemini missions had taken place during the previous 12 months. Gemini was the result of the work of over 4,000 contractors from 42 states. The comedian Bill Dana, a favourite of the pilots, joked that every part had been made by the lowest bidder. Gus Grissom, veteran Mercury pilot, said to his wife shortly before the launch, 'If there's a serious accident in the space program, it'll probably be Gemini, and it'll probably be me.' Even among the testosterone-fuelled astronauts Grissom stood out. He had to win at everything. He had alarmed his bosses at NASA by being a bit too realistic at a pre-launch press conference: 'If we die,' he said, 'we want people to accept it. We're in a risky business.' One of the hardest times for the astronauts' wives was the period immediately after a launch while they waited for news. They called it the Death Watch. When told that green dye would show the recovery crew where the capsule had splashed down, Wally Schirra's wife, Josephine, remarked drolly: 'Is that how we'll know where to throw the wreath?'

Grissom named his Gemini capsule *The Unsinkable Molly Brown*, much to NASA's annoyance. When they insisted he call it something else, he said how about *Titanic*? NASA as a whole was not known for its sense of humor. They seem to have settled on an abbreviated *Molly Brown*. After the first manned mission no other Gemini capsule was given a name.

Gemini 3 was also nicknamed the Gusmobile, because Grissom had had significant input in its design. The astronauts were often

involved in the design process, and New Nine astronaut Mike Collins said it was one of the wisest decisions NASA made. The capsule fit Grissom like a glove, but Grissom was only 5 feet 5 inches tall. When NASA realized that 14 out of

Gus Grissom and John Young

the 16 astronauts couldn't fit into it, the design was modified. Collins said that getting into the Gemini capsule was like getting into a Volkswagen Beetle. As it happens, the Volkswagen Beetle was Lindbergh's favourite car, which he loved precisely because it fit so snugly. He sometimes slept in his car, his legs sticking out the window. He used one of his shoes – he had large feet – as a pillow. The car can now be seen at the Lindbergh museum, in what was once his childhood home.

During the flight, Grissom ate a corned beef sandwich that he had secreted on board. Crumbs might have got into the machinery. Questions were asked in Congress. The medics were up in arms and said it negated all of their tests. But apart from that and some difficulties to do with the escape latch, the flight was a success. In fact it turned out to be more comfortable than had been anticipated. The Gemini simulator used for training on Earth had been torture; no one could bear to stay in it for more than a few hours, but in zero gravity the real experience turned out to be more pleasant.

3 June 1965. Gemini IV. Crew: Jim McDivitt and Ed White. Flight time: 4 days, 1 hour, 56 minutes, 12 seconds; 66 orbits. (For some reason the official designation of the flights went into Roman numerals after the first manned mission.)

A first attempt was made at a rendezvous in space. Gemini was sent in pursuit of a 'pod' that had, sometime earlier, been ejected away from the vehicle. One of the problems of rendezvous was the counter-intuitive need to move into a lower orbit in order to increase speed. The rendezvous attempt failed.

Back on Earth Jim McDivitt was asked – yet again – to respond to Titov. 'I did not see God looking into my space cabin window as I did not see God looking into my car's windshield on Earth,' McDivitt said, 'but I could recognize His work in the stars as well as when walking among the flowers in a garden. If you can be with God on Earth you can be with God in space as well.'

Ed White became the first American to walk in space. He was tethered to the capsule and had a camera slung around his neck. Part of his task outside the craft was to take pictures of the craft itself, which had suffered some minor damage. He took 39 photographs, to be assessed back on the ground to see if the cause of the damage could be determined. But he also took the opportunity to turn his camera away from the vehicle and take some photographs of Earth. At that moment he was 135 miles above the Nile Delta. After 20 minutes outside he was told to return. 'This is great! I don't want to come back inside,' he said. He had to be ordered back in. He said he had felt ecstatic during the walk, and that getting back inside the capsule was the saddest moment of his life.

Back on Earth White's photographs, developed overnight, were spread out on a table, and pored over by a small group of NASA employees. Richard Underwood, unsurprisingly, was immediately drawn to the photographs from the end of the roll that showed the Earth, not those that showed the outside of the spacecraft.

'Hey, Dick,' said Gilruth, 'all the action's up here.'

'Well, I don't think so Dr Gilruth. I think it's all down here.'

And Gilruth came and looked at the photographs and said, 'Those are just pictures of the Earth.'

'Yeah, but we're looking at things that no human being has ever seen before, parts of Africa and other places.'

There had been several attempts to fire Underwood but Gilruth was now persuaded, and from then on was a supporter of photography as part of the manned space programme, despite continuing opposition from other quarters. On the whole the engineers only wanted technical photographs of clear scientific or engineering value and nothing else. 'My recollection of those days,' said

Ed White and
Jim McDivitt

Underwood, 'is a constant battle with my friend Deke Slayton and the other engineers, scientists and astronauts.' Out of Project Apollo's multi-billion-dollar budget, Underwood fought to secure an annual budget of initially just $20,000. But Underwood now had an ally at the top, and Webb, too, became a fan. Thereafter Underwood had privileged access to the astronauts to ensure that they learned how to bring back the best possible photographs of the Earth, and later of the moon. White's were the best photos so far. Underwood said that the photographs that came back from Gemini IV were some of 'the most spectacular photographs ever to come from the space program'.

If he never entirely won over the engineers, Underwood convinced most of the astronauts. 'Your key to immortality,' he told them, 'is in the quality of the photographs and nothing else.' In the long term no one is going to care about all the data that has come back, those thick black log books will count for very little, what is going to have eternal value is the photographs. After Gemini IV the demand for photographs of the Earth from scientists escalated. Underwood later recalled a visit he had from someone from the Bureau of Commercial Fisheries who happened to be passing the Manned Space Center and had a sudden desire to look at some of the space photographs to see what was in them, particularly photographs of the Gulf of Mexico. He said to Underwood that in five minutes he had learned more about the movement of shrimp than he had in five years on the sea. The specialists were getting the point. And soon so, too, would the public.

21 August 1965. Gemini V. Crew: Gordo Cooper and Pete Conrad. Flight time: 7 days, 22 hours, 55 minutes, 14 seconds; 120 orbits.

New Nine astronaut Pete Conrad had gone through the earlier Mercury selection procedure and might have made the cut, as his crewmate Gordo Cooper had, had he not rebelled against what in Conrad's opinion had been an invasive and humiliating admissions procedure. When shown a blank card he handed it back and said, 'It's upside down.' Asked to deliver a stool, he gift-wrapped it and tied a red ribbon around the box. His application was rejected with the note: 'not suitable for long-duration flight', which was ironic given that Gemini V would be NASA's longest flight to date. NASA ended its research into the psychological effects of space flight before Project Mercury was at an end.

Cooper and Conrad inside their Gemini capsule

A second attempt at a rendezvous with a 'pod' ejected from the vehicle was made, again without success.

The long flight time meant that the crew had plenty of time to take some 'magnificent photographs' to please demanding geoscientists.

25 October 1965. Gemini VI. Crew: Wally Schirra and Tom Stafford. Flight time: 0 secs, and no orbits.

Gemini VI was to be an attempt at a rendezvous between two capsules. An Atlas rocket would first launch an Agena capsule into orbit, and then – launched atop a Titan II rocket – Schirra and Stafford would follow in their Gemini VI capsule. But as the crew was waiting to take off, the unmanned, famously unreliable Atlas rocket blew up six minutes after its launch.

There was a change of plan. Gemini VII would launch next, and a postponed Gemini VI, now designated VI-A, would follow some days later on an abbreviated flight. Instead of chasing an Agena, Gemini VI would now try and meet up with Gemini VII. The two vehicles were not designed to attach to each other but it would be test enough for now if they could be brought close together.

4 December 1965. Gemini VII. Crew: Frank Borman and Jim Lovell. Flight time: 13 days, 18 hours, 35 minutes, 1 second; 206 orbits.

Gemini VII was launched without incident. Gemini VI-A was due to launch nine days after Gemini VII on 13 December 1965. The second attempt to launch Gemini VI was a second failure, almost a disaster. The launch procedure suddenly shut down automatically. Schirra, the Commander, should have pushed the abort button, which would have blown the capsule apart from

Frank Borman and Jim Lovell back on Earth after almost two weeks in space

the booster rocket and parachuted the capsule and crew to safety far away from the launch pad, but Schirra remained cool, suspecting a false alarm. If the booster rocket had been deployed the mission would have to have been cancelled; as it was, it could be rescheduled. The next day an investigation found that a plug had shaken loose. It was also discovered that a plastic dust cover had been accidentally left in a fuel line, which would almost certainly have caused the launch to abort anyway: a reminder, if a reminder were needed, of how much could go wrong, and how tiny the fault could be.

16 December 1965. Gemini VI-A. Crew: Wally Schirra and Tom Stafford. Flight time: 1 day, 1 hour, 51 minutes, 24 seconds; 16 orbits.

The third attempt at a launch was successful. Because of the delays, the flight was abbreviated, but there was time enough for Gemini VI-A to get within 5 feet of Gemini VII. Each crew could see the other crew smiling. The first rendezvous had been a success of a kind. A vital operation that would have to be performed if men were to land on the moon had been shown to be possible. The American flights were coming thick and fast. The Russian programme meanwhile was suffering delays, partly due to design problems and partly because of diminished interest from the new Soviet leader.

Early during the Gemini VII flight one of the urine bags burst. Lovell said it was like spending two weeks in a latrine. The question astronauts get asked more than any other is what do they do if they need to urinate or defecate? For everything in space there was a procedure. The Chemical Urine Volume Measure System Operating Procedure was a document comprising 20 instructions: number two is 'place penis against receiver inlet check valve and roll latex onto penis' and number 4 is 'urinate'. The following 16 instructions detail what is to be done with the waste product. In zero gravity, the hardest part was sealing the bag before the contents escaped and started to float around. If any urine escaped the other astronaut would immediately know about it. Gordo Cooper said that during his Gemini flight some urine did escape and all they could do was push the droplets together to make a single ball and keep an eye on it. Astronauts who couldn't urinate properly in space were called wetbacks. One astronaut said that the best way to defecate in space was not to. But if you had to, there was a blue bag. Basically it stuck to the nether parts. A tube of germicide had to be mixed into whatever had been deposited into the bag. In the confined space of a Gemini capsule any kind of physical activity was almost impossible. Having the other astronaut right next to you ensured that you were only going to go if you absolutely had to, but during a two-week mission such a time would inevitably come. The urine was regularly dumped into space. *National*

Tom Stafford and Wally Schirra

Geographic reproduced a colour photograph of urine expelled from Gemini VII. The droplets immediately froze in space and glittered like diamonds. Schirra said it was one of the most amazing sights of the mission. He called it Constellation Urion.

The last three days of Gemini VII's almost two-week mission had been practically unbearable. But at least now there was no doubt that the human body could tolerate being in space for long periods without any serious consequences, and long enough certainly to get to the moon and back. Borman and Lovell had been in space for more than the length of any of the future Apollo missions.

In the early days of manned flight space was expected to produce a lot of surprises, which was why pilot-engineers had been chosen as the first explorers of space, not scientists: pilots already knew how to react calmly and swiftly to the unexpected. At that time, some scientists had thought that even a few seconds of weightlessness would impair bodily functions, that astronauts might not be able to swallow properly, that nutrients might not get to the stomach or be assimilated. Heart and lungs might get confused. Mike Collins said that when Alan Shepard came back after his 15 minutes in space NASA physicians moved the decimal point. Minutes might be OK but not days in space. After days in space the body's autonomic system might forget how to work, blood would pool in the lower parts of the body, not reach the brain, astronauts would pass out. Astronauts took a jaundiced view of the contribution made by the medics: 'The truth of the matter,' said Collins in his typical dry fashion, 'is that the space program would be precisely where it is today had medical participation in it been zero, or perhaps it would even have been a little bit ahead.' In fact weightlessness does cause some problems, but they are not serious. The flapper valves in veins that stop blood falling down into lower extremities give up after a while in space. Prolonged weightlessness can make you feel light-headed when you return, before the flapper valves kick back into action. Bone density reduces and muscles atrophy. Some astronauts built themselves up in preparation, but, perhaps surprisingly, NASA had no formal fitness programme for astronauts. There is a clichéd idea that the astronauts were perfect physical specimens with genius IQs. In fact their average IQ was around the 135 mark, smart but not genius level. They were fit but nothing like the athletes of today. As far as NASA was concerned, Apollo astronauts could do as much or as little exercise as they liked. Neil Armstrong, first man on the moon, did no exercise. He once told a friend: 'I believe that every human being has a finite number of heartbeats. I don't intend to waste any of mine running around doing exercises.' In the USSR, however, training was rigorous. Film footage exists showing cosmonauts in training; leaping off high diving-boards, running bare-chested and in shorts through birch forests.

The Gemini VII crew returned with photographs of the Earth showing rapidly changing tropical weather patterns. The information contained in just four photographs taken from space would have required the cooperation of thousands of weather stations on the ground.

16 March 1966. Gemini VIII. Crew: Neil Armstrong and Dave Scott. Flight time: 10 hours, 41 minutes, 26 seconds; 6 orbits.

The two previous flights had shown, finally, that two craft could be brought close to each other in space; now it was time to show that two craft could dock together in space.

We know how one of the threads of this story will end. Neil Armstrong will become the first human being to walk on the moon. Armstrong had made his first flight with his father in a Tin Goose in 1936. He had a pilot's license when he was still a schoolboy. He had been selected to take part in the air force's project MISS (Man in Space Soonest), cancelled in 1958 and replaced by Project Mercury. MISS had relied on the X-15 space plane, one of the air force's X-series of experimental aircraft used to test out new technologies to their extreme. There were only ever 12 X-15 pilots, just as there would only be 12 astronauts who walked on the moon: Armstrong belonged to both groups. Armstrong was also a consultant on X-20 Dyno-Soar projected, also cancelled after Kennedy had insisted NASA focus on the moon. The X-15 had been the world's first sub-orbital manned spacecraft. It made its maiden flight on 8 June 1959, launched from a B-52 bomber at an altitude of 8 miles; so high up there is no air to lift the craft by its wings. The plane became, briefly, a rocket. For less than two minutes before cutting out, a rocket engine fired and accelerated the plane to a speed of around 4,000 mph. The pilot was at this point weightless. It glided back into the atmosphere, steadied by jets, where it again behaved like a normal aircraft with normal controls.

In the 1960s the United States Air Force still defined space as anywhere beyond 50 miles above the surface of the Earth. Any air force pilot who reached that altitude qualified for astronaut wings. According to the United States Air Force definition, 13 flights by eight pilots made it into space. Some of the eight were civilian pilots and they were not awarded their wings until 2005, when the qualifying rules were modified. Some flights reached altitudes beyond the Kármán line (100 kms, 61 miles), the international

Dave Scott and Neil Armstrong

designation of where space begins. In 1963 Joseph A. Walker was the first X-15 pilot to cross this arbitrary barrier. He reached an altitude of 67 miles, and so joined an exclusive group of human beings that then included only a handful of NASA astronauts and Soviet cosmonauts. Armstrong reached an altitude of almost 40 miles in an X-15, travelling at around 4,000 miles an hour. An X-15 flight retains the world speed record for manned powered aircraft, 4,520 mph. Robert Poole in his book *Earthrise* says that spectacular photos were taken from X-15s during its five-year period of service and that there were no photos like them until the space shuttle. It seems that all the photographs were taken over America, and though some of them had brief news value in America, apparently they went unnoticed by the rest of the world.

Research on X-planes had begun in 1946 and for a time it looked as if space planes were the obvious, logical step after aircraft, and surely would have been if not for the moon race. The most famous of the series was probably the first, the Bell X-1 in which Chuck Yeager broke the sound barrier. Chuck Yeager was made even more famous as the hero of *The Right Stuff* (1979), Tom Wolfe's account of how the Mercury Seven fought behind the scenes to be recognized as test pilots in space rather than just some disposable human component. Chuck Yeager was scornful of Project Mercury: 'Spam in a can,' he called it. He had reason to be bitter: even though many contemporaries thought he was the best pilot of all of them, he had been excluded from the astronaut programme because he had only had a high school education. Some astronauts agreed that Mercury pilots were no more than human cannonballs: the risks were outside their control, but on Gemini a new kind of piloting skill came into play, about to be tested on Gemini VIII: the ability to dock one craft to another while both are in orbit.

'The concept of modern air power,' von Braun wrote, 'was a product of thousands of second lieutenants who learned to fly and who familiarized themselves with the new ocean of air that was the challenge of their day.' With Gemini came new challenges: 'We must get men out there in freely maneuverable craft and we must let them log hour after hour in space. Soon they will become acquainted with their ships and their environment and they will begin to look about, seek new maneuvers, new ideas, and new concepts.' Von Braun might just as well have been writing about the barnstorming days of early aviation. This was why the astronauts adored him: he understood them; and he was a pilot too, one of their tribe.

Von Braun was less well liked elsewhere in NASA. It must have rankled that apart from the astronauts von Braun was the only other celebrity. He had initially been cut out of Mercury, and Bob Gilruth, the director of the Manned Space Center (MSC), had chosen the air force's Titan II rockets over von Braun's Saturn rockets. Von Braun was naturally peeved. 'Remember now, we're entering enemy territory,' was what one of the team from von Braun's Marshall Center remembered being told when they went on a visit to the MSC. Gilruth was suspicious of von Braun, both because of his past – when drunk Gilruth had been heard to complain of 'that damned Nazi' – and for what he perceived as his opportunism. He said to Chris Kraft, who had been director of Mercury and was now putting together what would be known as Mission Control: 'Von Braun doesn't care what flag he fights for.' Von Braun was undoubtedly an opportunist, but the accusation of disloyalty doesn't hold water. Von Braun's biographer, Michael Neufeld, argues that 'any portrayal of [von Braun] as willing to do anything to advance his self-interest or ambition is simply incompatible with the fact that he regarded loyalty as more important than financial gain.' If anything von Braun could be accused of being too loyal.

Armstrong and Scott were so insulated inside their capsule that they felt nothing when the Atlas-Agena combination was launched barely a mile away. Soon after, their Titan II-Gemini combination took off in pursuit. The launches were a success. Armstrong and Scott in their Gemini capsule, and the Agena vehicle they were required to chase and catch, were both put into orbit. At 185 miles above the surface of the Earth, Armstrong steered Gemini VIII so that it met up with and docked with the Agena vehicle. And then things started to go wrong. As soon as the two vehicles were locked together the new configuration began spinning. Armstrong had no choice but to undock, but then Gemini started to spin even faster. The crew was in danger of losing consciousness, and the craft risked breaking up, but quick thinking and untold hours in a simulator paid off and Armstrong was able to bring Gemini under control just in time. The mission was aborted and an emergency landing procedure kicked into action for the first time in the US space programme. Because the mission was ending unexpectedly, three days early, the capsule landed in the Pacific rather than the Atlantic as had been planned. NASA's ability to second-guess possible changes of plan came into its own. By the time they splashed down there was already a destroyer waiting to pick them up.

Among the astronauts there was gossip. The men could be surprisingly bitchy. The fault must have been Armstrong's, some of them implied. NASA's management, however, was impressed with Armstrong's ability to remain cool under great pressure. On this, and another occasion to come, Armstrong's piloting abilities were tested to his considerable limit.

Bitchiness hid an astronaut's greatest fear: making a mistake in front of colleagues. A few weeks before the Gemini VIII flight, New Nine astronaut Elliot See crashed a NASA jet trainer, T-38, as it came in to land. In the plane with him was Group 3 astronaut Charlie Bassett. They both died. Together they had been the crew assigned to Gemini IX. The weather had been bad but a subsequent investigation headed by Alan Shepard ascribed the fatal accident to pilot error. Deke Slayton had been heard to say that 'Elliot See flew like an old woman,' and was particularly angry that See's poor judgement, as he saw it, had killed Charlie Bassett too. For the astronauts, nothing afterwards was ever as dangerous as the training at their Edwards Air Force Base had been; the chances of being killed were endless. In the 1950s an average of one test pilot died every week. In 1952, 62 pilots died in a nine-month period. Out of the Group 3 intake of 14 astronauts, four died in accidents in T-38 training jets. In addition to Elliot See and Char-lie Bassett, Group 3 astronaut Theodore Freeman died before he had even been selected for a NASA flight. Fellow Group 3 astronaut Clifton Williams died in 1967, after he had been selected for Gemini X.

NASA protocol dictated that the widows were to be informed of their husbands' death only by an astronaut, not by friends or neighbours. Many of the astronauts and their families lived close, sometimes next door, to each other, so the first people on the scene were naturally the wives of other astronauts, sent by their husbands but told not to say anything, just to wait until the designated male turned up. NASA was meant to be a civilian operation but many of its customs and practices mimicked those of the military. It was military practice that the widow of an officer who had died in action should move out of the camp as swiftly and anonymously as possible. When Marilyn See turned up at the Manned Spacecraft Center to pick up her husband's belongings she was barred from entering.

See and Bassett's flight was reassigned to Tom Stafford and Ed Cernan with a launch date set of 17 May 1966. In his memoir, Cernan describes what had become the typical Gemini boarding experience: 'We rode the clattering elevator up, watching the shiny metal skin of the tall rocket creep down past us ... We went through the little door into the White Room, the clean area which was the domain of Guenther Wendt, the Peenemünde refugee who was now chief of the closeout crew. Everyone in there, including our backups, Jim [Lovell] and Buzz [Aldrin], wore long white coats and white caps. They looked like morticians.' The astronauts called Wendt, who had a thick accent and wore thick glasses, 'the czar of the launch pad'. It was his task, as it had been during Mercury and would be during Apollo, to perform the last of the closing-out duties.

Cernan's wife, parents and sister were at that moment at Mass: 'God was up to date on my plans.' Cernan had asked if he could wear his religious medallion on the flight. The doctors said no but Deke Slayton overruled them. Once inside the tiny Gemini capsule, perched on the Titan II booster, the crew waited for the launch of the Agena target vehicle. Soon after it was launched the unmanned Atlas-Agena ensemble crashed into the ocean. The launch of Gemini IX was aborted.

3 June 1966. Gemini IX-A. Crew: Tom Stafford and Ed Cernan. Flight time: 3 days, 20 minutes, 50 seconds; 47 orbits.

A second attempt at a launch was successful, but once again there was a problem docking. This time the target vehicle was already rotating when they approached it. The docking attempt was abandoned.

On the last day of the mission, Cernan went on the first long spacewalk. NASA 'shrinks' had warned him that he might become disoriented 'or swamped by euphoria', as if he were in a headlong fall. 'Ridiculous,' he said. He experienced no sense of disorientation, falling, or of euphoria. Other astronauts had reported differently on all counts; have said that for the first moments it was impossible not to think that the Earth is where one will fall to, that at first there was a strong desire to hold on tight as if to a cliff that is itself falling, that only gradually did the realization come that even if you let go there was nowhere to fall to. To be in orbit is to be in a state of perpetually falling. It is what is happening right now to all of us here on Earth, even if we are unaware of it experientially.

Cernan had more to worry about than the risk of euphoria. He was wearing 14 layers of clothing: long johns; a nylon comfort layer; a black neoprene-coated pressure suit; a restraint layer of Dacron and Teflon link net that was like chain mail and intended 'to maintain the shape of the pressure suit'; seven layers of aluminiumized mylar 'with spaces between each layer for thermal protection'; a layer to protect against dust-sized meteors; a white nylon outer layer; and finally heat-resistant leg coveralls. He said you could take a blowtorch to him and he would not have felt it. After his space suit had been pumped up to the prescribed pressure it was so stiff he said it was like wearing 'a rusty suit of armor'. From the start, the suit 'took on a life of its own'. Outside the craft in the harsh environment of outer space, the sky was pitch-black but the sun blindingly bright. Inside the pressurized clothing his body was burning up huge amounts of energy carrying out even the simplest task. Weightlessness is all very well if you just want to float around aimlessly, but try to perform the smallest deliberate action and weightlessness becomes a hindrance. The suit's cooling system struggled to cope, and in the

Tom Stafford
and Ed Cernan

blinding light his facemask soon fogged up. After two hours and seven minutes, his EVA at an end, Cernan expended so much energy squeezing himself back into the capsule that he became exhausted, and his heart rate soared dangerously high. Stafford was so shocked at the colour Cernan had turned that he sprayed him with water, knowingly going against procedure and running the risk of an electrical short.

Richard Underwood's informal position within the NASA structure allowed him a degree of freedom that was unique within what was otherwise a strictly hierarchical organization. The astronauts trusted him, and told him that they were too wired to sleep during their designated sleep periods; that they put on a pretence of sleeping to satisfy the control centre, and spent the time 'Earthgazing', looking out the windows at views no one had ever seen before. From Gemini IX onwards there was supposedly a photographic timetable for each flight, but Underwood perpetrated a subterfuge that allowed the astronauts some leeway. Inside the adapted Hasselblad used on all Gemini flights, Underwood hid a piece of paper on which he outlined a schedule of surreptitious sleep-time photography, a list of times when the craft would be over some point on the globe of interest to some lobbying geoscientist. Underwood had access to weather satellites, and would give the astronauts a heads up when the conditions were good enough to take a photograph. He called it photography for insomniacs. The astronauts were more than happy to play along; it gave them a rare sense of independence and the opportunity to be more than just a man in a can; to be artists. The flight controllers were sometimes baffled afterwards about when some of the photographs could have been taken. Despite NASA's attempts to formalize everything, Underwood's relationship with the crews would remain informal. Not that he couldn't also be strict. He would shout at any astronaut he thought was doing a bad job.

NASA had come to appreciate the fact that having the crew take their own photographs helped keep the security services at bay – surveillance cameras would have attracted the interest of the CIA. Only Underwood truly understood the value of 'pictures that recorded it the way the astronauts saw it'.

When Underwood looked at the images that came back from Gemini IX, on some there was what appeared to be a UFO, a huge dark object. He decided to keep quiet about it. It was only a decade later that he worked out that the object was a passing Soviet Proton spacecraft, a very large vehicle. At one moment, over southern Africa, the two craft had come within four miles of each other.

18 July 1966. Gemini X. Crew: John Young and Mike Collins. Flight time: 2 days, 22 hours, 46 minutes, 39 seconds; 43 orbits.

This time the docking was a success and the combined Gemini-Agena craft climbed to a higher orbit, 475 miles above the surface of the Earth, a new record. A significant step on the journey to the moon had been taken. The crew was higher above the Earth than any human had ever been, grazing the edge of the Van Allen belt. Scientists had wondered how large a dose of radiation they might experience. In the event it turned out to be much lower than had been feared.

John Young and Mike Collins

During his 90-minute spacewalk, Collins noted that the stars don't twinkle in outer space but are steady. The sky was unrelieved blackness, with no shades of blue. All was quiet. He had so many tasks to perform there wasn't much time to look around, but he did have time to notice the colours out there: 'subtleties ... clearly discernible to the supersensitive instrument we call the eye, [that] are unfortunately not captured by the rather crude emulsion of the film'. One of his assigned tasks during the EVA was to photograph not 'the grandeur of the universe' or the Earth seen from 475 miles up, but, 'believe it or not', a titanium plate about 8 inches square divided into four sections, each a different colour. The photograph was to be used back on Earth as a kind of colour test card. If he wanted to take a few tourist photographs, when he had a moment, that was up to him.

When he tried to get back into the capsule, he radioed through to Young that he had a problem. 'What's the matter, babe?' Young asked. The problem was that his eyes were watering in the sunlight, so much so that he couldn't see out of his visor, couldn't easily see where he was going. During the struggle to get back inside he lost his Hasselblad. All that they would have in the way of visual evidence of the mission would be the film from one movie camera: 'an uninterrupted sequence of black sky'. As Collins ruefully observed, 'in the second half of the twentieth century, an event must be seen to be believed'.

At 38,000 feet, the first parachute, termed a drogue – or what Collins called 'the little rascal' – unfurled; it was only 6 feet across, but it slowed the capsule down enough that at 10,000 feet the main chute – 58 feet across – could then be safely deployed. The capsule swayed crazily from side to side as it headed for splashdown.

A few days after splashdown, Collins submitted his travel expenses: three days at $8 a day.

Collins wrote later that he had felt 'God-like' as he stood erect in his 'sideways chariot, cruising the night sky'. Too bad, he said, that he hadn't been given longer to let it all sink in. There was so much work to do he was hardly conscious of the Earth below him: 'Work, work, work! A guy should be told to go out on the end of his string and simply gaze around – what guru gets to meditate for a whole Earth's worth?' Did he see God? 'I didn't even have time to look for Him,' he said.

Collins had been frustrated after his Gemini flight by his inability to record in words what he felt about being in space. 'John Magee would have known how to do it,' Collins wrote in

John Magee

his memoir. Magee, a Spitfire pilot, died aged just 19 in a mid-air collision over Lincolnshire in 1941. A farmer said that he had seen him push back the canopy, and stand on the plane ready to jump. But the crash had occurred at only 1,400 feet. His parachute never opened. Magee had been born in China, his mother British, his father an American missionary. In 1940 he won a scholarship to Yale but chose instead to join the Canadian Air Force. In his short life he wrote just a handful of poems, the most famous of them 'High Flight'. Collins' wife, Pat, had typed out the poem on a small card which she had put into the small bag of personal belongings her husband was permitted to take on the mission – in NASA-speak his PPK (personal preference kit).

> Oh! I have slipped the surly bonds of earth,
> And danced the skies on laughter-silvered wings;
> Sunward I've climbed, and joined the tumbling mirth
> Of sun-split clouds, —and done a hundred things
> You have not dreamed of —Wheeled and soared and swung
> High in the sunlit silence. Hov'ring there

I've chased the shouting wind along, and flung
My eager craft through footless halls of air. . .
Up, up the long, delirious, burning blue
I've topped the wind-swept heights with easy grace
Where never lark or even eagle flew—
And, while with silent lifting mind I've trod
The high untrespassed sanctity of space,
Put out my hand, and touched the face of God.

The first and last lines of 'High Flight' were inscribed on Magee's headstone. 'All that from the cockpit of a Spitfire,' said Collins. 'I cry that he was killed.'

12 **September 1966.** Gemini XI. Crew: Pete Conrad and Dick Gordon. Flight time: 2 days, 23 hours, 17 minutes, 9 seconds; 44 orbits.

Another successful docking with Agena was accomplished. The crew returned with some spectacular photographs of the Persian Gulf. Somehow, during his second spacewalk, Gordon managed to fall asleep outside the craft. At the same time, inside the craft, Conrad was also asleep.

Dick Gordon and Pete Conrad

11 November 1966. Gemini XII. Crew: Jim Lovell and Buzz Aldrin. Flight time: 3 days, 22 hours, 34 minutes, 31 seconds; 59 orbits.

With the death of Elliot See and Charlie Bassett, Buzz Aldrin, who had been very close to Bassett, made it onto the last Gemini mission. And if he hadn't made it onto the Gemini mission, he almost certainly wouldn't have been included in the first mission to land on the moon.

As they were homing in on the Agena target, the radar failed. There and then, using pencil and paper, Aldrin calculated the correct trajectory. The rendezvous and docking was otherwise a success. The two vehicles remained together for almost two days. Aldrin's EVA was the most successful of all the Gemini space-walks, the longest – five hours and 30 minutes – and the most trouble-free. A handrail had been added to the outside of the craft: a simple addition that made all the difference. Aldrin proved that it was possible to work in space for an extended period without ill effects. Cernan, who had almost died during his EVA, thought that Aldrin took too much credit. Aldrin had apparently been disdainful of Cernan's brute force approach. Cernan said Aldrin went round as if he had solved all the problems himself. At times astronauts sound like badly behaved teenagers. During his EVA Aldrin photographed the star field, retrieved a micrometeorite collector and did other chores. By the end of the Gemini missions, NASA was satisfied that rendezvous, docking and EVA had been fully tested.

The Soviets had planned to make four more manned Vokshod flights but all of them were cancelled. Gemini had been a complete reinvention of Mercury, whereas Vokshod had basically recycled and modified capsules left over from the Vostok programme. Leonid Brezhnev's relative lack of interest in the space programme in his first months in office, combined with Vokshod's overextended ambition, had led to its downfall. Sergei Korolev had turned his attention to what would be the Soviets' own reinvention of their manned spacecraft, Soyuz. The first unmanned Soyuz craft was launched a little under two weeks after the last Gemini mission was over. The Soviets had fallen behind, but the failure of Vokshod had not been the only impediment. Early in 1966 Sergei Korolev had died aged 59, his health compromised after the years he had spent in the Gulag. He had gone in for routine surgery to have a tumour removed from his intestinal tract and bled to death. Eight years after Korolev's death Oberth wrote in a letter to a TASS correspondent: 'Unfortunately I do not know the names of the people I respect who have created those powerful rockets and the first spaceship.' Pre-eminent among them had been Korolev. The Nobel Prize committee had at one time asked Khrushchev if the Chief Designer's name might be put forward. 'The creator of the USSR's new technology was the entire nation,' Khrushchev replied. Now, finally, after his death, Korolev's identity was revealed to the Soviet people, and so to the world.

Korolev's star had begun to fade after the death of Khrushchev, and his giant N1 booster, planned to be even more powerful than von Braun's Saturn V, was 'mired in technical and financial difficulties'. In the race between boosters the Soviets were definitely behind. Just six weeks after Korolev's death, von Braun's Saturn IB was successfully launched. It could put a payload of 16 tons into Earth orbit compared to the 11 tons of Saturn I. The two-stage rocket was a marriage of the first stage of Saturn I and the third stage of Saturn V. The three-stage Saturn V had been designed to put a payload of 120 tons into orbit, but for now there were problems with the second stage.

Korolev's successor was his deputy Vasily Mishin. He had neither the charisma nor the influence needed to head a project facing unrealistic deadlines.

Korolev's Soyuz rocket would not have been able to get men to the moon, but it would prove in future years to be the most frequently used launch vehicle in the world. After the Space Shuttle was abandoned, Soyuz was, and still is, the only route to the International Space Station (ISS). Another of Korolev's designs, the R-11, would later be known as the SCUD missile.

Korolev had written a column for *Pravda*, 'Talks with a Chief Engineer', under the pseudonym Professor K. Sergeev. In his 'New Year Cosmic Greeting' on 1 January 1964 he wrote: 'I can't help wanting to exclaim, "So much has been done, so many steps taken!" At the same time I must say, "How little has been achieved so far and how much we still have to bring about!"' As we continue, haltingly, to take our first steps out into the cosmos, his words may well always be true no matter how far we travel into the infinite universe.

The Soviet space programme had not entirely stalled during 1966. Two weeks after Korolev's death the Soviets' Luna 9 soft-landed on the moon, another impressive first. When Johnson had asked von Braun what America might do to beat the Soviets in space, a soft landing on the moon had been one of his suggestions. Yet another historic moment occurred on 3 February when Luna 9 transmitted photographic data from the surface of the moon to the Earth. The panoramic images were intercepted – it is likely that the USSR had intended that they should be – and decoded at Jodrell Bank in England. They were reproduced worldwide. At the beginning of March the Soviets' Venera 3 crash-landed on Mars, the first spacecraft ever to reach the surface of another planet. At the end of March the Soviet space probe Luna 10 became the first spacecraft to orbit the moon, orbiting 460 times before its batteries died.

America made its first soft landing on the moon at the end of

May the same year. Surveyor 1 took thousands of photographs of the moon's surface that were then transmitted back to Earth and scrutinized for possible landing sites. It was crucial that whenever a manned lander did approach the moon's surface NASA had worked out both how to land reliably, and that the chosen landing site was not boulder-strewn or too close to the edge of a crater. Surveyor could have taken colour photographs and it could have taken colour photographs of the Earth, but 'some guys with three PhDs apiece' argued against it, Richard Underwood said. They claimed they'd get more technical detail from black-and-white film.

Numerous unmanned missions were undertaken in the next couple of years – a further seven Surveyor craft, and, in addition, programmes like Mariner and Lunar Orbiter – all with the objective of identifying the best place to land a manned craft when the time came. On 10 August 1966 Lunar Orbiter 1 became the first American spacecraft to orbit the moon. The craft had only enough power to take 211 pictures in medium resolution, but Underwood hoped that at least one of the photographs would show not the moon but the Earth. The photographs would be developed and scanned on board the craft and then transmitted back to Earth in electromagnetic form before being reconstituted, a single photo taking over 20 minutes to transmit. A decision had not been reached even when the craft was in orbit around the moon. There were arguments right up to the last minute, Underwood being told that it was a waste of film, that such a photograph would have no scientific value, and so on. The usual. In the end a senior figure at the meeting, the vice president of Boeing, said, 'To hell with it. It is a public service. It might be tremendous.'

To take a photo of the Earth, the probe would have to be reoriented and take the Earth photo first, before the moon photos were taken. There was great risk involved, and if it went wrong Boeing – Lunar Orbiter's manufacturers – would lose its bonus.

Word spread around the centre that the picture had been

taken. Senator Joseph Karth, chairman of the congressional committee on space sciences, was soon on the phone to Lee Scherer, NASA's programme manager: 'What's all this about you taking a picture of the Earth?' Scherer was defensive, and started to reel off any reasons that popped into his head that seemed to justify the taking of the photograph: 'Well, I don't give a damn why you did it,' Karth cut in, 'but me and 200 million other Americans thank you.'

NASA's Public Affairs Office led with the Earth photograph, but described it in NASA's usual technical style: 'the purpose of the photograph was to obtain data, long of interest to scientists, on the appearance of the Earth's terminator as viewed from …' Not only is the description terminally boring, it's not even true that that was the purpose of the photograph. When the photographs from the mission were released those of the moon attracted more attention. Admittedly, without high resolution, and in black and white, the Earth looks just like any other heavenly body, another of the zillions of other spheres whizzing about out there in outer space. As it turned out, repositioning the probe actually improved the photos of the surface of the moon. Seen obliquely there was more detail. Images of nine possible landing sites were identified. Boeing got a 75 per cent bonus. The subsequent official report summarizing all five Orbiter missions did not mention the

The first photograph of the Earth seen from the moon.
Taken by Lunar Orbiter 1 on 23 August 1966

photograph of the Earth at all. The main significance of the photograph had been missed. Though it was in black and white, and though much of the Earth was in shadow, here for the first time in human history was an image of the whole Earth.

CHAPTER TWO

Each Apollo astronaut had been allotted a particular area of expertise. New Nine astronaut Mike Collins was entrusted with pressure suits and spacewalks. He was also assigned one of the Apollo Command Modules for the 'care and feeding of'. The design and construction of the Command Module had been contracted out to the engineering firm North American. Collins' particular module, 014, was currently undergoing final assembly at North American's plant. Collins was tasked with getting to know the machine intimately; not just the workings, but also what was going to go in it: all kinds of test equipment, medical experiments – including a wired-up frog, and a collapsible bike.

Command Module 012 had been assigned to the first manned Apollo flight, Apollo 1. An early test version, 002, had been put into orbit at the end of January 1966. There were to be many such test launches in the weeks and months that followed. Gemini had been complicated, but Apollo was something else altogether. In the command module 'there were three hundred of one type of switch alone'.

The astronauts had been surprised when North American rather than McDonnell had first won the contract to build the Command Module. Collins was critical of the company. It had become clear to him that they had a long way to go before they matched the professionalism of McDonnell, the main contractor that had worked on Mercury and Gemini. The McDonnell Aircraft Company had welcomed the design input of the astronauts. North American did not. The Apollo engineers gained a reputation for being arrogant. They didn't want to know anything about what had happened on Mercury or Gemini: 'a veil came down, the eyes became slightly glassy, and one was informed – generally in a cool and faintly supercilious tone – that it was simply not done that way on Apollo'. And when they weren't arrogant, some of the Apollo engineers seemed naive; or worse, over-confident. Collins said that it was as if nothing had been learned from Mercury and Gemini, everything had to begin afresh.

Nothing major appeared to be going wrong with Collins' Command Module but he was frustrated by a litany of minor imperfections and irritations. Every day there were slippages in the timetable. There were more serious problems elsewhere: a service module (017) ruptured during a test because the wrong kind of aluminium had been used. It was rumoured that there were problems with the construction of the second stage of Saturn V.

Apollo 1's flight module (012) – supposedly fully finished and flight-worthy – was delivered at the end of 1966. On 27 January 1967 the crew of Apollo 1 – Gus Grissom, Ed White and Roger Chaffee – clambered aboard 012, which had been placed into position on top of a Saturn IB booster rocket at Pad 34 at Cape Kennedy. The rocket wasn't fuelled. It was to be just a routine check of the systems before the manned flight scheduled for 21 February.

Inside the capsule Chaffee complained of the smell, which he said was like sour buttermilk. Then there were radio glitches between the different buildings. Grissom was furious: 'How are we going to get to the moon if we can't talk between two or three buildings? I can't hear a thing you're saying. Jesus Christ!' And then, after five hours, Grissom radioed the message: 'We've got a fire in the capsule.' And then White shouted: 'Hey, we're burning up in here.' And then there were screams, and then there was the sound of static.

All three astronauts died within minutes. Some said NASA had lost its finest pilot in Gus Grissom. He might well have been first on the moon if he had lived. Apollo 1 was to have been Roger Chaffee's first mission. Ed White's wife, Pat, never recovered. She committed suicide in 1983, after a number of failed attempts over the years. Grissom had once told his wife, Betty, 'If I die, have a party.' Betty sued North American for $10 million. She was eventually paid $350,000 in an out-of-court settlement, as were the wives of the other crew members.

The Armstrongs had been close to the Whites, and were neighbours. When a fire had broken out at the Armstrongs' home in 1964, Ed White had helped save Armstrong's son from the burning building. 'Ed was able to help me save the situation, but I was not in a position to be able to help him,' Armstrong said after the Apollo fire, or what would forever after be known simply as The Fire.

Those who heard the radio transmission said the screams haunted them ever afterwards. Those who tried to open the escape hatch said it was the smell of burning they never forgot. And somehow the fact that the fire had happened on the ground made it so much worse.

On 5 April 1967, not four months after the fire, an investigation delivered to Congress a 19lb document several thousand pages long. Over 2,000 people had been on the investigation team. Mike Collins' 014 Command Module was taken apart. Over 1,400 errors were detected, and 10 possible causes of the disaster were identified, all of them electrical failures. Inside the capsule was 'a jungle of wire that had been invaded over and over again by workmen changing, and snipping, and adding, and splicing, until the whole thing was simply one big potential short circuit'.

The fire had most likely been caused by a short setting fire to a piece of Velcro, but a single cause was never identified. That there was a superfluity of possible causes was all the more damning. It had been a serious mistake to make the environment 100 per cent oxygen; there was too much combustible material in the capsule; worse, there were no proper procedures; too many changes had been approved without proper checks being made afterwards to ensure that the changes had been properly effected.

The accident report criticized the management style of North American, and soon accusations were made, at congressional hearings and in the press, of corruption. It turned out that North American hadn't even won the competition – though the difference in the scores NASA had given them were small, and the pressure at the time to make a decision had been immense. What became clear from the report was how much power Webb had.

Under intense questioning by Congress, Webb and others prevaricated. In the press NASA was said to stand for Never a Straight Answer. The congressional hearings became more and more acrimonious. At one point New Nine astronaut Frank Borman shouted at them to stop the witch hunt. NASA's management style also came in for criticism. Webb was devastated. He had thought that NASA's management structure could become a model for all types of business, and even be used to alleviate poverty and homelessness. He had talked about 'Space Age management'. His vision was in ruins. He saw now that he had got too close to Joe Shea, the manager of the Apollo Spacecraft Program Office (ASPO), who was in overall charge of Command Module progress. Webb realized that Shea and Mueller had been in too much of a rush. The management structure would need changing again. Shea was eased out, and George Low, who had been deputy director of the Manned Space Center, was brought in as the new manager of ASPO. Shea, a devout Catholic, had been critical of North American himself in the past, and was weighed down with guilt, feeling he should have done more to avert the disaster.

Most damaging of all to Webb's authority was the revelation during the investigation that Mueller had, late in 1965, commissioned a report into safety procedures, but the subsequent report, named the Phillips Report, had been kept from him. When he was ambushed at a congressional committee meeting, and told that the fire risk had already been assessed, it was clear Webb had no idea what report the congressmen were talking about.

Apollo had got off to a very bad start. There was a moment when it looked as if Project Apollo would be cancelled before it had really begun. There were strong factions at work in America at the time that would have liked to have seen Apollo brought low. By the mid-1960s there were 36,000 people working on Apollo within NASA, but there were 10 times as many working on Apollo in the private sector. The counter-culture that defined itself around ideas of scepticism and anti-authoritarianism was on the rise; to many Americans it would have come as no surprise to learn that companies like North American were corrupt, that the very same companies which were making the armaments that were fuelling the war in Vietnam were also absorbing most of NASA's vast budget, their tax dollars.

Apollo was stalled for 20 months. During the hiatus, Group 3 astronaut Gene Cernan told Deke Slayton that he wanted to serve in Vietnam. Slayton told him he was free to go but that there was no guarantee they'd have a place for him when he got back. Cernan chose to stay.

It became clear that it wasn't only the Command Module that was faulty. Nor was North American the only problematic contractor. When the Lunar Module was delivered by the Grumman company in June 1967, it didn't fit inside its protective shroud. It was judged not fit to fly. Rocco Petrone, NASA's Chief of Launch Operation, laid into the manufacturers: 'What kind of two-bit garbage are you running up in Bethpage?' With hindsight Apollo's ability to turn itself around in the period after the accident was one of its greatest achievements. Perhaps the accident had saved the project from some worse disaster further down the line. NASA and its contractors rose to the challenge. Out of acrimony came cooperation. Project Apollo was able to grow up. Cernan later wrote about how impressed he was by Grumman, that the workers there were so wedded to Apollo that they kissed each Lunar Module before sending it from the factory to NASA. George Low, as the new manager of ASPO, took responsibility for the redesign of the Command Module. They made 1,341 alterations at a cost of $75 million. Everything was made non-flammable. The 100 per cent oxygen environment inside the capsule would only kick in after the launch; before then the astronauts were to breathe a less flammable mixture of nitrogen and oxygen.

After the death of Sergei Korolev, the Soviet space programme was having its own problems. The first manned Soyuz mission was launched on 23 April 1967, just under three months after the Apollo 1 fire. The pilot, cosmonaut Vladimir Komarov, had told friends and colleagues that he thought the flight was doomed. Even though hundreds of faults had been reported, the decision to fly had been taken at the highest political levels, probably in an attempt to take advantage of the Apollo fire. Gagarin offered to fly in Komarov's stead, knowing that the authorities wouldn't allow the mission to go ahead if there was a chance a national hero might be killed; but Komarov refused to take the chance of putting Gagarin's life at risk. The launch took place as scheduled, and at first everything went without a hitch. But some way into the flight the spacecraft failed and, for 26 hours, span out of control. Komarov had time to bid his wife Valentina goodbye by videophone. In the usual secretive way of the Soviet programme she hadn't known about the mission in advance. At an American listening post in Istanbul, Komarov was heard talking about the future of their children and how his wife should best arrange financial matters. The future premier Alexey Kosygin was in tears when he came on line. It is said that during his final moments, as he plummeted towards the Russian steppes, still conscious, Komarov cursed Russian bureaucrats with his final b reaths, though that might be later mythologizing. Gagarin apparently never recovered. He lived another year, dying when the jet fighter he was piloting spun out of control.

In tribute to the first crew, the name Apollo 1 was not reassigned, and for mysterious, or perhaps merely idiosyncratic reasons, the intended missions Apollo 2 and 3 were scrapped. The next Apollo mission was called Apollo 4. Originally the launch had been anticipated to take place late in 1966. But after The Fire, and because of continuing problems with the manufacture of Saturn V's second stage, the launch of Apollo 4 was put back.

In July 1967 the DODGE (Department of Defense Gravitational Experiment) military satellite was launched. The director of DODGE predicted an educational use for satellites; almost no one foresaw – apart perhaps from von Braun – the power of the satellite to change life on Earth utterly. Curiously, the evidence was already there for anyone to see. Just a few days earlier the BBC had broadcast a live TV show called *Our World*, conceived by BBC executive Aubrey Singer, joining much of the world together for the first time by airwaves. Artists from 19 countries performed, including Picasso and Maria Callas. During the closing segment from the UK, The Beatles, dressed in gorgeous hippy gear and surrounded by flowers, sang 'All You Need is Love' for the first time. The broadcast had been made possible only because of the existence of the latest geosynchronous satellites. A satellite put into orbit at 23,000 miles falls around the Earth at the same rate as the Earth spins on its axis, with the result that the satellite remains at the same point above the Earth's surface. The first geosynchronous satellite, Intelsat 1, known as Early Bird, had been launched in December 1965, and was used to broadcast the live splashdown of Gemini VI-A. Intelsats 1, 2 and 3, and NASA's ATS-1 – the latter launched at the end of 1966 and used to broadcast weather data – were all brought into play to broadcast *Our World* to an audience, variously estimated, of between 400 and 700 million viewers.

Less than a week after the BBC broadcast, the DODGE satellite transmitted back to Earth the first colour images of the full Earth. The best of them, taken from altitudes around 20,000 miles above the Gulf of Mexico, showing the movement of a hurricane, were subsequently published in the *Washington Post*. At the time few newspapers had the technology to publish in colour, and the *Washington Post* only did so rarely. The photographs weren't of particularly high quality; the colours were muted, the focus blurry. Again, the world failed to notice the momentousness of the achievement.

The Apollo 4 launch had been rescheduled for mid-October, but even then there were further delays. Over a period of 17 days fuel-pumps failed, computers crashed. 'Can we ever get this baby to go?' Rocco Petrone, director of launch operations, asked despairingly, 'Can we ever get all the green lights to come on at the same time?' Then, on 9 November, success. The launch was the first full test of the rocket Chris Kraft called von Braun's masterpiece. 'It's not a noise,' Collins said of the launch of Saturn V. 'It's a presence. From the tip of toes to the top of head, this machine suddenly reaches out and grabs you, and shakes, and as it crackles and roars, suddenly you realize the meaning of 7.5 million lbs of thrust.' Saturn V was capable of hoisting a massive 120-ton payload into Earth orbit (compared to the 11 tons of a Saturn I, or 16 tons of Saturn IB), or taking a 45-ton payload into outer space. In the CBS television studio von Braun was heard to shout, 'Go baby, go!' Though the studio was three miles away its windows were blown in by the sound waves. The rocket's power, and what would prove to be its reliability, would in large part contribute to NASA's coming triumphs.

Much of modern technology works because each tiny part is perfectly engineered. It was what von Braun understood in his bones. Hundreds of thousands of parts needed to work, not only together but each reliably in itself: 'When *all* of the elements of a complicated installation are practically trouble-free, the installation in its entirety also becomes practically trouble-free,' he once said. The rocket was a mark of the success of the team as a team, and of von Braun as its leader. Von Braun would say from time to time that for all the power of teamwork, 'most significant advances in science and technology have not emanated from teams but from singular efforts of dedicated and solitary individuals'. Saturn was an example, as The Spirit of St Louis had been, where both together had been necessary.

A camera mounted on the capsule on top of the rocket captured 700 images. Apollo 4 returned with the first colour negatives of the full Earth – the whole Earth seen in full illumination, the counterpart of the full moon – taken from 11,000 miles above its surface. After splashdown, the camera was retrieved and the negatives developed. In their first press release about the mission NASA didn't mention the photos. The photographs were later made available publicly but yet again didn't make much impact.

In early November ATS-3 had been launched and it, too, sent back some of the first photographs, again in muted colour, of a fully illuminated Earth. The images were released by NASA a week after the photos from Apollo 4. The press reported that the photos were of great interest from a meteorological point of view. It had still not sunk in that we were looking at the answer to an ancient question: what did the Earth look like seen from the outside? On the evidence of these photographs, a kind of drab blue-and-white sphere.

At the end of January 1968 the Apollo 5 mission, using a Saturn IB rocket as the booster, put a Lunar Module into orbit for the first time. At the beginning of April the Apollo 6 mission was a second, unmanned, test of the Saturn V. Apollo 6 was not as successful a mission as the first test, Apollo 4, had been. Two of the engines failed after launch and the Saturn V rocket became highly un- stable, shaking violently for a time; an effect called pogoing. At one point the rocket flipped over and was pointing Earthwards. The controllers had their fingers on the abort button. The rocket eventually levelled out. Several instruments in the command capsule had been shaken into radio silence, and part of the metal shroud that protected the delicate Lunar Module fell off. The poor performance of Saturn V on its second outing might have attract-ed a great deal of attention, but on that same day, 4 April 1968, Martin Luther King was assassinated. Riots erupted across the country. During the next few days, 39 people died, 34 of whom were black.

In what space historian Piers Bizony describes as 'probably one of the cleverest and fastest (and least known) engineering achievements in history', von Braun and his team fixed the pogo-ing problem in less than a month. Ground tests seemed to show that the fix worked.

Three months later George Low suggested a change to the Apollo sequence of missions. Apollo 8 had initially been sched-uled to be a second test of the Lunar Module in Earth orbit. Why not use that flight to send a crew to the moon instead ? Webb's first reaction was, 'Are you out of your mind ?' But the more he thought about it, the more he was persuaded. Although the first test of the Lunar Module in Earth orbit had been a success, the Lunar

Module itself was still thought to be too heavy to risk a manned moon landing. A second test at this stage would be a waste of a mission, and JFK's deadline was in danger of being missed. Why not turn the delay into an advantage?

11 October 1968. Apollo 7. Launch vehicle: Saturn IB. Crew: Wally Schirra, Donn Eisele, Walt Cunningham. Length of mission: 10 days, 20 hours, 9 minutes, 3 seconds.

As a rehearsal for the first manned lunar mission, Apollo 7 took place entirely in Earth orbit. It has come to be known as the most bad-tempered of all the Apollo missions. The Commander, Wally Schirra, was on a short fuse and his mood seems to have infected his crewmates, Donn Eisele and Walt Cunningham. Schirra cancelled a planned TV broadcast at the last minute; he argued with the NASA flight controllers, calling one of them an idiot, which put him in very bad odour; and he refused to carry out what he called 'some crazy test we've never heard of before'. The astronauts said they didn't like the food, and then to cap it all they each developed head colds and refused to wear their helmets during re-entry. Chris Kraft, the director of flight operations,

Donn Eisele, Wally Schirra and Walt Cunningham

generously put Schirra's irritableness down to fear, and a delayed reaction to the death in the Apollo 1 fire of his friend and next-door neighbor Gus Grissom. Deke Slayton, in charge of flight crew operations, was less generous. None of the crew ever flew for NASA again.

Four days after the crew of Apollo 7 was safely returned to Earth, cosmonaut Georgy Beregovoy brought his Soyuz 3 spacecraft into rendezvous with an unmanned Soyuz 2. The planned docking failed. The facts coolly presented decades later would seem to show that the Soviets were now some way behind the Americans in the race to the moon. At the time, it wasn't so clear. The CIA reported to NASA that the Soviets were intending a manned fly-by of the moon before the end of the year.

Several months earlier, the Soviets had launched the first in a series of Zond missions that attempted to loop around the moon. The flights took advantage of the so-called free return trajectory which uses the moon's own gravitational field to sling the craft back to Earth again. The Zonds were stripped-down versions of Sergei Korolev's Soyuz. They were launched on a Proton rocket, designed by Korolev's rival Vladimir Chelomei. The Proton wasn't powerful enough to put a craft into orbit around the moon, and certainly not to land astronauts on the moon, but it might be capable of taking a craft on this looping trajectory around the moon. At a pinch the Proton might just about be able to send a Zond craft and two cosmonauts around the moon and back to Earth again. It was the best the Soviets could hope for.

The Soviets' first – unmanned – attempt at a fly-by of the moon was a success until re-entry, when the craft's guidance system failed. Three days after the launch of Apollo 7, a second attempt brought a cargo that included tortoises, insects and bacteria to within 1,200 miles of the surface of the moon. The tortoises lost

10 per cent of their body weight but returned otherwise unharmed. It is quite possible that somewhere in the former USSR the first living creatures to fly past the moon are still alive. Also on board was a tape machine that played and broadcast a recording of a cosmonaut talking. When NASA picked up the signal, it was, for a moment or two, sent into a panic.

The Soviets had scheduled a further unmanned fly-by for November, which they hoped to follow up with a manned fly-by the following month, but that mission, as the first mission had, developed problems on the return flight. The cabin depressurized, killing its cargo of assorted living creatures, and then when the craft re-entered the Earth's atmosphere the parachutes failed to deploy. The manned mission planned for December was cancelled.

What the Soviet programme desperately needed was the power of Korolev's massive N1 rocket, but Korolev was dead and the rocket was behind schedule. NASA was unaware that the Soviets were struggling. Von Braun still thought they were going to beat them: 'All our information indicates that the Russian program is richer than ours . . . I'm convinced that, unless something dramatic happens, the Russians are going to fly rings around us.'

CHAPTER THREE

21 December 1968. Apollo 8. Crew: Frank Borman (Commander), Jim Lovell (Command Module Pilot), Bill Anders (Lunar Module Pilot). Duration of mission: 6 days, 3 hours, 42 minutes.

At 2.30am on 21 December 1968 the crew are woken up by Deke Slayton. Borman says that he hasn't slept much. After a final medical check-up the crew eat the astronauts' traditional last meal before a launch of steak wrapped in bacon with eggs, orange juice and coffee without milk; euphemistically – and perhaps optimistically – described as a 'low-residue' meal. Anders has privately decided that he is going to see if he can get through the entire

Apollo 8 crew at breakfast, *left to right*:
Frank Borman, Jim Lovell and Bill Anders

Frank Borman, Jim
Lovell and Bill Anders

mission without defecating at all. Joining them at breakfast is
the back-up crew: Neil Armstrong, Buzz Aldrin and Fred Haise.
George Low, whose idea it was to make Apollo 8 a shot at the
moon, is also there. After breakfast the crew clamber into their
suits. None of them will be leaving the craft on this mission so
the suits aren't an absolute requirement, but are worn in case of
sudden loss of pressure during the launch, and to give a measure
of protection should fire break out. When they are on their way
they can take them off. Finally, as if in some futuristic coronation
service, their helmets are lowered and snapped into place. The
astronauts become mere observers of a world that has become
detached from them; reduced to what can be seen through the
visor. The world outside can no longer be smelled, felt, or tasted,
and all that any of them can hear is the swish of their yellow
rubber galoshes as they drag along the ground, and the hiss of
oxygen coming into their suits. It is an eight-mile ride to the

launch pad. Saturn V is still being fuelled up. The rocket emits a fog as if it is breathing.

Saturn V is actually three rockets stacked on top of each other: Stage 1 is the tallest at 138 feet, the second stage is 81 feet tall and the final stage 58 feet. On top of the three stages is a section called the Adaptor that on future missions will enshroud the Lunar Module (LM). Above the Adaptor is the Service Module (SM), which houses the oxygen, the propellant and remaining rocket engine. It is also where the food is stored and electricity generated. Above it is the Command Module (CM), the astronauts' home for the next few days. In the elevator that takes them up to the CM they feel the rocket humming and vibrating next to them. Chunks of ice slide and fall away from the rocket's skin, as 'her cryogenic lifeblood', liquid oxygen and liquid hydrogen, boil and bubble in her guts. At a height of 320 feet they get out onto an arm that leads into the White Room which in turn leads into the CM, a conical space just under 13 feet at its widest, and under 10½ feet tall. Above them, rising another 40 feet, is the escape rocket – the rocket's apex that ends in a small pointed cap – attached to

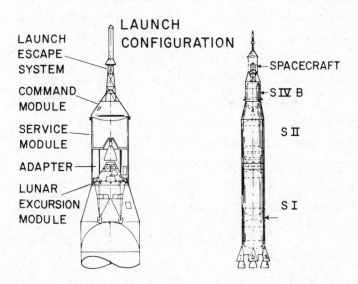

Apollo 8 seen from above

the CM, and able to eject them to safety if something goes wrong. The whole ensemble, often referred to as Apollo 8, the name of the mission, weighs 3,300 tons, 90 per cent of which is fuel. At 363 feet tall, it looks like some modernist iteration of a Gothic cathedral spire.

In the White Room, as ever, Guenther Wendt is there to greet them. As he leaves, he says what he always says in farewell: Godspeed. It falls to back-up astronaut Fred Haise to make last-minute checks. As the crew enter the Command Module he squeezes his way out past them and shakes their hands, closing the massive hatch behind him as he leaves.

The canvas seats are positively spacious compared to those on Mercury and Gemini. Even so, Lovell, who is in the middle, barely has room to move. He thinks, 'My God, they're serious!' The crew begin to go through their checklist. Anders feels surprisingly calm, almost bored. He looks at the window cover – they cannot see outside yet – and watches a hornet trapped between the glass and the cover. He wonders if it might be about to start building a nest.

It is still possible for the mission to be aborted. If anything goes seriously wrong Borman can activate the escape rocket, or else – if there is time – the crew can clamber into a small gondola suspended on a wire, and be whizzed down into a concrete tunnel, onto a 40-footslide, and be deposited into a rubber room, shock-proofed to withstand even the explosion of Saturn V above them. The room is big enough to accommodate up to 20 people, with supplies to last at least three days.

With 15 minutes to go until takeoff, medics at the control centre notice that the astronauts' heart rates are beginning to rise. At T minus five minutes the White Room and access arm swing away. There is no way out now except via the escape rocket. The crew can hear fuel still pouring into the rocket's tanks. At 10 seconds before takeoff the tanks are finally full and there is silence. At T minus 8.9 seconds, valves open. The five F-1 engines of the first stage of Saturn V come alive: 15,000 gallons of kerosene are being pumped into the engines every second, along with 24,000 gallons of liquid oxygen. The pumps are at greatly different temperatures. That the fuels are mixed carefully is an understatement. Flames scorch the concrete launch pad. At zero the rocket shudders. Only now, after 43 tons of kerosene and liquid oxygen have been converted into 6.4 million lbs of thrust, does the rocket slowly rise above the ground. Millions of parts of this massive machine, made to a precision within millionths of an inch, work together in response to the collective intent of hundreds of thousands of human beings.

Around the base of the rocket, sand blown onto the launch pad from nearby beaches is vitrified: what was opaque is made transparent. Standing three miles away, Anne Morrow Lindbergh turns from the rocket to watch some birds take to the air.

The crew hears what sounds like a distant rumble of thunder rolling towards them. Seconds later they are being shaken around, and much more violently than they had anticipated. They feel the many quick and jerky corrections that are being made by the rocket's computer. It has been an immense challenge to build a computer small enough to fit inside a spacecraft. The task was accomplished by the Electronic Research Center (ERC) set up in Cambridge, Massachusetts, in 1964 expressly to solve this problem. Now that its work is done the centre will, in the next few months, be closed down; its lasting legacy the constant drive to make ever-smaller computers.

This is the first manned test of Saturn V. Atlas and Titan rockets each had around 50 test flights before they took men into space. Saturn V had two, and one of them had not been successful. The rocket suffered what the programme director described as 'several important technical failures and malfunctions'. The pogoing problem was supposedly fixed, but now that the crew is being buffeted about it must be hard not to wonder if the problem might return at any moment. It is so noisy inside the capsule the crew cannot hear themselves speak. No number of simulated flights could have prepared them for the experience they are now going through.

It takes 12 seconds for Apollo 8 to creep up and away from the gantry before rapidly accelerating skywards. Frank Borman's wife, Susan, thinks it is like seeing the Empire State Building taking off.

Under two and three-quarter minutes later the first stage has burned through 500,000 gallons of fuel. Apollo 8 is 42 miles above the surface of the turning Earth, travelling at 6,000 miles per hour. The massive first stage has fulfilled its function and is sloughed off. Bolts are blown and it falls away, heading towards the Atlantic Ocean.

The second stage fires up its five J-2 rockets. They burn through a further 600,000 gallons of fuel, and accelerate what is left of Apollo 8 to 15,000 miles per hour and to an altitude of 60 miles.

After three and a quarter minutes in to the flight there is a loud explosion above the astronauts' heads as the escape capsule is jettisoned. Though they know this will happen, it still comes as a surprise to the crew, perhaps like that moment for the rest of us when the wheels go up as our plane takes off. With the escape rocket now gone, for the first time the windows are uncovered and light floods into the capsule.

After eight minutes in to the flight the rocket starts shaking again, quite violently. The astronauts can hear the metal skin of the rocket straining and creaking. The rocket is pogoing. The problem, it seems, was not entirely fixed. But a minute later the second-stage engines shut down and the pogoing stops. The second stage has now served its purpose and it, too, is cast off and sent plummeting to Earth. Already, 97 per cent of the ensemble that left the launch pad has been shed or burned up.

For a few seconds the crew is weightless and all is quiet. Odd objects begin to appear and float around the Command Module, a screw, a piece of string, reminders that the machine is made by fallible human beings who have left things behind.

The crew is waiting for the third stage to fire, and though only a few seconds have passed since the second stage shut down, it feels like forever before the single J-2 engine of the third stage ignites – and, because they are being accelerated, gravity returns.

At 11¾ minutes into the flight the third stage shuts down, and gravity departs once more. They have been put into orbit 100 miles above the surface of the Earth, travelling at 17,432 mph. For the next two and a half hours they will stay in orbit with their heads pointing towards the Earth.

There is much to do. Commander Frank Borman warns his crew: 'I don't want to see you looking out the window.' No time for Earthgazing.

It is Bill Anders' first experience of being in orbit around the Earth, and he is itching to see the Earth below, but during that two and a half hours he only manages to catch two brief glimpses: a thunderstorm at night over Australia, and he just has time to identify the location of his childhood home town of San Diego, shrouded in fog.

At precisely two hours, 27 minutes and 22 seconds after takeoff, Mike Collins at Mission Control – the first Capsule Communicator (CapCom) on duty – radios through: 'You are go for TLI.' NASA's cool liturgy steadies the moment. The timing of this so-called trans-lunar injection is crucial. The third stage fires up again. The craft is accelerated to 25,000 mph, out of Earth orbit and onto a trajectory that has been carefully calculated to take them to the moon. The many launch processes have been successfully activated one after another, like a line of falling dominoes: manual switches have been thrown; computer and other types of automated control have kicked in at precise moments. Sometimes it has been the acceleration of the rocket itself that has caused some sequence of operations to unfold: tapered pins have fallen out just when they needed to; metal rings have dropped away because of some inertial change. Everything has happened as it was meant to do, and now the first human beings ever to leave the pull of the Earth are on their way to the moon.

Pete Conrad and Dick Gordon in Gemini XI rose higher above the Earth than any other human beings in history, but with every passing second that record now recedes into insignificance. Through the round window of the Command Module, the Apollo 8 crew can see the Earth receding rapidly. From their perspective the Earth is positioned with Antarctica at the top, except that in space there is no up and down. After a while the whole Earth – 8,000 miles across – can be taken in in a single glance. It is the first time in history that human beings have seen the entire Earth with their own eyes, for what it is: a sphere in space. Even those back on the ground at Mission Control are shaken by the images of the round Earth being beamed back to them. 'We were the first humans to see the world in its majestic totality,' Borman will write, 'an intensely emotional experience for each of us. This must be what God sees.' Anders thinks: where have I seen this before? And remembers that it was three months earlier, at the premiere of Kubrick's *2001: A Space Odyssey*. After a while the Earth appears not to shrink any further, and now it feels as if their craft is hardly moving at all.

Though the third stage has fallen away, they can see it out of one of the Command Module windows, as if it is following them.

Borman asks Mission Control if they can slightly deviate from their course in order to open up some greater distance between the CM and the spent booster rocket. From the ground Collins tells them to use the Earth to align the craft. Borman replies, 'OK, as soon as I can find the Earth.' There is astonished laughter back on the ground. The Apollo 8 astronauts are not only the first humans to see the whole Earth but the first to lose sight of it altogether.

When they reach seven hours into the flight, the craft is put into what is called barbecue mode. The Command Module, together with the Service Module (sometimes referred to as the CSM), rotates steadily in order to keep the outside of the craft at an even temperature, otherwise one side would rise to 250 degrees Fahrenheit and the other minus 250 degrees. Inside it is a pleasant 70 degrees. As the CSM rotates, the Earth and moon appear successively in the crew's windows.

It is Anders' first proper experience of zero gravity. During training they had flown in so-called reduced-gravity planes, taken up into a great parabolic arc. At 90,000 feet the plane levelled out and for a few seconds the astronauts experienced zero gravity. They called them vomit comets. But this is nothing like that. Anders unbuckles his harness and stays exactly where he is, floating above his couch like an Indian mystic. At first his eyes bulge, and his face flushes and gets puffy. After a few hours the body adjusts and Anders feels as if he has always been weightless. Without his suit on, weightless, wearing just long johns, Anders experiences a sense of freedom as never before. And then for a moment he feels nauseous, and then the feeling passes.

Mission Control knows exactly where the Command Module is at all times, but if radio contact fails it will be Lovell's job as the navigator to guide them back home. One of his tasks is to make star readings at regular intervals throughout the journey, and to compare his calculated trajectory with the actual one. He uses what looks like a sextant – a compass in outer space is of course useless: there is no East or West, North or South. As at sea, Lovell uses the stars to sail by, though in space the constellations are harder to recognize; the star field is much richer than it is when seen from Earth.

When the time comes to try to go to sleep, they zip themselves into sleeping bags in an attempt to create the effect of blankets on a bed: 'Instinctively I feel I am lying on my back, not my stomach, but I am doing neither – all normal yardsticks have disappeared, and I am no more lying than I am standing or falling.' If their arms stray outside the hammock they drift up like a praying mantis's, and stay there. A pillow is no help. The head floats just above it, except not of course 'above', or even 'below' it. Nor are they lying with their arms floating up; they could just as well be said to be lying on their front, arms floating down. Or standing on their head. Anders finds it hard to sleep. In the silence of space, and without gravity to overcome, he hears his heart beating loudly and more forcefully than he is used to. Nor can Borman get to sleep. He radios NASA to ask if he can have permission to take a Seconal. NASA phones his wife Susan to ask if he usually has trouble sleeping. No, she says, and generally he doesn't like to take drugs. NASA gives him permission to take the pill anyway. The next day, Sunday, he has a headache and then begins to feel sick. He vomits. Worse, he gets diarrhea. The smell is revolting. Anders puts on his oxygen mask. The crew spends some time chasing down globules of floating vomit and faeces with paper towels. Anders says it is like going on a butterfly hunt. One globule is the size of a tennis ball, 'shimmering and pulsating in three dimensions ... in some kind of complex fluid vibrations made possible in zero gravity'. It splashes onto Lovell.

Borman doesn't want to tell Mission Control that he has been sick, but he is persuaded otherwise by his crewmates. He decides he doesn't want to speak to the ground directly and tapes a message. He says he is sure it is just a 24-hour virus. When the message is played back on the ground, for a while NASA considers aborting the mission. In the end medics on the ground decide that Borman suffered a bad reaction to the Seconal. In fact it has been the first experience of space sickness among the American astronauts. One in two of them will suffer from it.

On Sunday 22 December 1968 the crew of Apollo 8 makes its first broadcast. The Intelsat-3A satellite was rushed into service to cover the mission; its first transmission was made just a few hours earlier: the Pope celebrating Mass from St Peter's in Rome. Now the satellite transmits images of the Earth from 120,000 miles away. The images might have been clearer but the crew can't get the telephoto lens onto the TV camera, which anyway only broadcasts in black and white. The television audience must take it on trust that the white blob they are looking at is indeed the Earth.

Two days and seven hours into the flight, the Command Module moves into the gravitational pull of the moon, though the moon itself has hardly been in view at all during the journey. The angle of the trajectory has put them in the moon's shadow, or, when the moon might have been visible, the light of the sun was too bright.

But then, suddenly, on Monday 23 December there it is, on top of them, what looks like a great disk of black emptiness in space. The hairs on Anders' neck stand up. They have arrived. It took only two small in-flight corrections to get them here.

Now they are about to lose radio contact with the Earth for reasons entirely foreseen. They are about to pass to the side of the moon never visible from the Earth, the so-called dark side; though not dark because it is not illuminated but dark because we cannot see it from Earth, and dark because the astronauts will be out of radio contact. They will orbit the moon 10 times before returning home, putting them out of radio contact every couple of hours.

The Service Module's single engine fires for four minutes, just long enough to slow the craft down to 3,700 mph, the speed required to put the craft into lunar orbit: slower and it would crash into the moon's surface; any faster and it would shoot past the moon altogether. For the crew it is the longest four minutes of their lives. The craft settles into a circular orbit 69 miles above the moon's surface. Since they took off, 66 hours have passed. They are 234,000 miles from home.

'Oh, my God!' says Anders. 'What's wrong?' asks Borman. 'Look at that,' says Anders. Vast mountains have, without warning, appeared below them. The largest window is clouded up because the sealant has decomposed; Anders' side window is smeary, as if covered in oil; and the two so-called rendezvous windows are tiny – but, even so, they can see the moon's surface in extraordinary detail. Because the moon has no atmosphere, and the sunlight is so strong, the mountain peaks feel very close, as if their craft might clip one as it rushes by. They are the first human beings to see the far side of the moon at first hand. But then, for pioneers, everything is new. 'Alright, alright, come on,' Borman says, 'you're going to look at that for a long time.' Radio contact is resumed 45 minutes later, as they leave the far side of the moon.

If the moon takes them by surprise, so does their first experience of seeing the Earth rise over the horizon of the moon. At a press conference in November the crew was asked what part of the mission they were most looking forward to. Lovell said, 'To see Earth set and Earth rise.' The Earth has risen and set three times since they have gone into lunar orbit but they have been too busy to notice. (In fact the Earth doesn't rise over the horizon of the moon in the way that the sun appears to rise over the horizon of the turning Earth. The moon and Earth are locked together in a synchronized gravitational dance, the result of which is that the moon only ever shows one side to the Earth. The Earth only appears to rise over the moon's horizon because the craft is in orbit around the moon.)

Now, as they move to the dark side for the fourth time, they do notice, and Borman has his camera at the ready.

Borman: 'Oh, my God. Look at that picture over there.'
Anders: 'What is it?'
Borman: 'Here's the Earth coming up. Wow, that is pretty.'
Anders: 'Hey, don't take that, it's not scheduled.'
Borman: (Laughter). 'You got a color film, Jim?'
Anders to Lovell, laughing: 'Hand me that roll of color, quick, will you?'
Lovell, ignoring him and joining them at the window: 'Oh man, that's great!'
Anders to Lovell: 'Hurry. Quick.'
Anders snaps on the colour magazine.
Lovell impatiently: 'You got it? Take several of them. Here, give it to me.'
Borman: 'Calm down, Lovell.'
Lovell: 'Are you sure we got it now?'
Anders, sarcastically: 'Yeah. It'll come up again, I think.'

Frank Borman claims later that it was he who took the famous photograph later known as Earthrise, which is ironic given what he said about not looking out of the window; nor did he want to take along a TV camera, seeing it as a distraction. It will be overlooked for decades, but Borman took some black-and-white photographs of Earthrise before Anders snapped the colour magazine onto his camera. Borman had taken the first photograph of the Earth rising over the horizon of the moon but the famous photograph is in colour and it was Anders who took that. Colour will make all the difference.

Up to now, Anders has been so focused on photographing the

surface of the moon that he hasn't thought much about the Earth. But seeing Earthrise he realizes what it is that he has been missing whenever he looks at the moon. The Earth is the only thing out there that has any colour to it. The universe looks black and white; except for the Earth. Anders sees now that the moon is not as interesting as it first seemed. He has become oppressed by the moon's unrelenting sameness. Has he come 240,000 miles just to see something that looks like a dirty beach? Borman says the moon looks like the Earth might have done before the advent of life, or as it will after life is extinct. Earthrise makes the contrast clear: the beauty of the Earth and 'this ugly lunar surface'.

Earthrise happened when they were out of earshot of Mission Control, on the dark side of the moon. Curiously, despite their obvious excitement, they don't say anything to their colleagues back on the ground when the craft comes back into radio contact.

Later that day, as they approach their tenth and last orbit of the moon, the crew makes a final TV broadcast back to Earth. It is 8.11pm Houston time on Christmas Eve. At a pre-launch conference on 7 December 1968 a reporter questioned the Christmas timing of the flight – in fact, there was only a small window of opportunity each month when the launch could take place. Lovell cut him off mid-sentence: 'I can't think of a better religious aspect to the flight than to further explore the heavens.' Borman interjected with a secular justification: 'When you're finally up at the moon looking back at the Earth, all those differences and national traits are pretty well going to blend and you're going to get a concept that this is really one world and why the hell can't we learn to live together like decent people.'

Their final broadcast opens with another view of the Earth as a white blob. Much of the rest of the broadcast is taken up with showing images of the surface of the moon. In turn, each member of the crew gives his impression of the moon. Borman says that it is 'a vast, lonely, forbidding type of existence or expanse of nothing'; Lovell, too, talks of the lonely moon, essentially grey with no colour, but, unlike Anders, he finds the surface of the moon mesmerizing. He describes space as 'a vastness of black and white, absolutely no colour ... The loneliness out here is awe-inspiring. It makes us realize what you have back on Earth. The Earth is a grand oasis in the vastness of space.'

And then, towards the end of the 23-minute broadcast, Anders says, 'We are now approaching lunar sunrise. And for all the people back on Earth, the crew of Apollo 8 has a message we would like to send to you.'

No one knew what was coming. Back at Houston the flight controller Gene Krantz said he felt a chill. Some time before the mission, Julian Scheer, NASA's deputy administrator for public affairs, gave Borman some oblique advice: 'Look, Frank, we've determined that you'll be circling the moon on Christmas Eve and we've scheduled one of the television broadcasts from Apollo 8 around that time. We figure more people will be listening to your voice than that of any man in history. So we want you to say something appropriate.' Borman had asked a publicist he knew, Simon Bourgin, Science Policy Officer at the US Information Agency for advice. He in turn asked a journalist, Joe Laitin; a Christian who searched the New Testament without finding anything that struck him as suitable. He asked his wife, and she said: 'Why don't you begin at the beginning.' Joe Laitin suggested that after the broadcast there should be silence.

Now, at the end of their final broadcast to Earth, and in a soft voice, Anders begins to read:

In the beginning God created the heaven and the earth. And the earth was without form, and void; and darkness was upon the face of the deep. And the Spirit of God moved upon the face of the waters. And God said, Let there be light: and there was light. And God saw the light, that it was good: and God divided the light from the darkness.

And then Jim Lovell takes over:

And God called the light Day, and the darkness he called Night. And the evening and the morning were the first day. And God said, Let there be a firmament in the midst of the waters, and let it divide the waters from the waters. And God made the firmament, and divided the waters which were under the firmament from the waters which were above the firmament: and it was so. And God called the firmament Heaven. And the evening and the morning were the second day.

And then Frank Borman:

And God said, Let the waters under the heavens be gathered together unto one place, and let the dry land appear: and it was so. And God called the dry land Earth; and the gathering together of the waters called he Seas: and God saw that it was good.

'And from the crew of Apollo 8,' says Borman, 'we close with good night, good luck, a Merry Christmas, and God bless all of you – all of you on the good Earth.'

At Houston, there is silence. 'For those moments,' Flight Director Eugene Krantz said later, 'I felt the presence of creation and the Creator ... Tears were on my cheeks.' Krantz was a former airforce fighter pilot, rugged-looking, crew-cut, nicknamed General Savage, might have been a character from *Catch-22*, but inside 'was as sentimental as they come ... At the start of the day he would go into his office and listen to John Philip Sousa marches, to get his blood flowing. He was a devout Catholic and a man of strong beliefs, and at his core he believed in the exploration of space.'

The astronauts go off air just as they enter the dark side for the tenth and last time. While they are out of radio contact with Earth the so-called trans-Earth injection (TEI) will take place. The craft's single engine will fire just the right amount – carefully calculated to take into account the moon's uneven gravitational pull – to put the craft on a trajectory back home.

On Earth the astronauts' wives are able to listen in to the mission from home via a device called a squawk box. When Susan Borman hears the squawk box fall silent a tenth time, she feels that the silence now has a different quality. Some time before the launch she said to Chris Kraft: 'Hey Chris, I'd really appreciate it if you would level with me. I really, really want to know what you think their chances are of getting home.' 'Okay,' Chris replied, 'how's fifty-fifty?' She is sure TEI will fail. She is so sure her husband isn't coming back that she has started to prepare his memorial service. She is calm on the outside, but has turned to alcohol, and drinks in private. Now she is at home with a camera crew – at NASA's insistence – for company. They are filming her as she waits to hear if TEI goes smoothly, or not.

TEI takes place exactly as it is supposed to do. The crew finds the journey home boring. Anders calls it falling for 240,000 miles. Not that there aren't still hazards to negotiate. At one point Lovell accidentally wipes all the navigational information from the onboard computer. He has to recalculate their position using his special sextant and the stars. He sends his measurements to Mission Control for double-checking.

On Christmas Day, back on Earth, Marilyn Lovell opens the present her husband has left behind for her. A mink jacket. It is what she has always wanted. She wonders if he thought: 'If I don't make it back at least she'll have the fur.'

The next day, during the last hour of the journey, the Service Module is jettisoned. All that is left of the original rocket is the Command Module, little more than a human container, 'a steel cone less than a dozen feet high', the fulfilment of Jules Verne's vision of a manned cannon shell.

Entering the Earth's atmosphere is a particularly hazardous moment. The capsule, travelling at 25,000 miles an hour – 10 times faster than a high-speed rifle bullet – must somehow be slowed down. The incoming trajectory has been minutely calculated so that the craft will skim the atmosphere, like a pebble across the surface of a pond, in order to lose energy. Of the four unmanned attempts to do this, two failed. If the approach is too shallow the craft will shoot off into space, too steep, and even the heat shield will burn up like a meteorite.

To viewers on Earth the capsule looks even now like a meteorite. If the heat shield fails they will burn up. The ionized air around the capsule becomes impenetrable to radio signals for a few minutes. The g-forces make the astronauts feel as if they are more than six times heavier than normal. And after days of feeling as if they weighed nothing at all, the effect is all the harder to bear.

The parachutes deploy in sequence: two drogue parachutes each 16 feet in diameter to stabilize the capsule, and then three main parachutes each 80 feet in diameter.

And suddenly there they are, upside down in the Pacific Ocean, 10-foot waves outside. They are only 3 miles from where USS *Yorktown* has been waiting for their return. Everything that is not lashed down is violently thrown about. Seawater begins to come in. Borman throws up, again. When the first navy swimmer arrives and opens the hatch, he reels back from the smell as if he has 'been kicked in the head'.

Back at Mission Control a couple of dozen men (they are all men, average age 26) are cheering and lighting up cigars. Each of them has invested so much in making this happen it feels like their flight too. CapCom Mike Collins wants to cry, but instead claps a few colleagues on the back and leaves. It could have gone wrong in a thousand ways, but it hasn't. Almost miraculously it has all gone to plan.

Four days later, Charles Lindbergh sends the crew a telegram: 'THE GREATEST FEAT OF TEAMWORK IN THE HISTORY OF THE WORLD. YOU HAVE TURNED INTO REALITY THE DREAM OF ROBERT GODDARD.'

CHAPTER FOUR

Richard Underwood developed the photographs that came back from Apollo 8 himself, by hand. By machine it would have taken minutes, by hand it took about five hours. 'It was a labor of love,' Underwood said, '[a] tender loving process.' Though it was slow work, it was soon clear that the images were even better than he had dared anticipate. NASA image AS8-14-2383, what would become better known as 'Earthrise', was released to the press with the usual technical caption – barely descriptive, certainly devoid of poetry:

> The rising Earth is about five degrees above the lunar horizon in this telephoto view taken from the Apollo 8 spacecraft near 110 degrees east longitude. The horizon, about 570 kms (350 statute miles) from the spacecraft, is near the eastern limb of the moon as viewed from Earth. Width of the view at the horizon is about 150 kms (95 statute miles). On Earth 240,000 statute miles away the sunset terminator crosses Africa. The crew took the photo around 10.40am Houston time on the morning of 24 December, and that would make it 15.40 GMT on the same day. The South Pole is in the white area near the left end of the terminator. North and South America are under the clouds.

NASA released the image as the Apollo crew had seen it. The press reproduced it in the way we are used to seeing everything from Earth, as if over a horizon. We humans believe the sky to be above us because that is all of the sky that we can see, and so though we might acknowledge intellectually and occasionally that the sky is as much below us, that is not the way we feel the world to be.

The day after the crew had publicly read from Genesis from space, Edward Fiske, in an article in the *New York Times,* wondered if Apollo 8 had made the need for some kind of synthesis between the sacred and secular urgent. He argued for a new version of Christianity, purged of the supernatural, something more in keeping with the space age. Even the 'secularly inclined' BBC aviation correspondent, Reginald Turnill, had said that the reading 'struck one instantly as stroke of genius'. Mike Collins said the same: 'It was impressive, I thought, a stroke of genius, to relate their primordial setting to the origin of the Earth, and to couch it in the beautiful seventeenth-century language of the King James I scholars.'

Not everyone agreed. Before they had landed, Madalyn Murray O'Hair was on air accusing the Apollo 8 crew of disrespecting those who do not believe in God. For good measure she added that they had also slandered those who followed religions other than Christianity. What they had done was unconstitutional. She would have been even more incensed if she had known that Borman had recorded a prayer while in space to be played back to his church during their Christmas Eve service. The operation even had its own codeword: Experiment P-1. O'Hair invited listeners to her radio show, American Atheist Radio Series, to write in and protest, and announced that she intended to sue NASA. By then she was one of the most prominent figures in American public life. NASA took her threat seriously. O'Hair said she was inundated with letters. She was shown photographed in front of what she said were 28,000 letters in support of her protest. Borman, however, said that he personally received 100,000 letters, all bar 34 in support of the Genesis reading, and many of them denouncing O'Hair. One letter described O'Hair as, 'Jezebel, Lady Macbeth, Lot's wife, Mary Queen of Scots, all wrapped into one foul-mouthed atheist'. When Loretta Lee Frye, a Detroit member of the Houston-based Apollo Prayer League, started a petition in support of religious readings in space, she received half a million signatures in three months. Harold Camping, president of Family Radio, said that the campaign in support of the Genesis readings was 'the largest voluntary commendation of an act by man that has ever occurred in our nation's history and perhaps in the history of the world'.

At the Congress meeting that followed every Apollo mission, Borman, a Roman Catholic, began to talk about the Genesis reading. And then, aware that the nine Supreme Court justices were facing him, he said: 'now that I see the gentlemen here in the front row, I am not sure we should have read the Bible at all'. The reference was clear. 'The whole chamber rocked with laughter and applause.'

Borman was asked to respond to Titov's comment about the absence of God in space, still a favourite with journalists. He said – a slight variant on the usual response – that he had not seen God either, 'but I saw his evidence'. Bill Anders said that during the Genesis reading the thought had come to him that they were 'trying to say something fundamental. This isn't just another space mission; it's a new beginning, for all of us.' But how novel or inclusive had it been to read from the Bible? The Earthrise photograph reinforced in a humanistic way what many would have found unpalatable about the Genesis reading. The *New York Times* said that the photograph was a 'humbling reminder of the world's insignificance'; the *LA Times* that it 'made introverts of us all'. Writing in the *New York Times*, the then fashionable poet, Archibald MacLeish, wondered if, in a matter of hours, human beings had shifted perspective, away from a conception of themselves as 'God-directed actors at the center of a noble drama', to one in which 'Man may at last become himself.' Like Fiske – whose article appeared in the same edition – MacLeish saw immediately what was philosophically (and perhaps theologically) remarkable about the experience of seeing ourselves from the outside. Ending his article with a last rhetorical flourish, MacLeish wrote: 'To see the Earth as it truly is, small and blue and beautiful in that eternal silence where it floats, is to see ourselves as riders on the Earth together, brothers on that bright loveliness in the eternal cold – brothers who know now they are truly brothers.' It was a point that would be made over and over again in the months and years that followed. On the Apollo 8 astronauts' post-flight world tour Borman said that we 'are first and foremost not Germans or Russians or Americans but Earthmen'. He said that 'raging nationalistic interests, famines, wars, pestilences don't show from that distance. From out there it really is "one world".' 'Would it not be ironical', said Wernher von Braun '– as well as instructive – if nations first learn to transcend their national interests many, many miles away from Mother Earth?' Movingly, the comments mirror the famous speech made by Martin Luther King from the pulpit of the Ebenezer Baptist Church in Atlanta, a year to the day before the Genesis reading from space and the taking of the Earthrise photograph: 'If we are to have peace on Earth ... we must develop a world perspective. As nations and individuals, we are interdependent. It really boils down to this: that all life is interrelated. We are all caught in an inescapable network of mutuality, tied into a single garment of destiny. Whatever affects one directly, affects all indirectly. We are made to live together because of the interrelated structure of reality.' King was assassinated four months later.

Ironically it was Pope Paul VI, addressing a crowd in St Peter's

Square soon after the Apollo 8 splashdown, who made one of the most perceptive secular observations about the mission. 'The stature of man,' he said, 'in prodigious confrontation with the cosmos emerges immensely small and immensely large.' The Earth appeared insignificant compared to the seeming infinity of space that surrounds it, and yet the colourful Earth looked so different from anything else out there that it also seemed to be special. For the first time in history, humans had seen their home from the outside, and yet this new perspective, paradoxically, was a loss of perspective. Bill Anders said that his faith faltered when he heard a broadcast of 'O Holy Night' disintegrate as they went around the dark side of the moon: 'If music could not survive an encounter with the universe, was there much hope for religious doctrine? And if the church could not claim to be universal, could it claim the authority to declare transcendent laws?' And yet man in space also seemed to highlight the need for religion, or something like it. Here, as Fiske so promptly realized, was an opportunity to forge a relationship between science and religion that was more open, braver and richer. Anne Morrow Lindbergh said that after Earthrise no one would ever look at the Earth the same way again: 'As Earthmen, we may have taken another step into adulthood.' And yet she wondered, too, if it might take centuries before we fully absorbed this new perspective: that there is no persepective to be had.

Soon after the Apollo 8 mission, the US Postal Service issued a commemorative stamp that reproduced the Earthrise photograph along with the words, 'In the beginning God.' It was the first time religious wording had appeared on a stamp since O'Hair's landmark judgment of 1963. Bill Anders said it was the stamp that everyone swooned over. Rather than finding a new synthesis, religion and atheism in America seemed to be dividing along old lines, mirroring the Cold War between God's America and Soviet Atheism.

Despite the avalanche of letters arriving at NASA in support of the Genesis reading, Deke Slayton was sufficiently disturbed by Madalyn O'Hair's threatened action that he forbade religious readings or prayers on upcoming Apollo missions.

CHAPTER FIVE

3 March 1969. Apollo 9. Crew: Jim McDivitt (Commander), Dave Scott (Command Module Pilot), Rusty Schweickart (Lunar Module Pilot). Duration of mission: 10 days, 1 hour, 54 seconds. Command Module: *Gumdrop*. Lunar Module: *Spider*.

The main goal of the mission was to test the procedures for the docking of the Command and Lunar Modules in Earth orbit before they were performed in lunar orbit during the next mission. When Schweickart developed space sickness, his spacewalk was postponed for fear that he might throw up inside his space helmet and suffocate. Several years later he would describe his EVA as a transcendental experience, but for now he kept quiet, worried that such talk might undermine his chances of another mission.

Photographs of the Earth at close quarters, developed soon after the craft returned home, generated some excitement. Von

Jim McDivitt,
Dave Scott and
Rusty Schweickart

Braun talked up the practical value of photographs of the Earth. They could be used, he suggested, to locate underground water, reduce disease in crops and forests, find fish, improve maps, detect illegal dumping of toxic waste, locate oil and minerals, measure ice and snow, soil fertility and salinity, and prevent famine.

Gene Cernan, Tom Stafford and John Young

18 May 1969. Apollo 10. Crew: Tom Stafford (Commander), John Young (Command Module Pilot), Gene Cernan (Lunar Module Pilot). Duration of flight: 8 days, 3 minutes, 23 seconds. Command Module: *Charlie Brown*. Lunar Module: *Snoopy*.

The main goal of the mission was to attempt to dock the Command Module and Lunar Module in lunar orbit. The crew came within 10 miles of the moon's surface, even closer than the Apollo 8 crew had. To come so far, and to get so close! The mission returned with some fine photographs of Earthrise, but everyone's focus was on the upcoming mission.

16 July 1969. Apollo 11. Crew: Neil Armstrong (Commander), Mike Collins (Command Module Pilot), Buzz Aldrin (Lunar Module Pilot). Duration of mission: 8 days, 3 hours, 18 minutes, 35 seconds. Command Module: *Columbia*. Lunar Module: *Eagle*.

Mike Collins,
Neil Armstrong
and Buzz Aldrin

President Nixon had asked if he might have dinner with the crew of Apollo 11 the night before the launch. Chuck Berry, the NASA doctor, forbade it, saying that the risk of infection was too great, which was odd given that the crew was not living in a germ-free environment, and odder still that the privilege had once been granted to the Lindberghs.

The next morning, in the elevator going up the side of the rocket to the White Room, Mike Collins wondered what was amiss, and then realized: there were no people. He was aware that a vast crowd had turned out at Cape Kennedy, but he said he already felt closer to the moon than to them. And yet if, in his mind, he was already on his way, he was also aware of the absurdity of it all: 'Here I am, a white male, age thirty-eight, height 5 feet 11 inches, weight 165 pounds, salary $17,000 per annum, resident of a Texas suburb, with black spot on my roses, state of mind unsettled, about to be shot off to the moon. Yes, to the moon.' He had decided that their chances of returning were about evens. Apollo's engineers were more optimistic: they estimated the astronauts' chances of being killed at one in 1,000. Everything they did was done with the survival of the astronaut paramount: perhaps the true gift of The Fire.

Once they were in the Command Module there was barely time left to shake hands with pad leader Guenther Wendt.

Collins clambered into the middle couch. Armstrong was on his left, Aldrin on his right. In front and around them were 57 instrument panels and 800 switches. The instruction manual was 330 pages long. The Mercury manual had been a mere 30 pages. Collins felt uncomfortable in his Apollo spacesuit, which was tight around the crotch. He had preferred the Gemini spacesuits, made by a different company. Fortunately it would feel more comfortable in zero gravity.

Among the million or so people who had turned out to watch the launch was Charles Lindbergh. Jim Lovell had been delegated to be his escort. During a conversation in which Lovell extolled the historic importance of what was about to be attempted, Lindbergh stopped him and said, 'You know, Apollo 8, to me that was the high point of the space program, because it was the first time humans traveled outside the pull of Earth's gravity ... You were the pioneers of this. Landing on the moon is just the icing on the cake.'

President Nixon wasn't at the launch, just in case it all went wrong. Nixon sent ex-President Johnson instead. A speech had been written in the event the astronauts did not return. The funeral service was to be modelled on the service for burial at sea. Johnson was in a foul temper: 'It was worse than I thought it would be,' he said afterwards, 'I hated it.' Johnson had apparently been in a bad mood ever since he'd lost the party nomination. In the end there had been something tragic about his Presidency. It had begun with 'a boldness of vision unprecedented since the Roosevelt era': Medicare, Medicaid and the Voting Rights Act were all part of his legacy, but so was Vietnam. By June 1968 America had committed 535,000 troops. As the war escalated so had Johnson's passion for space flight dimmed. The war was now costing $3 billion a month. America's spending was destabilizing the global economy. And still the war would see out another president.

During the launch, Aldrin's pulse climbed to a modest peak of 88, the lowest pulse rate of any astronaut during takeoff. 'What's there to be afraid of?' he said, 'When something goes wrong, that's when you should be afraid.' On the ground von Braun offered up a short prayer, 'Thy will be done.' Lindbergh later said that even from 3 miles away his 'chest was beaten and the ground shook as though bombs were falling nearby'. The astronauts' beloved nurse, Dee O'Hara, who probably knew them better than anyone other than their wives, began to cry. Herman Oberth, von Braun's one-time mentor, was there. He said that the launch was just as he had imagined 'only more marvelous'. Arthur C. Clarke said that it was the perfect last day of the Old World.

Apart from being thrown around a little, it was much less traumatic an experience than the launch of Gemini had been. Once they were safely underway, the crew could get out of their suits. Not easy. They were like three whales, Collins said, thrashing around in a small tank.

Early in the flight Collins had to perform a difficult manoeuvre. The Lunar Module (LM) had to be released from its protective shroud, reoriented through 180 degrees and brought into alignment with the Command and Service modules (CSM). The LM was housed between the third stage of the Apollo rocket and the CSM in a section called the Spacecraft Lunar Adaptor (SLA), a conical cover that protects the delicate landing module during launch. During the manoeuvre, the CSM and LM were entirely free of each other, hurtling together through space. NASA had wondered if they should be somehow tethered together during the reconfiguration, but realized it wasn't necessary: the apparent need was psychological rather than physical. Once the three rocket stages have fulfilled their function and fallen away, the four leaves of the SLA detach like petals to expose the Lunar Module. Collins had spent many hours back on Earth practising this manoeuvre in the

Apollo simulator. He would have put in eight hours a day if he could have done, but the simulators were always breaking down and the astronauts had had to take it in turns. John Young called the simulator the Great Train Wreck because of its odd shape seen from the outside. Armstrong had been involved in its design. Sometimes the simulations were so complex that the instructors knew there was no way out. They just wanted to see how the astronaut would react. Collins was the only member of crew trained to do the transposition and docking, and the manoeuvres kept him busy during the first hours of the journey. He worried that he had used up more fuel than he needed to, but otherwise the operation went smoothly. Now all he had to do was insert a plug to make an electrical connection between the CSM and LM.

At 28 hours and seven minutes into the voyage, Collins radioed a message to ground control: 'Houston, Apollo 11 . . . I've got the world in my window.' He said it was a sober, melancholy sight. Armstrong put up his thumb and blocked out the whole Earth. He said it made him feel not big and powerful but small and insignificant.

When they saw the moon for the first time – after a day of being in its shadow – Collins sensed that he and his crewmates were all feeling the same thing: that it was a scary-looking place. But no one said anything. Each mission caught the moon in a slightly differing light. The Apollo 8 crew said they saw the moon in shades of grey between black and white, the Apollo 10 crew that they saw browns too. The Apollo 11 crew saw a new shade: a 'cheery' rose colour that darkens through brown into black.

'You cats take it easy on the lunar surface,' Collins told his crewmates before he threw the switch that released the Lunar Module. 'If I hear you huffing and puffing, I'm going to start bitching at you.'

The Lunar Module – 'nothing more than a taut aluminum balloon, in some places only five-thousandths of an inch thick', easy enough to puncture with a sharp object – was on its way to the moon's surface. Building the LM had been one of the greatest challenges of the entire project. It is a clumsy-looking machine, like an insect. Volkswagen later ran a nine-second advertising film campaign: to a background of Sputnik-like space noises, an image of the LM on the moon is followed by VW's logo. An announcer says, 'It's ugly, but it gets you there.'

Collins was left alone to orbit the moon in the Command Module, *Columbia* – named after Jules Verne's *Columbiad*, shot to the moon out of a great cannon. Not that Collins ever felt particularly close to his Apollo 11 colleagues. Amiable strangers, was how he described them. He would like to have been closer, but – even among astronauts – Armstrong and Aldrin were a formidable pair. He liked Armstrong but didn't know how to get to him, didn't really know what to make of him. Armstrong kept everything to himself, and Aldrin's intense gaze was unsettling. He had the feeling that Aldrin was probing him, looking for weaknesses. Aldrin would have made 'a champion chess player', he said, 'always thinking several moves ahead. If you don't understand what he's talking about today, you will tomorrow or the next day.' Armstrong by contrast was very laid back. His application to NASA had arrived a week late. Strings had had to be pulled even to get him an interview. Collins said he had sent in his application before the ink had dried on the form. Neither Aldrin nor Armstrong had any small talk. All they ever wanted to talk about was technical stuff. Both were extraordinary intelligences, both very shy. Aldrin had a tendency not to say anything unless it was absolutely necessary, and then when he did, to be direct, which often got mistaken for rudeness. He might get excited about some small technical detail and talk about it all night long. He was aware that his brain had an exceptional computational talent. His doctoral thesis had been on manned orbital rendezvous, and he was obsessed with talking about trajectories. The astronauts nicknamed him Dr Rendezvous. Why would he not be proud of his abilities? But it was easy to mistake his pride for arrogance. He couldn't keep to a single subject and was always wondering how everything could be re-engineered or reconfigured in some way. He had the constitution of an ox and could slowly drink a whole bottle of whiskey and be none the worse for it the following day. Yet Aldrin was also emotionally sensitive. His father, Edwin Eugene Aldrin Sr, was 'distant and demanding', and had often been away while Buzz was growing up. Edwin had studied with Robert Goddard and, as an aviation consultant, knew Charles Lindbergh. Whenever Buzz came back from school, the first thing his father wanted to know was how Buzz had done in exams or at sport. If his son had come third, he wanted to know who had come first and second. 'Third place', Buzz wrote, 'doesn't hold quite the appeal to him that first place does.' Aldrin Jr graduated in third place from West Point. He flew 66 combat missions in Korea; Armstrong flew 78.

It was ironic, Collins thought, that the most sociable of the three should be the one left to orbit the moon alone. It was ironic, too, that if anything went wrong while they were on the moon – his greatest fear during the entire mission – he would be the one left working out from the Handbook how to take apart and rebuild some piece of technical equipment; he who couldn't even mend the latch on his screen door.

Collins had been asked countless times before the mission how he would cope being on his own. He said he liked being alone, that that was the essential experience of being a fighter pilot. Now, for 48 minutes every two hours, Collins would be cut off not just from his colleagues but from the entire world. He told his crewmates to keep talking to him, but soon he grew used to the experience and then to really like it: 'I am alone now, truly alone, and absolutely isolated from any known life. I am it. If a count were taken, the score would be three billion plus two over on the other side of the moon, and one plus God knows what on this side. I feel this powerfully – not as fear or loneliness – but as awareness, anticipation, satisfaction, confidence, almost exultation. I like the feeling.' That 'almost' is very Collins.

On a Pan American flight between Honolulu and Manila, Lindbergh wrote to Collins, who received the letter soon after he returned to Earth: 'Dear Colonel Collins ... What a fantastic experience it must have been – alone looking down on another celestial body, like a god of space! There is a quality of alone-ness that those who have not experienced it cannot know – to be alone and then to return to one's fellow men once more. You have experienced an aloneness unknown to man before. I believe you will find that it lets you think and sense with greater clarity ... As for me, in some ways I felt closer to you in orbit than to your fellow astronauts I watched walking on the surface of the moon.' He told Collins that, though he had observed every minute of the lunar walk, and though certainly it had been 'of indescribable interest', it was Collins' experience that seemed to him to be of 'greater pro-fundity'. In his Introduction to Collins' autobiography, Lindbergh was again drawn to Collins' experience of being alone:

Relatively inactive and unwatched, he had time for contempla-tion ... Here was human awareness floating though universal reaches, attached to our earth by such tenuous bonds as radio waves and star sights . . .

Only once before have I felt such extension as when I thought of Astronaut Collins. That was over the Atlantic Ocean on my nonstop flight with *The Spirit of St Louis*. I had been without sleep for more than two days and two nights, and my awareness seemed to be abandoning my body to expand on stellar scales. There were moments when I seemed so dis-connected from the world, my plane, my mind and heart-beat that they were completely unessential to my new existence.

Experiences of that flight combined with those of ensuing life have caused me to value all human accomplishments by their effect on the intangible quality we name 'awareness'.

The Lunar Module had been difficult to simulate on Earth. The Lunar Landing Training Vehicle, affectionately known as the flying bedstead (a name also given to earlier flying contraptions), was a notoriously tricky machine to handle. Armstrong had to bail out of one during a training flight in May the year before; another test pilot had bailed a few months earlier. Armstrong – like all test pilots, a constantly calculating machine – ejected two-fifths of a second before the vehicle crashed. Bob Gilruth pushed Deke Slayton to stop using it. It wasn't clear, anyway, how comparable the experience of landing it was to landing an actual Lunar Module in the moon's weak gravitational field. But Slayton argued that it was better to take the risk now rather than later above the surface of the moon.

After he had bailed and missed death by less than half a second, Armstrong went back to the office. Fellow astronaut Al Bean bumped into a group of astronauts huddled together in a corridor discussing the accident. 'That's bullshit!', Bean said to them, 'I just came out of the office and Neil's there at his desk . . . shuffling some papers.' Armstrong said later, 'I mean what are you going to do? It's one of those days when you lose a machine.'

Armstrong had put the odds of a successful landing at 50/50. At 40,000 feet, he turned the LM upside down in order to get a better look at the moon's surface. They were approaching at 3,000 mph, standing up – seats would have added too much weight – velcroed upright. The landing craft had been pared back, as *The Spirit of St Louis* had been just over 50 years earlier.

The Lunar Module's onboard computer had been programmed so that the entire descent could be made automatically, though control of the craft could also be shared or overridden. As the

LM got closer to the surface, the computer sent out an error code, 1202. Aldrin didn't know what it meant. Gene Krantz back at Mission Control didn't know what it meant either. He recalled seeing something like it before but couldn't remember when. To make matters worse there were also some radio communication problems. The situation was almost bad enough to justify aborting the landing. There was very little room for error, and only enough fuel for six and a half minutes of flight time.

After some discussion on the ground, Mission Control decided to ignore the error message. The computer was overloaded with information; alarms sounded a further four times as the descent continued. Armstrong had by now taken complete manual control of the landing. There was nothing more Mission Control could do. Slayton whispered to Kranz: 'I think we'd better be quiet.'

The craft proved to be hard to control manually, and responded sluggishly. The alarms, too, had been a distraction. Armstrong hadn't had time to look out for the landmarks he had familiarized himself with – poring over lunar maps for hours and hours – back on Earth; and Aldrin now had his attention completely directed towards the instrument panels. Armstrong realized that he had

The flying bedstead

overshot the designated landing site and was flying over an area strewn with boulders. With the gauge indicating that there was only a few seconds worth of fuel to spare, he went in search of a suitable landing area beyond the boulders. His pulse rate rose to 150 beats per minute. As he got closer to what looked like a good spot, dust that had probably not stirred for hundreds of thousands of years blew up from the surface of the moon. And then it was over; the craft settled and came to a standstill, did not disappear into feet of dust as some had feared it might. Later, it was calculated that there had, in fact, been enough fuel to have kept the craft aloft for another 25 seconds. The sloshing fuel had given a false meter reading, as it memorably once had for Lindbergh.

The *Eagle* had landed. Armstrong looked up out of the Lunar Module's small overhead window. Depending on the position of the sun, sometimes the sky over the moon would be unbelievably rich in stars, at other times, as now, it appeared black and empty. But not entirely empty; there was one brilliant shining object out there, the Earth: 'It's big and bright and beautiful,' Armstrong told Mission Control.

You'd think they'd have flung open the hatch itching to get out and explore, but there were procedures to follow. First, they had to go through a simulation of the next day's takeoff. They were somehow expected to sleep before the scheduled lunar walk in 10 hours' time. But even NASA now thought that 10 hours was too long to wait and moved the walk forward four hours.

Of course they could not get to sleep. They said it was the light from the Earth – three times brighter than moonlight – shining into the landing craft that kept them awake.

Buzz Aldrin had known that, if it happened, landing on the moon would be a transcendent moment in human history. He asked Dean Woodruff, the pastor of his local church (the Webster

Presbyterian Church in Webster, Texas) if he could think of some way that the significance of the event might be acknowledged. Could he 'come up with some symbol which meant a little more than what most people might be thinking of'? Woodruff suggested that Aldrin should mark the occasion by taking Holy Communion on the surface of the moon, and wrote to the church's General Assembly to find out if there were any theological impediments. Word came back that there were not. The bread and wine could be pre-consecrated.

Aldrin told Deke Slayton of his intention and said that he wanted the ceremony to be broadcast to the world. But Slayton, still mindful of Madalyn O'Hair's campaign and threatened legal action, told Aldrin that he might take Communion if he wished but he would have to keep it to himself. The most that Slayton would allow was the curious concession that Aldrin could recite a prayer on open mike on the way back. After the main event of the moon landing, Slayton thought that by then no one would be paying attention. Clearly the letters that were flooding into the Manned Space Center in support of the Apollo 8 Genesis reading carried less weight then O'Hair's protest.

Woodruff wrote a sermon, afterwards published as 'The Myth of Apollo 11: The Effects of the Lunar Landing on the Mythic Dimension of Man', which he delivered on 20 July, the morning before the lunar landing. He said the event would come to be seen as being more influential than the Copernican or Darwinian revolutions. He wrote that, flying high above its surface, we escaped Earth's bonds, both the literal bond of gravity and metaphorical bonds. Apollo 11 was a material manifestation, he said, of an ancient myth: that of 'magical flight'. The myth symbolized humankind's desire to transcend itself. 'Perhaps when those pioneers step up on another planet and view the earth from a physically transcendent stance, we can sense its symbolism and feel a new breath of freedom from our current claustrophobia and be awakened once again to the mythic dimension of man.'

At around 5.57pm Houston time, Aldrin prepared to celebrate Communion. He took the pre-consecrated wafers and wine from his personal preference kit, along with a mini flat-packed chalice. Aldrin used a small fold-down table underneath the keyboard of the abort guidance system as an altar. He turned on his mike and made a short statement: 'This is the LM pilot speaking. I'd just like to take this opportunity to ask every person listening in, whoever, and wherever they may be, to pause for a moment and contemplate the events of the past few hours, and to give thanks in his or her own way.' He was careful not to mention God. Armstrong – who described himself as a deist (though his mother was saddened that he was not more devout) – was unfazed: 'I had plenty of things to keep busy with. I just let him do his own thing.'

In the moon's gentle gravity the wine poured slowly out of the flask and curled gracefully against the side of the cup. Aldrin read silently from a small card on which he had printed words from the book of John:

I am the vine and you are the branches
Whoever remains in me and I in him will bear much fruit;
For you do nothing without me.

At home in Nassau Bay, his wife Joan was listening to old Duke Ellington records with only one ear tuned to the squawk box, but she heard her husband's message and approved. She didn't know that he was taking Communion at that particular moment but she liked his acknowledgement of a spiritual dimension to the experience of being on the moon beyond the technical. And yet, as Robert Poole acknowledges in *Earthrise*, 'It was not exactly an advertisement for the unity of mankind': one man celebrating Communion alone while another man tried to act as if nothing was happening. Aldrin later said that if he had the chance to do it all over again he would not have celebrated Communion. 'At the time I could think of no better way to acknowledge the enormity of the Apollo 11 experience than by giving thanks to God.' It had been meaningful to him, but the astronauts had come, he now said, 'in the name of all mankind'.

In the event, despite the revised schedule, the moonwalk was delayed. It had originally been planned for 10pm Houston time and then rescheduled for 6pm. Aldrin finally opened the hatch around 8pm.

Aldrin Sr had got involved in what became an embarrassing argument within the Manned Space Center about who should be first out of the Lunar Module. He encouraged Buzz to believe that he should be first on the moon, not Armstrong; and then began to pull strings on his son's behalf. Buzz appears not to have cared before his father intervened. When it was first announced that he was to be part of the first crew to land on the moon, Aldrin said to his wife that he would have preferred to have been on a later flight: 'I didn't want all the press and all the attention for the rest of my life for being on the first landing. Because that's all the press seems to care about.' But, seemingly because of the pressure from his father, he came to believe that he had been badly treated when he didn't get to be the very first man on the moon. Collins thought that Aldrin got so worked up over who should be first out of the Lunar Module that the joy he might otherwise have experienced from walking on the moon was spoiled.

Now, as he attempted to get out of the Lunar Module, Armstrong was forced to his knees in accidental obeisance. He had to crouch and gradually push his way through the door. Once he was outside he pulled a cord to release a TV camera so that the moments to come could be recorded.

The first task to be performed was to throw out the trash. (By the time Project Apollo was at an end, NASA would have put 118 tons of rubbish on the moon: redundant Ranger, Surveyor and

Lunar Orbiter probes, spent third stages of Saturn rockets that crashed there, LM descent stages.) Still on the ladder, Armstrong described 'for the benefit of scientists back on Earth' the surface of the moon: '[It] appears to be very, very fine grained . . . almost like a powder . . .' Once his feet touched the ground, he uttered the famous words: 'That's one small step for [a] man, one giant leap for mankind.' The missing 'a' is still argued over. One investigation claims to have uncovered the indefinite article hidden in static. There is something gloriously arbitrary about the statement's significance, as if everything else that Armstrong and Aldrin said to each other since landing on the moon counted for nothing compared to the first words spoken after human feet – unmediated by the craft, or a ladder, but, nevertheless, still mediated by a spacesuit – had touched the moon's surface.

The TV footage broadcast from the Command Module had been in colour, at least some of the time. The astronauts considered the onboard camera a nuisance and only worked out how to use it on the way out. Collins talked of 'camera clap-trap'. He said it was 'a bloody nuisance of an afterthought'. Armstrong and Aldrin didn't even know how to turn it on. The only PR training they had had was to be told, glibly, to put on a good show as the world would be watching.

There had been no time to develop a colour camera for use on the surface of the moon. At first, the engineers had not wanted the crew to take a TV camera on board the LM at all, because of the extra weight. Curiously it was the communications director, Ed Fendell, who argued against, saying at a meeting that there was no need to have a live broadcast from the moon. Kraft said, 'I can't believe what I'm hearing.' A row developed. 'We've been looking forward to this flight', Kraft shouted '– not just us, but the American taxpayers and in fact the whole world – since Kennedy put this challenge to us.' Even dead, Kennedy was still the final arbiter. And now, because they had left the decision so late, the world had to make do with black-and-white TV images of the first moonwalk. It had even been argued that the stills camera that was to be taken onto the moon's surface should have black-and-white film in it, because – an often repeated argument – black-and-white photographs showed greater detail. Then someone from the press office angrily asked what they thought a black-and-white photograph would look like on the front cover of *Life* magazine and the matter was closed. By the late 1960s photographs in magazines and newspapers were larger than they had been, and often reproduced in colour. *Time* magazine went into colour in 1968.

Around 600 million people back on Earth were watching, the largest audience in the history of television up to that time. Depending on which was closest to the spinning Earth, the TV signal was captured by one of three possible stations and relayed from there to Mission Control. From Mission Control the signal was bounced back into space to be picked up by satellite ATS-3, which then relayed the images to the world. We could see men on the moon only because getting them to the moon had brought about a worldwide telecommunications system. Those who saw the landing at the time will appreciate the observation made by Mark Armstrong, aged six. Searching the fuzzy TV images for his father, he said: 'Why can't I see him?'

Aldrin followed Armstrong out of the LM, pausing on the top of the steps, not to take in the grandeur of the scene, but to urinate inside his spacesuit. He may not have been first but he could at least mark the territory as his own.

The small planet pleased them. The moon felt almost intimate because of its strongly curved horizon, and there was something lovely, too, about the moon's light gravitational pull; just enough to give a sense again of up and down, yet light enough to turn the moon into a playground, perhaps not so different from the home planet of Antoine de Saint-Exupéry's Little Prince.

Armstrong and Aldrin had brought a flag with them.

Armstrong broke into a sweat as he struggled to push the flagpole into the ground. There was nothing between dust and solid rock.

Aldrin became aware of the paradox that they were both further from the Earth than any other human being and yet also the objects of the world's close attention. He knew that when he returned, he would be asked what it had been like on the moon. And he knew he would not be able to answer that question any better than he had the question everyone wanted an answer to after Gemini: what had it been like in space? When they saw the TV footage, later, back on Earth, Aldrin turned to Armstrong and said, 'Neil, we missed the whole thing.' Collins missed the whole thing too. Just as Armstrong was about to step onto the surface of the moon the link between the Command Module and the LM failed. The link was restored just as *Columbia* passed into the moon's dark side.

Before they got back into the LM, Armstrong and Aldrin left behind their boots, and their camera. Any weight that could be saved meant extra fuel. When Armstrong threw his backpack into the Lunar Module, Mission Control picked up the vibration on a seismometer on the moon's surface. 'You can't get away with anything any more, can you?' Armstrong told them. And he was right. It was the beginning of a new era of continual surveillance. The first moonwalk was over. It lasted barely two and a half hours, and because NASA was worried about how the spacesuits would hold up, they had walked not much more than 65 yards from the LM. Armstrong had been given only 10 minutes in which to collect rock samples.

The ascent engine fired up. The flag fell over. The Lunar Module rose above the moon's surface, leaving behind the descent engine and the ladder they had used to get down. On the ladder was a plaque – the so-called Goodwill Message from 73 world leaders – that read in part, 'We Came in Peace for all Mankind,' along with facsimile signatures from the crew and the President.

Nixon had wanted the plaque to read 'We Came in Peace Under God for all Mankind', but someone at NASA decided to omit the words Under God. O'Hair's influence had turned the moon landing into an almost entirely secular event; had turned the focus instead to world peace. No prayers were broadcast from the moon. President Nixon declared 21 July National Day of Participation, calling on every American to pray for the safe return of the crew. Madalyn Murray O'Hair nevertheless protested that among the messages of goodwill was one from the Pope.

The fragile LM quickly accelerated to 57,000 mph. It was much easier to escape the moon's soft gravitational pull than the Earth's.

Collins said that the moment he feared most during the whole mission was the possibility that he might have to return alone if anything went wrong with the lunar lander. He said that, afterwards, back on Earth, the moment he always revisited in his imagination was when he first caught sight of the *Eagle* making its way back to *Columbia*. It was the best sight of his life, he said. He took a photograph showing the LM, the moon and the Earth together all in a line. All of humanity, all life was in front of him.

Only after the LM had returned did Collins allow himself to think for the first time that perhaps they were going to make it. He wanted to kiss each of them on the forehead but thought better of it. Instead, he shook their hands.

He flipped a switch and what was left of the *Eagle* was jettisoned with a small bang. Collins was relieved, Armstrong and Aldrin sad to see it go.

Congratulations came in from all over the world. Collins thought they were a little precipitous. Couldn't they at least have waited until after they had left lunar orbit? Among the messages was one from Esther Goddard. Could Robert Goddard ever really have anticipated this moment, Collins wondered?

The crew made their last broadcast of the mission. Aldrin told the world that 'in reflecting on the events of the past several days, a verse from Psalms came to mind: "When I consider thy heavens, the work of thy fingers, the moon and the stars, which thou hast ordained; What is man, that thou are mindful of him?"' Collins wondered afterwards if they had missed an opportunity, wondered whether their messages had been just a little trivial. He tried to imagine what a crew made up of a philosopher, priest and poet might have made of what they had experienced. But he had come to the conclusion that, though his fanciful crew might have expressed themselves better, they might not have made it back – forgetting, perhaps because their minds were on higher things, to push the circuit breaker that enabled the parachutes to open.

Lindbergh turned down Nixon's invitation to greet the returning astronauts at sea. He said he did not want to be taken back to a 'life I am most anxious not to re-enter'. He meant, one in which he was the centre of attention.

After splashdown, the reporter Eric Sevareid turned to the CBS News anchorman Walter Cronkite and said, 'You get a feeling that people think of these men as not just superior men but different creatures. They are like people who have gone into another world and returned, and you sense they bear secrets that we will never entirely know.'

When NASA spread out the photographs that came back from the mission, among them were Earthrise photographs even better than those from Apollo 8, but the public and press would focus all their attention on the landing.

'Where are the photographs of Neil?' someone at the meeting asked. Armstrong had taken a fine portrait of Aldrin standing by the flag, but there was nothing like that of Armstrong, the first man on the moon. Underwood said Aldrin hadn't taken a photo of Armstrong because he was still mad at him for being the first out, but more likely it was because the camera, a Hasselblad of course, was hooked onto Armstrong's chest, so that he could more easily photograph features of the lunar surface and materials to be collected. It wasn't easy to take it off and hand it over. Apparently the swap *was* about to happen, but the President came on the line. Nixon told them, and the world, that it had been 'the greatest week since the Creation, that for one priceless moment in the whole history of man, all the people of the earth are truly one ...' And so, because of Nixon, there were no photos of Armstrong.

Someone at the meeting suggested that perhaps they could pretend that the photo of Aldrin was of Armstrong. Who would know? It wasn't as if you could see his face or anything. But Underwood said it wasn't worth the risk. 'There's some nine-year-old kid out there who's a space groupie, and he knows every aperture and wire and seam in a spacesuit. The day after you publish it, the *New York Times* is going to get a letter from that nine-year-old kid saying, No, you're wrong. That's Buzz Aldrin.' So nobody mentioned it. Underwood said that this was NASA policy for a lot of things. The only images of Armstrong that came back are of him in the shadows, working. Nobody in the press seemed to notice.

The day after the launch, the *New York Times* ran an article referring back to the article it had run on 12 January 1920, in which Robert Goddard had been accused of not understanding Newton's Third law. The *New York Times* had then ridiculed the idea of propulsive space travel. The article, published on 17 July 1969, ended: '*The Times* regrets the error.'

The President threw an extravagant party for the returning astronauts in Los Angeles. Guests at the Moon Ball included Wernher von Braun, Charles Lindbergh, Eddie Rickenbacker, Mrs Robert Goddard, Howard Hughes, Jackie and Aristotle Onassis, Fred Astaire and Joan Crawford. LBJ and Lady Bird Johnson sent their regrets. There were protesters outside the Moon Ball. One placard – ahead of its time – read: 'Fuck Mars'. Speaking on behalf of a sizeable minority, Picasso said, 'It means nothing to me. I have no opinion about it, and I don't care.' John Updike later voiced that same malaise in his novel *Rabbit Redux*: 'For the twentieth time that day the rocket blasts off, the numbers pouring backwards in tenths of seconds faster than the eye until zero is reached: then the white boiling beneath the tall kettle, the lifting so slow it seems certain to tip, the swift diminishment into a retreating speck, a jiggling star. The men dark along the bar murmur among themselves. They have not been lifted, they are left here.' The *New York Times* reported that bars in Harlem had been tuned to a baseball game, not to the lunar landing. 'The moonshot...was imposing,' the Stockholm *Expressen* allowed, 'but it also gives a horrible feeling to think that the USA can handle tremendous technical problems with such ease while it has considerably more difficultly coping with those of a complicated social, political, and human nature.' The year after the moon landing the poet and musician Gil Scott-Heron released his debut album 'Small Talk at 125th and Lenox'. The track 'Whitey on the Moon' – 'I can't pay no doctor's bills / But Whitey's on the moon' – nailed what to some was a major failing of the moon landing: that it had been just another exercise in white supremacy.

Collins could have gone on to command a later mission if he had wanted to, but he decided that this was to be his one shot. He said that after the experience he just couldn't get excited about anything the way he could before Apollo 11: 'I seem to be gripped by an earthly ennui which I don't relish, but which I seem powerless to prevent...not many things seem quite as vital to me any more. My threshold of measuring what is important has been raised; it takes a lot more to make me nervous or to make me blow my cool...That doesn't mean I have acquired a complete guru-like detachment. I still get irritated, and I still express irrational annoyance.' And the question that would make him most annoyed was the one everyone wanted an answer to: 'If one more fat cigar smoker blows smoke in my face and yells at me, "What was it really like up there?", I think I may bury my fist in his flabby gut; I have had it with the same question over and over again.'

Flying in space had, however, changed his perspective on himself. It wasn't obvious to others, he thought. Outwardly he seemed the same. And certainly he hadn't found God on the moon, but he cared more now about what really mattered to him. He had taken up painting, he wrote poetry:

> The moving line skims, sure and swift
> Green as a snake across the wall.
> A linear lie of circular progress, it tells us nothing,
> Except that man must keep his sensors saturated.

Collins wouldn't be the only astronaut who came back and turned to art.

When he had been a boy undergoing painful dental work, Collins had learned to imagine himself floating up at ceiling height looking down on himself. It was a way, he discovered, of removing himself from the pain. The boy he looked back on in pain was not him, just someone he was looking at. He had done the same thing as a pilot, and had taken his aerial perspective back with him to Earth. Now, when things were not going well on Earth, he would lift his mind out there into space, 'and look back at a midget Earth'. As a pilot, the remembered view from the clouds was comforting but as an astronaut, the view from the moon's orbit was 'even more supportive'.

'I travelled to the moon,' Buzz Aldrin said, 'but the most significant voyage of my life began when I returned.' He had what he called, 'a good old American nervous breakdown'. For a time he became depressed and turned to drink. His mother had died of a self-administered drug overdose the year before. Her father had also committed suicide. As a child he had read a story in which travellers to the moon had returned insane. The story had haunted his mission. Aldrin told his wife Joan that he was sick and tired of talking about the moon. How many times can you say it was mystical up there? But his father didn't let up on him: now that he had been to the moon, the world might be at his feet, if only he would assert himself more.

Aldrin described himself as 'introverted, supersensitive, a perfectionist, concerned about what people thought of me ... No wonder I was in trouble.' He thought he had become a better person for going through his breakdown: 'I got a chance to redo my life.' Joan described him as being a 'curious mixture of magnificent confidence, bordering on conceit, and humility'. She had hoped that after the moonwalk he might become more relaxed and open up. Joan told her husband she hoped their life would get back to normal, but Buzz said: 'Joan, I've been to the moon, and I'm never going to be allowed to live the way I once lived. Neither are you and neither are our kids.'

After the Moon Ball, the astronauts had gone on the 'Giant Step' tour of the world: 45 days, visiting heads of state in 23 countries. Aldrin found the experience particularly hard to bear. One night he and Joan both got drunk, and he told her that he and the rest the crew were 'fakes and fools' for having allowed themselves 'to be convinced by some strange concept of duty to be sent through all of these countries for the sake of propaganda, nothing more, nothing less'. He and Joan separated afterwards.

When he was asked if he had any regrets, Aldrin said that he wished he had looked out of the window more, Earthgazing.

Whatever happened to Neil Armstrong on the moon he mostly kept to himself. When he returned to Earth, he became even more reclusive than he had been before. He announced that he did not intend to fly again. Two years later he resigned from NASA. His longtime friend Robert Hotz, the editor of *Aviation Week*, said that he understood why: 'Hell, you're in this high-tension world of aerospace. You get out on the farm. You look at the mountains across the valley, which are several million years old and are going to be there through the life of the planet. You understand that you're a short-term phenomenon, like the mosquitoes that come in the spring and fall. You get a perspective on yourself. You're getting back to the fundamentals of the planet. Neil feels that way, because we've talked about it, and so do I.' After he returned to Earth, Armstrong and his wife Janet split up.

The chalice Aldrin had taken to the moon found its way back to his church, the Webster Presbyterian Church, Texas, not far from the Johnson Space Center at Houston. Each year on the Sunday closest to the anniversary of the lunar landing it is used to serve Communion wine. Almost any object that had been to the moon and returned to Earth seemed to have had conferred on it some mysterious and invisible quality that made the objects – Bibles and First Covers popular among them – somehow hallowed, or at least collectible. Collins had loaded his personal preference kit with small items: 'prayers, poems, medallions, coins, flags, envelopes, brooches, tie pins, insignia, cuff links, rings, and even one diaper pin'. Also included were 50 elephants carved from slivers of ivory, housed inside a hollow bean. In the 1990s Armstrong stopped signing autographs when he realized that they were being sold for profit, though of course this has only made his signatures all the rarer, and all the more valuable. In 2005 Armstrong intervened when he discovered that his barber had sold his hair-clippings to a collector for $3,000.

Less than two weeks after Apollo 11 splashed down, Madalyn O'Hair finally served her civil suit. Apparently the last straw had been the prayer Aldrin read publicly on the return journey. The action was directed against Thomas Paine in his role as NASA administrator. She sought a court order preventing him and NASA from allowing any further religious activity in space.

During 1969 NASA received 185,876 letters on the subject of religion in space, mostly in support of the Genesis reading.

To each letter NASA sent out a standard reply. Other organizations across America had received an estimated 3 million letters.

14 November 1969. Apollo 12. Crew: Pete Conrad (Commander), Dick Gordon (Command Module Pilot), Al Bean (Lunar Command Pilot). Duration of flight: 10 days, 4 hours, 36 minutes, 24 seconds. Command Module: *Yankee Clipper.* Lunar Module: *Intrepid.*

The crowd that turned out to watch the launch of Apollo 12 was a third the size of the crowd that had turned out for the launch of Apollo 11. Kennedy's goal had been achieved, a man (indeed two men) had been safely returned from the surface of the moon. Perhaps the public wondered what was the point of the missions that came after? Even at the White House interest had waned. A memo written to White House staff noted that at 5.52am Commander Pete Conrad would emerge from the Lunar Module and climb onto the moon's surface: '(You know ... the same old thing – the Armstrong-Aldrin bit.) (Ho-hum).'

During lift off, lightning struck the rocket twice. Pete Conrad's pulse rate didn't alter. At Mission Control there was concern that the lightning might have damaged the parachute mechanism, with potentially disastrous consequences for the crew's re-entry.

Pete Conrad, Dick Gordon and Al Bean

Since there was nothing that could be done, Mission Control decided to keep their concern to themselves. Inside the Command Module the computer screen had completely filled up with error messages. Conrad started to laugh. His crewmates joined in and they laughed their way into orbit. Apollo 12 became known as the most joyous of all the Apollo missions. The three astronauts were close friends; had matching gold Corvettes which they drove in convoy, and wore matching gold aviator sunglasses with their matching powder-blue NASA flight suits. Commander Pete Conrad was particularly well liked at the Manned Space Center. No one had a bad word to say of him.

As they orbited the moon, Al Bean thought it looked absurdly cartoon-like. Even close to it was so clearly spherical. He remembered being scared at the sight of the craters but telling himself he wouldn't be able to do his job if he was scared so he'd look back inside at the control screens until he had the courage to look outside again, and then he'd get scared again . . .

By this time a colour camera had been developed to use on the moon's surface, but 42 minutes after he left the Lunar Module to take the first moonwalk, Bean accidentally pointed the camera at the sun and burned out the video feed.

The Apollo 12 crew and their matching gold Corvettes

For Dick Gordon – as it had been for Collins – one of the highlights of the trip was welcoming his crewmates back on board after their return from the moon. Gordon beamed at them, telling them to get back in but not to mess anything up. Bean said that when he saw Gordon's welcoming smile he was filled with intense love for his crewmates. He had just walked on the moon, but for him the defining moment of the trip was that instant of love.

At a news conference after their return, fellow astronaut Pete Conrad was sitting there thinking how in some ways the whole thing had been curiously disappointing; not that different from the experience in the simulator beforehand, and then, as if telepathically, Bean turned to him and said: 'It's kind of like the song: Is that all there is?' He was referring to the uniquely dark and humorous popular song, written by Jerry Lieber and Mike Stoller, and first performed by Peggy Lee, that had been released in 1969, the year of the moon landing. Each verse recounts one of a series of disappointments seen from the perspective of the song's narrator: the circus act she saw as a girl; the fire that burned the family house down; her first love affair. She would kill herself if she didn't suspect that even death will turn out to be yet another disappointment.

For a time Bean was gripped by the 'earthly ennui' that Collins had also felt, and yet sometimes he'd just sit in the mall watching people and find the experience surreal and thrilling. Collins had turned to poetry; Bean took up painting. He painted only the moon, obsessively over and over again, mixing moon dust into the pigments. He said the best day of his life was when he figured out a technical detail to do with the use of a particular colour.

Charles Berry, the astronauts' doctor, said that 'No one who went into space wasn't changed by the experience ... I think some of them really don't see what happened to them.' Bean said that 'everyone who went to the moon came back more like they already were'. Several of the Apollo astronauts would articulate the same idea: that they had not been fundamentally changed by the experience of space so much as become more themselves, but perhaps to become more like yourself is to change. Perhaps it is what we mean by growing up.

Bean wasn't religious, but he wondered afterwards if the Earth as a whole was what the writers of Genesis had had in mind when they wrote about the Garden of Eden. He said that when we look through our telescopes there is nothing we can find out there so beautiful as the Earth that the Apollo astronauts who went to the moon looked back on. 'We've been given paradise to live in,' he said: 'I think about it every day.'

Pete Conrad insisted that for him going to the moon had been no big deal. When he was on the moon, he had felt that he was in the right place at the right time and then when he returned to Earth 'that just shut the door' on the experience. He didn't go out and look at the moon afterwards and reflect on his time there. He said no one believed him, but that it was nevertheless true. Some of his fellow astronauts said that that was just the 'Right Stuff' talking: that he was supposed to say something like that. But who can deny him his own experience? Anyway, like Collins he became tired of talking about it. After a while, when he was asked what it had been like, he'd just say: 'Super! Really enjoyed it.'

11 April 1970. Apollo 13. Crew: Jim Lovell (Commander), Jack Swigert (Command Module Pilot), Fred Haise (Lunar Module Pilot). Duration of flight: 5 days, 22 hours, 54 minutes, 41 seconds. Command Module: *Odyssey*. Lunar Module: *Aquarius*.

Earlier in the year the director of NASA, Thomas Paine, cancelled Apollo 20, a flight scheduled a few years into the future. With the costs of the Vietnam War escalating, and dwindling public interest, severe cuts were required of the Apollo programme. Public interest in Apollo had fallen after the first moon landing; it

Jim Lovell, Jack Swigert
and Fred Haise soon
after their return

fell even further after the second. There were no plans to broadcast the Apollo 13 mission live. 'New and unusual events have always excited our curiosity and captured our imagination,' von Braun once wrote. 'Our first day at school, first airplane flight – and first love ... With repetition and the passage of time, however, they lose freshness and become routine even if no less significant.' Mike Collins said something similar: 'Part of life's mystery depends on future possibility, and mystery is an elusive quality which evaporates when sampled too frequently, to be followed by boredom.' The Apollo missions had in the eyes of the public already become routine, but all that changed when, 56 hours into the flight, the crew heard a loud bang. At first they thought a meteoroid had struck the vehicle. Every moment of the drama that then followed was recorded live on TV. In fact an oxygen canister had exploded in the Service Module. NASA told the astronauts' wives that there was only a 10 per cent chance of them making it back alive. The combined Command and Service module lost its oxygen supply and its power after just two hours. The Lunar Module would have to be used as a 'lifeboat', providing oxygen, power and thrust. That they survived was in part due to the crew's sangfroid and inventiveness. At one point, guided by Mission Control, they were required to construct an air-filtration system out of cardboard and storage bags, grey tape and socks. The crew's survival was also a tribute to the trajectory experts at Mission Control – among them Katherine Johnson, the black mathematician who had been so admired by John Glenn during the Mercury days – and to the Houston computer. The moon's gravity was the only way the craft could be put on a homeward-bound trajectory. It meant that the crew had to complete their journey to the moon, moving further away from the Earth before they could start to make the journey back again. The craft arced the dark side of the moon in a mere 20 minutes, less than half the time it would have taken if, in preparation for landing, they had put themselves into a settled orbit. Haise and Swigert began to take photographs, much to Lovell's annoyance. 'Relax, Jim,' Haise said. 'You've been here before, and we haven't.'

During the return journey the course had to be altered twice using the LM thruster and a watch for timing.

In the dim lighting the crew would have had the best view of the Earth of any mission.

The President declared the day after their return to be a National Day of Prayer and Thanksgiving, but Commander Jim Lovell said that the mission 'in reality was a failure. When we got back, there were no accolades or trophies, and we weren't escorted anywhere by the vice president and other VIPs. NASA just wanted to forget about it and move on.' 'The only acknowledgment I received,' he said, 'was a handwritten letter from Lindbergh who congratulated me on a successful return.' NASA at that time seems to have been incapable of recognizing the public's need for a human dimension. Once the crew was returned safely public apathy in Apollo immediately resurfaced. The Apollo 13 mission only began to acquire its current mythological status after Ron Howard's film *Apollo 13* appeared in 1995.

In the summer of 1970, the US had only narrowly approved the building of a space station. The future Vice President, Senator Walter Mondale, had argued passionately against it: 'I believe it would be unconscionable to embark on a project of such staggering cost when many of our citizens are malnourished, when our rivers and lakes are polluted, and when our cities and rural areas are dying.' The vote was won by a slim majority, but the writing was on the wall. A few months later the Apollo programme was curtailed further: Apollo 18 and 19 were cancelled.

During 1970 NASA received a further 901,810 letters specifically about religion or the Genesis reading.

31 January 1971. Apollo 14. Crew: Al Shepard (Commander), Stu Roosa (Command Module Pilot), Ed Mitchell (Lunar Module Pilot). Duration of flight: 9 days, 1 minute, 58 seconds. Command Module: *Kitty Hawk*. Lunar Module: *Antares*.

The Apollo missions had become about moon rocks and scientific experiments. It was not enough to ignite public interest.

Stu Roosa, Al Shepard and Ed Mitchell

Perhaps the scientific case could have been made, but an internal struggle between engineers and scientists at NASA had not yet played out. The scientific argument hadn't even been successfully made to the astronauts. Al Shepard had made his disdain for geology apparent. A number of astronauts felt as he did. Mike Collins said that his 'curiosity about things geologic [was] easily quelled'. He thought that NASA geologists did their best to take the magic out of going to the moon. Other astronauts only affected not to be interested in geology, a Right Stuff front that masked genuine inquisitiveness.

In his seminal account of the moon missions, Andrew Chaikin called Apollo 14 the nadir. As if to rub salt into the wound, Apollo 14 became infamous for its flirtation with pseudo-science.

The day after takeoff, Ed Mitchell carried out an experiment; but not one that had been officially sanctioned, this was a secret experiment NASA knew nothing about. While he was floating in his sleeping bag, Mitchell took out a clipboard on which were written a list of random numbers. Each number was assigned one of the symbols typically used in ESP experiments: a circle, a square, wavy lines, a cross, or a star. He then picked a number and concentrated on it and its associated symbol for a few seconds. He repeated the process several times. Back on Earth test subjects were, at that very same moment, trying to decide what number and symbol Mitchell was conjuring up. Mitchell said that the results were promising and that he was neither encouraged nor chastised when NASA found out. Deke Slayton remarked, 'Hell, NASA doesn't know everything.' Von Braun had shown an interest in Mitchell's experiments. Mitchell claimed that von Braun had hinted at the possibility of using some of NASA's resources to pursue the experiments further, but that he'd left NASA before anything came of it.

As they approached the moon Stu Roosa was playing his personal tape: a choir was singing the hymn 'How Great Thou Art':

When I in awesome wonder
Consider all the works Thy hands have made
I see the stars, I hear the mighty thunder
Thy power throughout the universe displayed
Then sings my soul, my Saviour God to Thee;
How great Thou art, how great Thou art.

Like other Command Module Pilots, Roosa relished his time alone orbiting the moon. The darkness out there was palpable, like some damp substance, eerie but not terrifying. The moon made him feel big. It was the Earth that made him feel small. The moon was a dreamscape. Out of the long, sharp shadows he conjured up fabulous creatures, like a child with a magic lantern. He had an epiphany that everything was about light, that humans are creations of light not darkness. He knew now what it was to be utterly alone, and he knew he could bear it.

On the surface of the moon Al Shepard leaned backwards – a precarious movement in a spacesuit – to look at the Earth. He said he began to cry. He was the only astronaut who ever admitted to crying in space.

Ed Mitchell said that when he walked on the moon he felt immediately as if he had become a native of the moon, as if the landscape that had not changed for millions of years had all that time been waiting for them to arrive. The silence was startling. Without an atmosphere, outer space begins at the moon's surface. Even meteorites crash there silently. Inside their spacesuits and inside the lunar landing module they had brought sound to the moon. How different being there than on the Earth's surface, cocooned by the Earth's atmosphere!

When they were all back on board the Command Module, *Kitty Hawk*, climbing away from the moon, Ed Mitchell was filled with a profound longing, a kind of homesickness for the moon he was leaving and would never see again, like the sickness of love perhaps, or the troubadours' longing for their distant lady. 'It wasn't merely the view that was so powerful, it was the *idea*.' The next day, during the journey back to Earth, he looked at his home planet and something inside him changed, though at the time he was unaware what it was. He had looked at the Earth many times during his trip to the moon, and each time had been mesmerized, but this time was different. It would take him several years to assimilate what had happened to him. Out there, ten times as many stars are visible than from the most propitious vantage point on Earth, and are ten times brighter. 'There is a sense of being swaddled by the universe,' Mitchell said. In space the universe seems 'more intelligent than inanimate'. He was vividly conscious on his return journey both of the separateness of the stars and planetary bodies but simultaneously also was aware that he was 'an intimate part of the same process'. He said that after such an epiphany nothing could ever be the same again. Back on Earth he felt a strong 'desire to live life to the fullest, to acquire more knowledge, to abandon the economic treadmill'. Two years later he founded the Institute of Noetic Studies, an organization devoted to the study of consciousness. He became interested in Buddhist and Hindu writings. He met 'Native American Indians, Kahuna of Hawaii, shamans of South American tribes, and voodoo priests of Haiti,' many of whom 'spoke of kind and loving spiritual connections to all life as fundamental to their existence.'

When asked if going to the moon had changed him, Al Shepard grudgingly admitted: 'I suppose I'm a little nicer than I used to be.'

Stu Roosa had been a smokejumper with the US Forest Service. Among his personal belongings he took to the moon 500 seeds from five different types of tree. About 420 of the seeds proved to be viable back on Earth. Seedlings were planted across America in the mid-1970s. About 50 so-called moon trees have been identified, among them a Loblolly Pine growing at the Lowell

Elementary School in Boise, Idaho, a Douglas Fir at the US Veterans Hospital in Rosebury, Oregon, a Redwood in Friendly Plaza, Monterey, California, a Sycamore at the Goddard Space Flight Center in Maryland and two Sweetgums at the Forest Service Office in Tell City, Indiana.

Soon after he had returned to Earth, Ed Mitchell and his wife divorced. He said that what she really wanted was to be married to a shoe salesman. Out of the combined intake of Mercury Seven, the New Nine and the Fourteen, only seven couples stayed together. 'If you think going to the moon is hard, try staying home,' Gene Cernan's wife once remarked.

16 July 1971. Apollo 15. Crew: Dave Scott (Commander), Al Worden (Command Module Pilot), Jim Irwin (Lunar Module Pilot). Duration of flight: 12 days, 7 hours, 11 minutes, 53 seconds. Command Module: *Endeavour.* Lunar Module: *Falcon.*

NASA would describe Apollo 15 as the most successful of its manned missions, but that success did not translate into public interest. In 1969 the Space Task Group had produced a report: the Post-Apollo Space Program, Directions for the Future. There would be a base in lunar orbit by 1976, and a base on the moon itself by 1978. By 1980 there would be a space station orbiting the Earth with 50 personnel on board. A manned mission to Mars would be launched before 1981. Hundreds of humans would be living in space by 1985. It was a vision that von Braun had mapped out 20 years earlier in *Collier's* magazine and in his Mars novel. But by 1971 there was no government or public appetite for such grand schemes. With the long-term goals of space now uncertain, the point of the remaining Apollo missions was cast into doubt. Apollo 14 had proven that humans could live and work for extended periods of time on the moon without ill effect, but that was of limited value if there was to be no base there. The

Dave Scott,
Al Worden and
Jim Irwin

success of Apollo 15 was to be primarily geological. What was to prove popular with the public was the first outing on this mission of the Lunar Roving Vehicle (LRV), affectionately known as the Moon Buggy. Again, the concept can be traced back to von Braun in *Collier's*. And it was von Braun who had pushed to make the rover a reality, and his operation at Huntsville that had overseen its development.

On the journey out to the moon Jim Irwin said that the Earth shrank 'from a basketball to a baseball, a golf ball, and finally a marble ... the most beautiful marble you could imagine' – though presumably it had been the most beautiful basketball too. When they got to the moon they couldn't stop looking at its surface. Al Worden commented that, rather than appearing to be a forbidding place, as others have found it to be, to him it simply looked dead. 'I had journeyed all this way to explore the moon,' Worden said, 'and yet I felt I was discovering far more about our home planet, our Earth.' He said the experience had been mind-altering. Other astronauts had said much the same thing right from the first manned flight. Perhaps they were beginning to say what was expected of them, or were running out of ways of saying the same thing. Perhaps there truly was a common transformative experience, which most of them underwent and which was a struggle to process.

Dave Scott and Jim Irwin's time spent on the moon was the longest of any Apollo mission, which meant too that Worden as Command Module Pilot was alone for longer than any other Apollo astronaut. For three days he saw stars to the limit of his eyesight, and the Earth repeatedly rise and set over the surface of the moon. After a while he'd been around so many times he said he began to recognize features on the moon. How odd, he thought, that already I'm seeing things that are familiar to me. He imagined life spreading between the stars, 'timeless, always there, adapting, propagating, spurred by survival'.

Meanwhile, on the surface of the moon, Scott and Irwin saw shades of gold, another hue to add to the litany of moon colours. On their second moonwalk they came upon a rock, lighter coloured than the rest, and sitting on a pedestal as if it had been placed there 'to be admired'. The rock was almost as old as the primordial crust of the moon. On Earth the rock was labelled 15415, but became popularly known as the Genesis rock. Dating tests indicated that it was 4 billion years old, almost as old as the solar system itself. The Genesis rock was perhaps the single biggest scientific discovery of the Apollo missions. It helped to show how and when the moon, and indeed the Earth, had been formed. The mission also returned with a core sample that displayed 42 identifiable layers. The bottom layer had not been disturbed for half a billion years.

These and other discoveries from the final missions proved that the moon had never been truly volcanic. It had never lived as the Earth lives now. The moon was how the Earth had been, billions of years ago before it had grown an atmosphere. The Earth had first come to volcanic life before it could support biological life in all its variety. Many scientists wanted to believe that the moon had once had a volcanic age, that it had been more Earth-like in its past. But the missions proved once and for all that the moon had only had a brief and desultory volcanic period and had then died. Most volcanoes on Earth are young, around 100,000 years or less. The volcanoes on the moon are small, rare and between 3 and 4 billion years old.

When Irwin first saw the Genesis rock he 'sensed the beginning of some sort of deep change taking place' within him. He wanted to hold some kind of service near where they had made the discovery, but Scott put him off. They didn't have clearance from NASA. But, the day after, Irwin recited a line from Psalm 121 at the foot of the moon's Apennine Mountains: 'I will lift mine eyes unto the hills, from whence cometh my help,' a prayer not in desire of anything in particular, he said, but in gratitude for what was. He became the first, and to date only, astronaut publicly to quote from the Bible while on the moon. Irwin had hoped that Madalyn O'Hair would object. Her action against NASA had gone all the way to the Supreme Court, but on 8 March, four months before the Apollo 15 launch, the court ruled that the relevant principle was the astronaut's own right to free speech. So long as it was clear that the astronauts spoke in a personal capacity, there was no case to answer. O'Hair's biographer Anne Seaman wondered if the near-disaster of Apollo 13 had influenced the judges. O'Hair said nothing in response to Irwin's reading. Perhaps by then she realized that the battle was lost, or that there were other more pressing or controversial causes that demanded her attention. O'Hair had always been expert at choosing to attack what would bring her the most publicity. She may simply have become as uninterested in Apollo as much of the rest of the world had.

Worden was as relieved as his fellow Command Module Pilots before him had been to welcome his crewmates back on board. 'I'd kept our home clean and tidy for them. But now, as I opened the hatches between the spacecraft I saw two grimy faces,' he later wrote, as if he were a mother welcoming her children back, too long out at play, with a clip around the ears. 'Their spacesuits were dirty, and I could smell the moon dust in the air. It was a new, peculiar odor to me, dry and gunpowdery. I kept the hatch closed as much as possible . . . hoping the dust would not spread.' Scott and Irwin had so exhausted themselves during their final moonwalk that their potassium levels had plummeted. By the time they were back in the Command Module, neither of them had slept for 23 hours, and both had developed heart problems. If they had been on Earth they would have been treated as if they were having heart attacks. NASA's doctor Charles Berry said that, effectively, in their 100 per cent oxygen environment they were already in an intensive care unit and receiving the best possible treatment; even better for their hearts, they were in zero gravity.

On the way home, Worden made the first deep-space EVA of the Apollo mission. There would be only two others. He had a unique viewpoint, the first person in history to see the entire Earth when he turned his head one way and the entire moon when he looked the other way. The moon was still close enough that he could see its craters clearly, and the frozen patterns of ancient lava flows. And then in the other direction there was the dynamic, vibrant Earth. *National Geographic* magazine had complained that the number of photographs coming back from the Apollo missions was falling off. There was also an increasing demand for photographs from scientists. After a lot of petitioning from Underwood's department a camera had been attached to the outside of the craft. Part of Worden's task during his EVA was to retrieve the camera and film from outside. The camera had jammed. The mission returned with just a few fuzzy photographs. Worden had wanted to take a camera out on his spacewalk with him, but NASA had ruled against it, saying that he'd be too busy. At one point Irwin poked his head out of the capsule to make sure everything was OK. When Worden turned to look at him he saw himself reflected in Irwin's helmet, and behind Irwin a moon as large as the craft itself. It could have been the most famous photo of the space programme, Worden said.

As the capsule approached the Atlantic only two of the three main parachutes opened, and one of those looked as if it might fail at any moment. It was a hard landing.

Worden said the weightlessness of space was like a homecoming. It felt so natural: 'As if I had been that way before or belonged in space.' When they were back on Earth, for a while they were not able to walk easily. They would push at objects, expecting them to move effortlessly away as they would have done in outer space. Nurse Dee O'Hara said that she often witnessed the astronauts' frustration, in their first days back, at the Earth's limitations: 'They have something, a sort of wild look, I would say, as if they had fallen in love with a mystery up there, sort of as if they haven't gotten their feet back on the ground, as if they regret having come back to us . . . a rage at having to come back to Earth.'

During the two-week debriefing period that followed their return, Apollo 15 astronaut Al Worden started jotting down impressions of the trip in his hotel room. He said the words seemed to be coming from somewhere else. Most of what he wrote wasn't even complete sentences. He organized the fragments and published them as a collection of poetry.

A spacewalk
Is like
Being let out
At night
For a swim
By Moby Dick

Worden said his poems were 'about as good as you might expect from a pilot', but perhaps the real value of the art being produced by returning astronauts was that it was being made at all. Worden felt that at 39 years old he had been reborn. He said, as a number of the Apollo astronauts had, that he had a new perspective on every aspect of life.

In the recovery ship, the day after he had returned to Earth, Irwin knew 'his soul had been stirred. He was a nuts-and-bolts man who had come back to something he had never anticipated: the seed of spiritual awakening.' As Al Bean had, he began to take delight in the simplest activities. Even to sit in a chair was a vivid experience. Two months later he was baptized at Nassau Bay Baptist Church, and soon after that founded a Baptist ministry called High Flight, the title of the pilot John Magee's poem. He became a powerful speaker. He said that when he was walking on the moon he had felt the power of God as he had never felt it before, and when they had first come upon the Genesis rock, he heard the voice of God speaking to him. He said it had been a literal revelation. After he had read the Psalm at the foot of the moon's Apennine mountains, his love of mountains deepened. In 1973 he went to Mount Ararat in search of the remains of Noah's Ark.

Irwin was the first of the 12 moonwalkers to die, of a heart attack aged 61. Whether or not the heart attack was a result of the stress of the long moonwalks is not known for certain.

16 April 1972. Apollo 16. Crew: John Young (Commander), Ken Mattingley (Command Module Pilot), Charlie Duke (Lunar Module Pilot). Duration of mission: 11 days, 1 hour, 51 minutes, 5 seconds. Command Module: *Casper*. Lunar Module: *Orion*.

As the appetite for a long-term future in space dwindled, so did the scope of the last missions feel diminished. The main purpose of Apollo 16 was to collect even older rocks than had been collected on the previous mission.

Ken Mattingley,
John Young and
Charlie Duke

Before the launch, Charlie Duke had a dream that was so vivid he said it was one of the most real experiences of his life. He dreamed that he and Young were on the moon and spotted rover tracks. They followed the tracks and saw another rover in front of them with two astronauts aboard who looked just like they did, except that it was suddenly clear to him that the astronauts in front of them had been there for thousands of years.

On the return journey Ken Mattingley went on a spacewalk to retrieve film canisters from the camera fixed to the outside of the craft. This time the camera had operated as it should. Mattingley said that his EVA training had prepared him for everything except for the experience itself. He looked around him and saw the moon in one direction and a crescent-shaped Earth in the other, and yet what he was most aware of was the feeling that everywhere else he looked there was nothing at all.

Charlie Duke went on to become a Brigadier General in the US Air Force, and active in prison missionary work. He said that though he had walked on the moon, his walk with God would last forever. His wife, Dotty, once got into an argument about Ed Mitchell's experience in space: 'But it's not the same God,' she said.

John Young was one of only three Apollo astronauts – the other two were Jim Lovell and Gene Cernan – who went to the moon twice. Back on Earth, Young had come to the same conclusion that several other Apollo astronauts had already reached. 'We worry about the wrong things,' he said, 'like the price of a gallon of gas, rather than the Earth as a whole.'

7 December 1972. Apollo 17. Crew: Gene Cernan (Commander), Ron Evans (Command Module Pilot), Jack Schmitt (Lunar Module Pilot). Duration of mission: 12 days, 13 hours, 51 minutes, 59 seconds. Command Module: *America*. Lunar Module: *Challenger*.

The last Apollo mission was the first launch to take place at night. In 1963, Valentina Tereshkova had felt the urge to bow to her rocket; now, less than 10 years later, Apollo 17 – bathed in the light of 74 xenon spotlights – also seemed to demand some form of obeisance. The astronomer and writer Carl Sagan said the launch was like a religious experience, as did the social philosopher William Irwin Thompson, who was also at the launch, and who compared the takeoff to the elevation of the Host.

Jack Schmitt was the first and only scientist to make it onto an Apollo mission. Gene Cernan was typical of the other astronauts in being critical of the push that had been made by scientists to be a part of the Apollo programme: 'It seemed to us that NASA was caving in to the scientific community, bargaining for dollars and support by promising a ride to some guy toting test tubes.' In fact Schmitt was a geologist and probably didn't spend much time 'toting test tubes'. The freemasonry of the pilot-engineers was being invaded. They thought that the safety of the missions would be compromised by having a scientist on board. Just as well no one had pushed for Collins' poet. Cernan made his reservations known to Deke Slayton, who told him he had a choice: either take the flight with Schmitt as Lunar Module

Pilot or step aside. Cernan of course decided to make the best of what he considered was not an ideal situation. In his own mind Cernan believed that he came to embrace Schmitt as a colleague, but his praise remained, at best, lukewarm. In his memoir, written decades later, Cernan describes Schmitt as: 'a genius type of pebble-pusher' who 'had a passion for thinking'. When Slayton told Schmitt he had made the grade, none of the astronauts congratulated him. A friend took him to Pizza Hut to celebrate. Even the astronauts' wives closed ranks against him. Schmitt acquired the nickname Dr Rock among his colleagues.

Schmitt suggested the mission make an attempt to land on the dark side of the moon, using a lunar orbiting satellite to maintain communication. The idea was rejected. Cernan said he thought it was crazy. And yet just such a mission may well have captured the public's imagination. Even though Cernan and Schmitt would visit places on the moon no other astronauts had been to before, as far as the public was concerned it was all just more of the same. *Life* magazine ended its contract with the astronauts.

The launches can't ever have become exactly routine, but Ron Evans somehow managed to fall asleep while they were waiting for takeoff, snoring loudly.

Jack Schmitt, Gene Cernan (seated) and Ron Evans

By the time the craft left Earth orbit it was midday. The Earth was full and highly illuminated. Cernan called out to his colleagues to take a look. 'I can't see a thing except the Earth,' Schmitt said. 'That's what I'm telling you to look at,' Cernan replied. 'Okay there's the old Earth,' Schmitt said, returning to his instruments. 'What else could I have expected from a scientist?' said Cernan. Even though he subsequently said that it was typical of Schmitt's droll sense of humor, Cernan didn't seem to realize that he was being teased. Schmitt was, in fact, in charge of photography. Underwood had told him that if he remembered to take a photograph about five hours into the flight, the chances were that the photograph would become a classic. At five hours and six minutes after takeoff, 28,000 miles from the Earth, Schmitt took one of the most famous photographs ever taken, the first detailed, high-quality, colour photograph of the whole Earth in full illumination.

A couple of days later they saw another sight only the crew of Apollo 16 had seen before. All other crews had approached the moon in its shadow. It had appeared suddenly, when they were almost on top of it; a menacing dark absence in the stars, or as a 'ghostly earthlit sphere'. Now, the moon revealed itself as a vast crescent bathed in sunlight. This moon – seen in detail and sunlight – was entirely different from the moon that had greeted Cernan on his earlier Apollo 10 mission. The experience made Cernan think that everything – the Earth, the moon, the stars, space – was all too wonderful to have been the result of chance.

On the moon's surface Cernan again tried to draw Schmitt's attention to the view of the Earth, but Schmitt wouldn't be distracted from the task at hand:

'Oh, man – Hey Jack, just stop. You owe yourself thirty
seconds to look over the South Massif and look at the Earth.'
'What? The *earth*?!'

Schmitt dwarfed by a large moon rock

'Just look up there?'

'Aaah! You seen one earth, you've seen them all.'

While Cernan tried to convey the emotion of his experience to Mission Control – 'Gosh, it's beautiful out here,' – Schmitt drily reported that the soil looked 'like a vesicular, very light-colored porphyry of some kind; it's about ten or fifteen per cent vesicles'. On the way back to Earth, Ron Evans performed the third and last deep-space EVA, an experience that he said, as others had before him, made him ecstatic. Apollo 17's splashdown on 19 December 1972 didn't even make the news. Nixon said that it was impossible not to see God's hand in the successful return of the 24 astronauts. A few days later, he cancelled the Apollo programme altogether.

In 1969 the first test of Korolev's N1 rocket had been a failure. The second test, in July the same year, had been a disaster. The rocket blew up 12 seconds into its flight. Material was spread over several miles, some of it landing in a children's playground. The following month Zond 7 was launched. Because the Proton rocket had proved to be unreliable, it was not a manned mission as had at one time been planned. As it happened the launch and re-entry were both successful. Inside the craft was a mannequin in the likeness of Gagarin, filled with biological matter, plant life and insects. The mission came back with many colour photographs of the whole Earth, more than three years ahead of the Blue Marble photograph, but the photographs were not of a high quality and made little impact. Underwood said that photography was never a high priority for the Soviets, and that they had little understanding of quality control: they used lousy film; they plucked reels out of a bin as if it were a tub filled with nuts and bolts; there was no consistency in the emulsions; some of the stock was out of date.

On 19 April 1971 the Soviets launched the first space station, Salyut 1. They may have almost given up on the race to the moon but at least they could still be first in other ways. In June that year three cosmonauts boarded the space station, and lived there for almost 24 days. The US would not better that record for two years. But the Soviet mission ended in tragedy when the crew all died on the way back due to a leak in their capsule. Georgy Dobrovsky, Vladislav Volkov and Viktor Patsayev are the only human beings who have died outside the Earth's atmosphere. Later that year the Soviet's unmanned Luna 16 mission returned with rock samples from the moon, raising the question: why send humans at all?

The Soviets made a third attempt at launching their N1 rocket on 26 June 1971 and a fourth on 23 November 1972. Both flights failed at some point. A planned fifth launch never took place and the N1 was cancelled. The N1 would have been the largest rocket ever built, if it had ever flown successfully. To this day, that accolade goes to von Braun's Saturn V.

In 1974 the Soviets quietly pulled the plug on their manned moon programme too. The US had spent $25 billion, the USSR perhaps half that. The manned space era ended with a whimper.

INTERLUDE

INTERLUDE

What does it mean to grow up? We can grow to be better selves – more the self that we always were, more our self – or into perversions of what we are. At best we might hope to grow into seriousness and complexity while retaining the innocence and playfulness of the child. Or else we might grow, as a tree does tortured by the wind, into a twisted version of what we are, arrested at some young age, five perhaps, forever stamping our feet. Hitler as a grown man, at the height of his powers, would sometimes roll on the floor in anger and chew the carpet; literally he did this, there were witnesses. Despite our strongest desires to think the best of everyone, there is evil and there are facts.

In memory we see generations past as if through both a microscope and a telescope. There are details, but too much detail and the overall effect is lost. But if we spy on our subject through a telescope only, the individual gets lost in the landscape of history.

The physical condition of being an individual human being is one of looking outwards, not back on one's self. What we see of our own self is little more than a headless body; as whole bodies are what other people look like. There are mirrors of course, and selfies, but those perspectives always come as a shock for being out of the ordinary. We see other people far more clearly than we see ourselves. We each constantly give ourselves away to others without ever knowing it. We do not know what we look like, and we do not know what we are. About ourselves we miss almost everything. And so it is, too, for humankind collectively on Earth. Until recently, we had only been able to imagine what our Earth home might look like from the outside.

The Ancient Greeks knew that the Earth was circular in some way: a disk, a cylinder or a sphere, but a Greek word that means spherical also means circular, and so who it was that first described the Earth as a sphere is still a matter of academic conjecture. It may have been Pythagoras in the sixth century BC, but everything we know about Pythagoras comes from secondary sources after his time. Plato wrote *Phaedo* 150 years later; an account of the last hours of Socrates' life, before he downed the fatal hemlock concoction. Socrates described to those gathered around his deathbed the Earth as he imagined it must look seen from the vantage point of the afterlife. After death he would see the Earth as it really is; but, this being Plato, it would be an idealized, Platonic Earth not the physical Earth we know when we are alive. 'The true Earth', Socrates said, is like one of 'those balls made of twelve pieces of leather, variegated, a patchwork of different colours'. The Earth – or at least this ideal of the Earth – is a vast dodecahedron positioned in the middle of the heavens, shining with colours more vibrant than any we know in life. The Roman General Scipio appears to his dreaming grandson, in a short section in Cicero's vast *De re publica* (*On the Commonwealth*) written 400 years later. He tells him that from his vantage point – 'a high place full of stars, shining and splendid' – Rome looks insignificant, just a small region of the Earth, and that the Earth, too, is insignificant compared to the stars. By the time another century had passed, the astronomer Ptolemy *knew* from scientific considerations, by observation and measurement, that the Earth must be a sphere. But knowing is different from seeing. Where was the visual evidence? To see for ourselves we would have to rise high above the surface of the Earth. There were balloons and then airplanes, yet even in the 1950s still no one knew what the Earth would look like seen from outer space. In 1952 Wernher von Braun got it wrong in significant ways: 'The continents will stand out in shades of grey and brown bordering the brilliant blue of the seas,' he wrote. 'One polar ice cap will show as blinding white, too brilliant to look at with the naked eye.' To see for ourselves we would have to escape the gravitational pull of the Earth. And then there were rockets. Between 1968 and 1972, 24 Apollo astronauts saw for themselves. For the rest of us, they returned with words and images.

PART THREE

CHAPTER ONE

Richard Underwood was keen to get the photographs that came back from the Apollo 17 mission developed as quickly as possible. He took some of them home to show his children, sure in his own mind that they were going to be influential. As he began to lay them out on the table, his son said: 'Uh oh, Dad's going to show us some more of that junk from space.'

NASA photograph AS17-148-22726, popularly called the 'Blue Marble', was released to the public on Christmas Eve 1972, four years to the day after Earthrise had been taken. NASA's press office made much of the image, though in the mission's official report the photograph isn't mentioned at all. Nor did the astronauts refer to it when – as was traditional after each mission – they reported to Congress.

Jack Schmitt had taken the photograph with Antarctica at the top, but NASA released it the 'right' way up. By continually orienting photographs of the Earth with Antarctica at the bottom we are encouraged to hold on to an old paradigm of what the Earth is. The enlightenment that came to most of those who had seen it for themselves, was to experience the Earth as a globe *falling* through space. We can understand it intellectually but to truly get it – that we live here on the surface of a sphere, rock and water bounded about by a skin of atmosphere, forever plunging through black space – that is something else. Can a photograph alone provoke that Zen-like leap of understanding? Certainly a photograph with Antarctica at the 'bottom' does not help. The Blue Marble showed a full (or, to be precise, an almost entirely full) Earth seen in full sunlight. Because it had been taken relatively close to the Earth there was also a lot of detail in the photograph. It was centred on Madagascar, which meant that the world got to see for the first time how truly enormous Africa is. As it is projected on most maps, Africa appears much smaller than its true size – as do all countries around the Equator. Even now most people do not realize quite how vast Africa is. The landmasses of all of China, the United States, India and Japan added together, plus nearly all of Europe, fit within Africa's coastline. Appropriately enough, one of the first people to see the photograph was Victor Hasselblad, who happened to be visiting the Manned Space Center at the time. It was his modified camera that had gone to the moon on all the Apollo missions, his camera that had taken Earthrise and the Blue Marble photographs.

In 1948 the physicist Fred Hoyle had predicted that 'Once a photograph of the Earth, taken from the outside, is available ... a new idea as powerful as any in history will be let loose.' What he hadn't predicted was that the photograph would need to be of the highest quality and in colour. Two such photographs spanned the Apollo era: Earthrise, taken during the first manned mission to the moon, and the Blue Marble, taken during the last. They have become two of the most reproduced images of all time. In *Life* magazine's 2003 publication: *100 Photographs That Changed the World* wilderness photographer Galen Rowell called Earthrise 'the most influential environmental photograph ever taken'.

In 1966, a 28-year-old rock promoter named Stewart Brand started a campaign to persuade NASA to release what was rumoured to – but did in fact not then – exist: a colour photograph of the whole

Earth. Brand was a member of Ken Kesey's 1960s commune 'The Merry Pranksters', based at Kesey's homes in California and Oregon, and a contributor to Paul Krassner's *The Realist*, the magazine for which Madalyn O'Hair also wrote. The idea for the campaign had come to Brand on a February afternoon in San Francisco:

I had taken a mild dose of LSD on an otherwise boring afternoon and sat, wrapped in a blanket, gazing at the San Francisco skyline. As I stared at the city's high-rises, I realized they were not really parallel, but diverged slightly at the top because of the curve of the earth. I started thinking that the curve of the earth must be more dramatic the higher one went. I could see that it was curved, think it, and finally feel it. I imagined going farther and farther into orbit and soon realized that the sight of the entire planet, seen at once, would be quite dramatic and would make a point that Buckminster Fuller was always ranting about: that people act as if the earth is flat, when in reality it is spherical and extremely finite, and until we learn to treat it as a finite thing, we will never get civilization right. I herded my trembling thoughts together as the winds blew and time passed. And I figured a photograph – a color photograph – would help make that happen. There it would be for all to see, the earth complete, tiny, adrift, and no one would ever perceive things the same way.

Buckminster Fuller – who was then in his early seventies, and at the height of his fame as a philosopher and inventor – agreed to help Brand achieve his objective. In 1951 Fuller had come up with the phrase 'spaceship Earth', a way to try and shock us into enlightenment using words. All of us on Earth are astronauts, he told us, all voyagers through space. To raise money for his campaign, Brand sold 25-cent badges bearing the legend: 'Why haven't we seen the whole Earth yet?' The badges were distributed widely; some were mailed to NASA officials, though no one there has owned up to receiving one. It is not clear how much Brand's campaign influenced NASA, but a few months after it had begun Lunar Orbiter had been repositioned so that it could take a single black-and-white photograph of the whole Earth, the first ever. And by the end of 1967 the ATS-3 and DODGE satellites, and a camera placed on Apollo 4, had taken the first colour images of the whole Earth. In the style of Warhol's epic movie *Empire* (1964) – eight hours of long takes of the Empire State Building – Brand screened 24 hours' worth of footage from the ATS-3, a film he titled *Full Earth*.

In September 1968 the first edition of Stewart Brand's *Whole Earth Catalog* was published, a counter-culture magazine that promoted the idea of communal living and a self-sufficient lifestyle. It listed clothing, books, tools, machinery, anything that might enable or sustain creativity or self-sufficiency. The catalogue was a response to a world that had become wasteful, nowhere more so than in wealthy America. The magazine encouraged a movement back to the land and a do-it-yourself attitude that was encapsulated in the catalogue's motto and first sentence, 'We are as gods, and may as well get used to it.' In America at that time, an estimated 10 million people were living in communes. Steve Jobs later described the catalogue as 'one of the bibles of my generation. It was sort of like Google in paperback form.' On the cover Brand used the photograph of the whole Earth that had been taken by ATS-3 the previous November. Subsequent editions of the magazine followed at roughly three-monthly intervals. 'Earthrise' was used on the cover of the 1969 spring issue. 'It gave the sense that Earth's an island,' Brand said of the photograph, 'surrounded by a lot of inhospitable space. And so graphic, this little blue, white, green and brown jewel-like icon among a quite featureless black vacuum.' The last issue of the catalogue came out in 1972, first to last neatly spanning the Apollo years. On the cover of the last issue Brand used a photograph taken by Apollo 4. It showed the Earth in partial shadow. He wondered later if the shadow might have been off-putting to people, frightening even. A thousand copies were printed of the first issue of the catalogue. The last issue was published by Penguin in an edition of a million copies. Brand immediately launched a new magazine, *CoEvolution Quarterly*, a platform from which to explore environmental concerns. The magazine published essays from various perspectives by a diverse group of writers that included Ursula LeGuin, Wendell Berry, Gregory Bateson, Eric Drexler, Lynn Margulis, Paul Hawken and Kevin Kelly (future editor of *Wired* magazine).

Only 24 human beings have ever seen the Earth from space, the 24 Apollo astronauts who went to the moon and were safely returned to Earth again. Only they have ever seen the Earth as if from the perspective of an alien. Only they have seen the Earth as it truly is, a sphere falling through space. Nearly all of them were struck by how fragile the Earth looks. From space the Earth's atmosphere appears as a thin blue line around the circumference of the planet, a hazy penumbra of a mere 50 miles clinging to a disk 8,000 miles across. 'It is a pity,' Mike Collins wrote, 'that my eyes have seen more than my brain has been able to assimilate.' And yet of all the astronauts who left Earth orbit, he came closest to describing in words the sense and the feeling of what he had experienced: 'The moon is so scarred,' he wrote, 'so desolate, so monotonous, that I cannot recall its tortured surface without thinking of the infinite variety the delightful planet earth offers: misty waterfalls, pine forests, rose gardens, blues and greens and reds and whites that are entirely missing on the gray-tan moon.' On his way back home, he looked out of the window and tried to find the Earth:

The little planet is so small out there in the vastness that at first I couldn't even locate it. And when I did, a tingling of awe spread over me. There it was, shining like a jewel in a black sky. I looked at it in wonderment, suddenly aware of how its uniqueness is stamped in every atom of my body . . . I looked away for a moment and, poof, it was gone. I couldn't find it again without searching closely.

At that point I made my discovery. Suddenly I knew what a tiny, fragile thing Earth is.

Apollo 17 Jack Schmitt did not agree. 'I think the pictures make it look a lot more fragile than it is,' he said. 'The Earth is very resilient . . . I know what blows it's taken.'

So which is it, fragile or robust?

The ecology movement had had a mystical and spiritual component to it from its origins in the late nineteenth century. 'The Earth is holy . . . We are here, part in the creation,' wrote the American botanist Liberty Hyde Bailey, a writer who had been influenced by the American transcendentalist Henry Thoreau, and who was one of the first scientists to promote the newly rediscovered works of Gregor Mendel. In 1926 the Russian geologist Vladimir Vernadsky – a proponent of cosmism – wrote about the Earth using the holistic terms geosphere, biosphere and noosphere. The geosphere described the totality of the world's inanimate matter, the biosphere the world's living forms, and the noosphere human knowledge. The palaeontologist Loren Eiseley (1907–77) – also influenced by Thoreau – went further and wrote of the whole Earth as something living that could repair itself: 'Like the body of an animal, the world is destroyed in one part, but renewed in another.' Eiseley wrote of the need for anthropologists to take time not just to observe and speculate but to dream. The idea of the

Earth as a giant ecosystem was popularized by the American biologist Eugene Odum (1913–2002) in the 1950s. In later editions of his *Fundamentals of Ecology* (1953) he reproduced the Earthrise photograph, which also hung as a poster on his office wall. In 1969 the American microbiologist René Dubos (1901–82), credited with coining the environmental slogan, 'Think globally, act locally' (although it was actually first coined by Jacques Ellul, who wrote about the tyranny of technology), said that seeing the Earthrise photograph had made him realize 'that the Earth is a living organism'. Like Eiseley he thought that the Earth was resilient and adaptable, and had evolved the ability to heal itself. He wrote that humankind was not pitted against the Earth but was part of the Earth's system: 'Earth and man are two complementary components of a system, which might be called cybernetic, since each shapes the other in a continuous act of creation.' Dubos called his ideas a 'theology of the Earth'. There were 'sacred relationships between mankind and the physical attributes of the Earth,' ... 'A truly ecological view of the world has religious overtones.' Stewart Brand, who had trained as an ecologist, said that ecology as a science is pretty boring, but 'ecology as a movement, as a religion, is tremendously exciting'.

In the 1960s NASA employed the biologist James Lovelock to come up with experiments that might answer the question: is there life on Mars? Lovelock thought it unlikely. He suggested testing the planet's atmosphere as a way of deciding the question, but his idea was rejected. In 1965, when it was discovered that Mars' atmosphere is mainly made up of carbon dioxide, Lovelock's hypothesis that it was a dead planet with a dead atmosphere was strengthened. Lovelock intuited that Earth was home to life in part from seeing the Earthrise and Blue Marble photographs. Scientists in the past had been able only to guess that the Earth looked different from other heavenly bodies; for the first time, here was visual evidence that it did in fact look different. In the seventeenth century the French astronomer Adrien Auzout had conjectured that the moon was probably not home to life. In a thought experiment, Auzout had tried to imagine what the Earth must look like seen from the moon and had guessed that it would appear much more vibrant than the moon seen from Earth. In the 1920s Vernadsky wrote that 'the face of the Earth viewed from celestial space represents a unique appearance, different from all heavenly bodies'. In his science-fiction novel *Star Maker* (1936) the British philosopher Olaf Stapledon describes the Earth as having 'the intricacy and harmony of a living thing'. It took high-quality colour photographs of the Earth seen from space to confirm what they had intuited: that from space the Earth looks different from any other heavenly body we know of. In black and white the difference is hardly apparent, but in colour we are forced to ask why the Earth looks different

Lovelock formalized his thinking into a testable theory that become known as Gaia: 'The entire range of living matter of Earth, from whales to viruses, and from oaks to algae, could be regarded as a single entity, capable of manipulating the Earth's atmosphere to suit its overall needs and endowed with faculties and powers far beyond those of its constituent parts.' The Earth, together with its atmosphere, is evolving, not moving to an equilibrium state. In that sense the totality of life on Earth behaves as if it too were some living entity. The Earth looks different because it is alive. And, as with any system that is alive and evolving, we cannot know what the Earth is evolving into; we have no idea where *we* fit in. We can get no true perspective. Lovelock wondered if, as a species, we might eventually come to see ourselves as part of some 'Gaian nervous system and a brain which can consciously anticipate environmental change'.

The principles of Lovelock's Gaia theory were first published in Carl Sagan's journal *Icarus* and in Stewart Brand's *CoEvolution Quarterly*. The theory was at first ridiculed and then heavily criticized (notably by a young Richard Dawkins) for not being properly scientific. In *Earthrise*, Robert Poole writes that when Lovelock's book *Gaia* was published in 1979, it was resisted by the orthodox scientific community with 'almost religious intolerance', fulfilling the first two stages of Schopenhauer's insight that 'all truth passes through three stages: first it is ridiculed, second it is violently opposed, third it is accepted as self-evident'. Today, to most of us, it seems almost obvious that our planet is itself a living system.

In 1972 the cultural historian William Irwin Thompson had created The Lindisfarne Association to encourage a new relationship between the humanities and science and technology: 'to foster a global ecology of consciousness'. Speakers at the conferences included Stewart Brand, the economist E. F. Schumacher, the poet Kathleen Raine and the biologists Stuart Kauffman, Lynn Margulis and James Lovelock. At the 'Planetary Culture' conference held in 1974, Brand talked about his campaign to persuade NASA to take a colour picture of the whole Earth. Apollo 9 astronaut Rusty Schweickart had also been invited to speak.

Schweickart was one of the more intellectual of the astronauts. He chose to take recordings of Vaughan Williams' late cantata *Hodie* (This Day), the story of Christ's Nativity, and Alan Hovhaness' Second Symphony 'Mysterious Mountain' with him into space. Even the most avid devotee of classical music might not come across either piece during a lifetime's listening. In his programme note Hovhaness wrote that:

Mountains are symbols, like pyramids, of man's attempt to know God. Mountains are symbolic meeting places between the mundane and spiritual world. To some, the Mysterious Mountain may be the phantom peak, unmeasured, thought to he higher than Everest, as seen from great distances by fliers in Tibet. To some, it may be the solitary mountain, the tower of strength over a countryside – Fujiyama, Ararat, Monadnock, Shasta, or Grand Teton.

Dave Scott hid the music until near the end of the mission. He said Schweickart never forgave him. To read, Schweickart took works by Elizabeth Barrett Browning and Thornton Wilder. After returning to Earth from his mission in Earth orbit, Schweickart was drawn to Zen Buddhism. Pete Conrad once described Schweickart as their token hippie. Schweickart grew his hair, not very long but long compared to the usual crewcut, and a beard. His wife, Clare, played the ukulele and sang folk songs. They were part of a book group that read consciousness-raising texts like George Gilder's *Sexual Suicide*. In 1972 he took the Maharishi Mahesh Yogi on a tour of the Manned Space Center. The Maharishi had become famous for introducing Transcendental Meditation to the West, and for being taken up by The Beatles. He could 'deliver *samdhi*, a holy state of expanded consciousness, without going to all the trouble of fasting and endless prayer'. Afterwards, at their home, the Maharishi walked barefoot across the Schweickarts' lawn and gave Clare her own personal mantra.

Several years had passed after his return from space before Schweickart had been able to work out what had happened to him during his EVA. He had decided not to talk about it at the time. It wouldn't have been very 'Right Stuff', and might have jeopardized his chances of being chosen to go on another mission. He wasn't chosen again anyway; perhaps for no particular reason, though on the whole Deke Slayton was suspicious of intellectuals, even more so of hippies. In 1972 Schweickart gave an interview to *Time* magazine during which he told the interviewer: 'I completely lost my identity as an American astronaut. I felt a part of everyone and everything sweeping past me below.'

National boundaries became meaningless and arbitrary, but also the boundary between self and not-self. 'There's a difference [afterwards] in that relationship between you and the planet and you and all those other forms of life on that planet, because you've had that kind of experience ... And all through this I've used the word *you* because it's not me, it's not Dave Scott, it's not ... it's you, it's we. It's Life that's had that experience.'

At the Lindisfarne conference Schweickart decided to speak spontaneously. He found himself talking of his experience in space in the present tense and second person. A member of the audience said it was like listening to 'a long, pauseless, prayer'. Brand later said that Schweickart 'seemed amazed at what he was saying, amazed at the gathering he was attending, amazed – still – at the events which led him to drift bodily free between Earth and Universe. Remember the Star Child at the end of *2001*? Like that?' Schweickart ended his address with a poem by E. E. Cummings: 'i thank You God for most this amazing day.' The poem ends with the lines, '(now the ears of my ears awake and/now the eyes of my eyes are opened)'.

Schweickart might well have been the perfect astronaut to experience going to the moon, but it was not to be. Among the astronauts who did go to the moon, the closest experience to Schweickart's is probably that of Ed Mitchell, who, after he had returned to Earth, devoted the rest of his life to studying consciousness, in part in order to try and understand what had happened to him out there. Mitchell had been brought up in New Mexico – that curious region of curious energies: home to many and various Native American tribes, where the first rockets and the first atom bombs were tested, and where UFOs are spotted with seemingly greater frequency than most other places on the planet. Mitchell remembered the luminous glow that filled the sky after an atom bomb had been detonated. He had no memories of the early rocket tests from White Sands, but he had heard the stories:

> Just a mile or so down the road from where I was raised lived a man who would loom large in my imagination ... each day as I walked to school along the white gravel road, I would pass the quiet home where a mad scientist was said to live. He was quite literally, a rocket scientist. He was also America's first, his name was Robert Goddard. This was deep in the bleakness of World War II, and across an ocean this man's German successor Wernher von Braun, was busy designing the V-2.

Mitchell felt as happy studying consciousness as he had in his youth studying technology. He approached both in the 'same secular manner', learning about the non-physical world as he had once learned about 'the dynamics of an airplane or the model of an atom'. The approach was the same even if the paths went in opposite direction. Scientists are led inexorably back to the beginnings of things, ultimately to the beginning of the universe, mystics inexorably back to the self. But the 'larger purpose of both science and religion', Mitchell said, is the same: our human desire, our secret hope, is 'to find our place in the vast scheme of things'. He said it was the larger purpose of every mission into space.

Mitchell identified what he had experienced in space as a kind of *samdhi*. It was the same experience that Schweickart had had in low orbit: a sudden understanding that separateness is an illusion and that 'an essential unity is the benchmark of reality'. After his experience Mitchell said he became more sensitive to his body and to his emotions. He had had 'a grand epiphany' and understood that the universe was 'in some way conscious'. He said he 'couldn't honestly call it a "religious experience"'. Despite the allusions he makes to religion, he was looking for a secular rather than a religious explanation for what he had experienced. 'Life itself is a mystical experience of consciousness,' he wrote. 'It's just that we have grown used to it through the millennia.' The ecstasy he had experienced was somehow a natural response of his body 'to the overwhelming sense of unity' of the universe. For a moment his ego had dissolved: 'You develop an instant global consciousness; a people orientation, an intense dissatisfaction with the state of the world, and a compulsion to do something about it. From out there ... international politics look so petty. You want to grab a politician by the scruff of the neck and drag him a quarter of a million miles and say, "Look at that!"' He thought that all astronauts had gone through the experience, but that 'if you happen to be closed off and are happy with [your] former belief system, you reject the information and nothing happens. Or if it happens to be too challenging or threatening, then it'll be consciously rejected, thrown out. But otherwise, it can be absorbed and expanded into [your] belief system, and you have a different view.' If there is an existing predisposition there already, those who have had the experience are convinced of its reality. An atheist might have the experience too, but can describe it only in terms of sensation and feeling. 'The mystic reels in the mysteries of the ineffable, while the scientist chafes at the lack of specificity.' He said that what he had experienced was equivalent to the Christian peace that passeth all understanding, or what theologian Paul Tillich called a 'union with the ground of our being'. After the experience of ecstasy comes a feeling of being cut off, which Mitchell likens in Christian theology to the Fall. The intensity of the experience necessarily fades, but he still got an intimation of it – as Collins also described – whenever he flew in a commercial aircraft. He said that 'space flight is one of the more powerful experiences that humans can have, and the technological event of breaking the bonds of Earth is far more important than the technology that went into it'. He came to believe that 'ageless wisdom based on integrity, tolerance, and goodness is still pertinent to the modern experience'. We have not yet grown into our brains, he said. The question remains: can we grow into our brains in time? Are we capable as a species of growing up? Schweickart, too, wondered if humans could keep up with the speed of technological progress: 'The Earth was at "a balance point in its evolution",' he said, 'teetering between potential greatness and colossal collapse.' After lives spent at the cutting edge of technology, von Braun and Lindbergh had come to similar conclusions.

CHAPTER TWO

After the war, Lindbergh inspected air force facilities around the country as consultant to the Secretary of the Air Force. He was also part of a committee that oversaw the development of long-range ballistic missiles. He flew on simulated atomic-bomb missions and quickly realized that a new kind of precision bomber was needed. He was a member, too, of the top-secret organization CHORE (Chicago Ordnance Survey), set up to evaluate the history of weaponry. The work was so secret the invitation to join had been delivered by hand. At one CHORE dinner the scientist sitting next to him told him what it had been like to be at the explosion of the first atomic bomb at Almagordo, part of the White Sands Proving Ground. 'Now, when I press this button,' Enrico Fermi – the creator of the world's first nuclear reactor – had told those assembled, 'there is a chance in ten thousand it will be the end of the world.' At 20 seconds after 5.29am on 16 July 1945, 'and before the astonished military observers could react, Enrico pushed the button'. Anne Morrow Lindbergh had been at the same dinner. She wrote in her diary afterwards: 'On the whole how ordinary these people are when they come out of the tunnel of their specialty! What children they seem in the field of living & feeling & being aware. How impractical, too, some of them. Like artists – but inarticulate ones. Like musicians, childlike – simple – wrapped in the cocoon of their own world.'

At the end of 1945 Lindbergh had been invited to address the 42nd Aviation Anniversary Dinner of the Aero Club of Washington. It was by then a rare public appearance. At first he had planned to talk about rocket science, but he changed his mind because 'such questions became dwarfed by the basic problems of how to keep aircraft from destroying the civilization which creates them'. In 1949 at the 46th Aviation Anniversary Dinner of the Aero Club of Washington – another rare public outing – he talked about man's need to 'balance science with other qualities of life, qualities of body and spirit as well as those of mind – qualities he cannot develop when he lets mechanics and luxury insulate him too greatly from the Earth to which he was born'.

Lindbergh became increasingly disillusioned with aviation: 'As my hours in the air increased in number, I lost the keenness of that early vision.' Flying was now just about pointing the plane in the right direction. 'The pure joy of flight as an art' was giving way 'to the pure efficiency of flight as a science'.

In his youth, science had been more important to him than either man or God: 'The one I took for granted; the other was too intangible for me to understand.' He would never have any patience for organized religion, but as he got older Lindbergh began to see science and religion not as inhabiting separate domains but feeding into each other: 'The shape of God we cannot measure, weigh, or clock, but we can conceive a reality without a form. The growing knowledge of science clarifies man's intuition of the mystical. The farther we penetrate the unknown, the vaster and more marvelous it becomes.' During the war he had taken the New Testament with him on his flying missions: 'That is my choice. It would not have been a decade ago; but the more I learn and the more I read, the less competition it has.' He wrote that after the war, 'No peace will last, which is not based on Christian principles, on justice, on compassion allied with strength, and on a sense of the dignity of man.' He became interested in Eastern religions. His favourite poet was Lao Tze, whom he often quoted. He liked to garden, and listen to birdsong: 'I realized that if I had to choose, I would rather have birds than planes.'

During the second half of his life Lindbergh gave himself over to Nature and to saving endangered species. He had come to the conclusion that neither aeronautics nor astronautics had been a boon to the human race. Aircraft had brought people together in peace, but in war they had been used to kill in a way that seemed 'to have little or no relationship to evolution's selectivity'. Missiles had become the rockets that allowed humans to explore space, but they were also the deadly weapons that had 'made our civilization subject to extermination within hours'. He blamed himself for having 'helped to change the environment of our lives'. Flying had brought many benefits to society, he continued to believe, but the Earth was being damaged because of it.

Lindbergh became an environmentalist before the term was in common usage. He was involved in the setting-up of the National Park Service and was active in The Nature Conservancy. He worked for the World Wildlife Fund, writing reports for its parent organization the Union for Conservation of Nature and Natural Resources. His efforts helped to bring attention to the ivory trade. He was involved in projects to save the blue whale and the great finback. When Aldabra, an island in the Indian Ocean, was being considered as an airbase, he wrote to protest. He spoke in Alaska on behalf of the Arctic wolf, and almost immediately protective legislation was passed. 'I don't want history to record my generation', he said of his conservation work, 'as being responsible for the extermination of any form of life.' But his was a life in which every word was weighed by the press: 'Where the hell was he,' the journalist Max Lerner wrote in response to this declaration, 'when Hitler was trying to exterminate an entire race of human beings?'

Thoreau's words 'in wilderness is the preservation of the World' resonated with Lindbergh. 'Real freedom,' he wrote, 'lies in wildness not in civilization.' He said that he believed 'there is wisdom in the primitive lying at greater depths than the intellect has plumbed, a wisdom from which civilized man can learn ... It is ... wisdom of instinct, intuition, and genetic memory, held by the subconscious rather than the conscious mind, too subtle and elusive to be more than partially comprised within limits of rationality.' Not that he didn't also recognize that 'the Garden of Eden is behind us and there is no going back to innocence; we can only go forward'. He worked in Africa with the famous family of anthropologists the Leakeys. He lived for a time with the Masai, and later with the Tasaday, a cave-dwelling tribe living in the mountainous rainforests of Mindanao, one of the southern islands of the Philippines. Lindbergh had been introduced to the tribe by Manuel Elizaide, a crony of the Philippine dictator, Ferdinand Marcos. A one-time hard-drinking playboy, Elizaide seemed to have turned his life around when he first came across the Tasaday – he said by chance – and joined them in their struggle against developers. Elizaide created an organization called Panamin (Private Association for National Minorities), and Lindbergh was encouraged to join the board. The Tasaday had apparently never before seen evidence of life in the world outside. Lindbergh was one of the first outsiders to make contact with the tribe. Aged 70, he jumped from a helicopter to land among them. Anne wrote to their daughter Ansy: 'I must say Father has really done it this time!' After two weeks Lindbergh, along with 44 anthropologists, ran out of supplies. An air force helicopter had to be sent in to rescue them. 'I don't know anyone better equipped to survive in the jungle than your father,' Anne wrote.

In midlife he had made a decision that he did not want to fly in a rocket: 'If I had not decided against specializing in fields of missiles and space, as I did, I might have been orbiting in a satellite instead of looking up at one from the framework of a jungle. Then the cramped and weightless interior of a rocket head would be familiar to me, and the gravity-bound expanses of East Africa the strange.' (Might he have become an astronaut if he had chosen to? It doesn't seem very likely but perhaps someone at NASA might have said: 'I'd like to see how the old boy does.') Astronauts had travelled through space; living with the Tasaday, Lindbergh felt as if he had travelled through time, moving in a moment from so-called civilization to 'a stone-age cave-dwelling culture. I felt I might have been on a visit to my ancestors a hundred thousand years ago.' They said to him: 'We do not know what to ask for because we do not know what we want.' They wished to stay forever in their caves as they had since time began. Lindbergh said he had never seen a happier people. Here among these tribal peoples, he encountered 'the juxtaposition of apparently opposed principles of nature – the importance and unimportance of the individual'. It is said that afterwards Lindbergh took to rolling naked in the mud at his home on Long Island Sound.

The Tasaday were soon drawn beyond the forest – as all human beings are, curious. By then Elizaide had fled the country with millions of dollars that had been raised by Panamin for the tribe's protection. It also emerged that he may have bribed the tribe to say that they had had no contact with the outside world.

On a number of occasions during the 1970s Lindbergh also stayed – at Anne's suggestion – with Benedictine nuns at the Regina Laudis Priory in Bethlehem, Connecticut. Charles wrote to one of the sisters to tell her how much he had appreciated the experience – 'the welcome, the singing, the sense of earth, the spiritual atmosphere, and with these qualities, broadness of viewpoint and sense of humor' were unlike any he had 'encountered before in a religious organization'.

Lindbergh spent much of whatever spare time he had reading and writing about conservation. He wrote his first piece of popular journalism on the subject in 1972 for *Reader's Digest*. In the article – 'Is Civilization Progress?' – he observed that 'Life seems to me to have more validity than any other standard as a measure for the progress of mankind ... On every continent and in almost every country, the crisis for wildlife is acute ... Man can stop the extermination if he has the desire to do so. To what extent he has this desire will, I think, be the measure of his greatness – whether he places more value on his own material accomplishments or on God's miracle of life.'

In his last published work, *Autobiography of Values* (1979), Lindbergh writes of man's despoliation of the Earth, but in the past tense, as if from the distant perspective of an alien recalling an Earth that no longer exists: 'Every day, increasing numbers of bulldozers and trucks tore into mountains, slashed through trees, leaving far greater scars on the Earth's surface than those created by bombs. Gases from civil vehicles polluted our atmosphere. Waste from civil factories poisoned our rivers, lakes, and seas. Civil aircraft made every spot on Earth open to the ravages of commerce ... What was the prospect for mankind?' Lindbergh had come to the realization that we live 'in a vicious circle, where the machine, which depended on modern man for its invention, has made modern man dependent on its constant improvement for his security – even for his life'. Humankind was caught in a bind of its own devising: on the one hand, 'in worshipping science man gains power but loses the quality of life', while on the other 'without a highly developed science, modern man lacks the power to survive'. He believed that the damage caused by technology might only be repaired, not by blindly rushing to create new technology, but by changing our relationship to technology and to the planet; by finding a new perspective.

'The human future,' Lindbergh wrote, 'depends on our ability to combine the knowledge of science with the wisdom of wildness.' The word 'balance' was forever on his tongue: between science and spirituality, nature and technology, body and spirit. He had, over the years, developed a sophisticated spiritual side.

Anne and Charles
Lindbergh in 1968

His life ultimately somehow managed to embrace the reductive and the transcendental. He said that the longer he lived, the more limited he believed rationality to be, a surprising statement from someone who had created so many problems for himself out of a too-firm grasp on purely rational argument. Lindbergh remained a virtual recluse for the rest of his life, declining all public invitations to talk.

'Your father is very busy with his many lives,' Anne Morrow Lindbergh wrote to her elder daughter in 1968. It is not clear if Anne knew quite how many lives her husband was leading by that time. Several times every year for the last decade Charles had been visiting his three other families spread several hundred miles apart across the borders of Switzerland and Germany.

In 1957, at a dinner party in Munich, his tall, beautiful research assistant, who seems only to have been known as Valeska, had introduced Charles to her friends, sisters Brigitte and Marietta Hesshaimer. Both were disabled, the result of bone tuberculosis suffered when they were girls. Lindbergh had already been having an affair with Valeska – 22 years his junior. He is rumoured to have fathered two children with her. Now he fell for Brigitte, a 30-year-old hat maker, 25 years his junior. Together they had three children: Dyrk born in 1958, Astrid in 1960 and David in 1967. He told Brigitte that if she ever told anyone their secret he would not return. To the children he was 'Careu Kent'. They knew he visited from America. They thought he was a writer because whenever they saw him he was carrying papers, and was always writing. They said he was loving and playful, even though Charles had no German and the children very little English. At some point he also fell for and took up with Marietta, without, apparently, any acrimony between the sisters. He and Marietta reportedly had two children together, Vargo born in 1962 and Christof in 1966. Anne once said, 'You know, my husband doesn't think a family exists unless there are twelve children.' Could she have known how close she had come to the truth?

Charles visited his other families several times a year. He would pick up a rental car, a pale-blue VW bug, and drive first to Montagnola, a small Swiss town overlooking Lake Lugano, where Valeska lived. From there he drove 130 miles to the Swiss town of Sierre, where Marietta lived. He bought her a house in the 1970s. Brigitte lived in Munich, a 375-mile drive from Sierre. He never stayed with any family for more than a week.

Charles and Anne's relationship had gone through a number of transformations over the years. The first years of their marriage had been idyllic. In the years after the murder of Charlie their relationship became increasingly fraught. When Anne delivered the manuscript of her account of the second survey for Pan American that she and Charles had undertaken in the months after the birth of their second child, the publishers had had the good idea of asking Antoine de Saint-Exupéry to write a short introduction. He sent them nine insightful pages. Anne was delighted. She felt validated as a writer in a way she had never done with Charles, despite Charles's support and encouragement. When *Listen! The Wind* was published in 1938 Anne won a second National Book Award. Antoine de Saint-Exupéry's resonatingly titled *Wind, Sand and Stars* came out in America the following year. Anne thought it said everything she had wanted to say about flying and time and relationships. The author was in America to promote the book and Anne was eager to meet him. Charles and Anne invited Saint-Exupéry to dinner and to stay the night. Charles was supposed to pick him up from his hotel in New York and drive him to where they were staying in Long Island Sound. At the last moment Charles was called away and Anne went to collect him instead. Although she described him as stooped and balding, Anne was immediately drawn to him. Within moments they found themselves talking about Rilke. Summer lightning was how Anne described their meeting. She wrote afterwards in her diary that she had talked too much, but she had felt alive and full of joy. When Charles eventually returned, he and Saint-Exupéry fell into a conversation about the place of the machine in modern life. Charles spoke no French and Saint-Exupéry very little English. Anne was their conduit. Saint-Exupéry said that he was optimistic that man would ultimately get the better of his machines and use them 'as a tool for greater spiritual ends'. He said that the modern world of machines was more foreign than America had been to the early English who had found themselves transplanted there, and it had taken them 300 years even to begin to develop a culture. He thought the signs of a spiritual revival were already evident. Charles was less optimistic.

The two pilots were similar in a number of ways: both had been close to their mothers, both were over 6 feet tall, both had flown the night mail, both were shy, both took notes and thrashed out philosophical ideas while in the air, both were interested in science. Unlike Lindbergh, Saint-Exupéry was unafraid to express emotion. It is not clear that the relationship between Anne and Saint-Exupéry was ever anything other than platonic, but Charles sensed their closeness and was jealous.

Between January 1941 and April 1942, Saint-Exupéry briefly settled in New York with his wife, doing what he could to persuade America to enter the war. Charles and Anne, meanwhile, were doing the opposite. After he left New York and returned to

France, Anne never saw Saint-Exupéry again. In 1944, when she was told that he was missing presumed dead, she said she felt as she had done when she learned of the death of her first child and then a few years later of the death of her sister Elizabeth. 'I felt incredibly alone,' she wrote in her diary, '[he was] as a sun or moon or stars which light earth, which make the whole earth and life more beautiful. Now the earth is unlit and it is no longer so beautiful. I go about stumbling and without joy.' She wondered if he had forgiven them for their stand over the war; whether he had forgiven her for *The Wave of the Future*. It is said that Saint-Exupéry's Little Prince was in part based on the Lindberghs' third son, Land, born in 1937.

By the early 1950s Anne had decided that if their marriage had not improved by the time the youngest of their six children – Reeve, born in 1945 – reached the age of ten she would leave her husband. Anne longed to have time to continue writing. 'Isn't it possible for a woman to be a woman and yet produce something tangible beside children?' she wrote to a friend. But whatever the complexities and complications of that relationship, neither of her parents ever felt truly alive, Reeve wrote years later, unless the other one was there. It wasn't possible to understand Anne, she said, without a deep understanding of Charles, and vice versa.

Anne confided to her diary that she was envious of Charles's success with *The Spirit of St Louis*. He was invading her territory, undermining her confidence. 'Too much of our life – our pain – our marriage – has gone into the maw of that book . . . And it is *His* book no matter how much of me is in it . . . He has written HIS book & I have never written *mine*. I know this. And I also know that it is chiefly my own traits of character – my cowardice – *my* inhibitions – *my* laziness – *my* lack of centeredness & sureness – *my* unhappiness & gropings – that have kept me from writing it.'

It had taken Charles 14 years to complete his masterpiece. He began to write it in Paris in 1938. He finished it in 1952 at one of their homes, Scott's Cove off Long Island Sound. The book was published in 1953 and won him a Pulitzer. When Anne read the manuscript, she said tears rushed to her eyes and throat. She tried to work out why exactly, and came to the conclusion that there was 'something in the directness – simplicity – innocence of that boy arriving after that terrific flight – completely unaware of the world interest – the wild crowds below. The rush of the crowds to the plane is symbolic of life rushing at him – a new life – new responsibilities – he was completely unaware of & unprepared for. I feel for him – mingled excitement & apprehension – a little of what one feels when a child is born & you look at his fresh untouched . . . little face.'

Anne was tired of playing the wife of the hero. She was tired of not standing up for her own life. She had been in therapy for a couple of years, much against Charles's wishes. She destroyed most of the diaries she wrote during that period. She was tired, too, of Charles's rages and his sermonizing, tired of his long absences. They were together for less than half the year. One Christmas, alone with the children, they all played a game, trying to guess where Charles might be: one child suggested China, another Japan. India and Alaska were other guesses. Anne said 'in the air', but the children all agreed that wasn't fair, so she guessed Germany.

Anne spent a lot of time alone in her room crying. She developed pneumonia. 'I must accept,' she wrote to her sister, 'the fact that my husband is as completely different from me as he can be – gets his stimulus differently, his contacts with people differently, his refreshment differently.' Eventually – facing the fact, even accepting the fact of their intrinsic difference – her resilience returned. She found confidantes, who may also have been lovers, and she started writing again. Her elder daughter, Anne (known as Ansy) later said that her sense was that Charles never knew, or just chose not to know, about the putative lovers: 'he knew that mother loved him and would never leave him. And that was all he needed to know.'

At home Charles was a martinet, controlling, even when he did not mean to be. He couldn't help it. He said he was fallible, but he didn't behave as if he was, and rarely backed down, never apologized. There were rules and chores, and endless lists. He taught his children to take calculated risks, but could not see the contradiction between the freedom implied by risk-taking and his own iron grip on the family. He talked of things being 'on your record' forever. He also had a soft side; Reeve said that she only had to say, 'Oh, father,' and he was liable to relent. Sometimes, when he was in a particularly good mood, he gave them back-rubs, bear hugs and piggy-back rides. There were no kisses. Sometimes he teased them to tears. He read to them every night. He taught his children to fish, sail and swim. Ansy said that even during play he was controlling and sucked all the joy out if it. The house shook when he was angry. His instinct was first to say no; no to almost everything. No TV, no sweets, no comics, no church. He was against chaos, obscenity, chatter, Pop Art, Mothers' Day and psychotherapy. He liked practical jokes. The children thought the pranks were hilarious. Anne did not. The children called him father, never anything more affectionate. They thought he was God. They did not find it surprising when a child at school told one of them that their father had discovered America. He had little to do with their friends' parents. The Lindberghs were their own tribe, 'close-knit, self-enclosed, and self-defining'. The children found it hard at times 'to separate individual identity from family identity'. Each child reacted differently, either withdrawing, being deferential or rebelling. When Lindbergh left home for a long trip abroad, Reeve said that there was at first a feeling of deflation that soon turned into a feeling as if of holiday. But by the time he returned it was as if the family woke up again, and now, for a little while at least, there was 'unbridled joy'. Anne's mother, Betty, wrote in her diary, 'He must control everything, every act in the household.' She said of the children that 'they are all apprehensive, never knowing when their father will fall upon them. The atmosphere – the tension in the house is so terrible – that when Charles goes off for a day or two – everyone sings!'

Charles used silence as a kind of tyranny. The children had each been made to swear never to talk of their eldest brother Charlie. Occasionally someone would turn up at the door claiming to be the Lindbergh baby grown up. They called them the Pretenders. No one at home ever referred to Lindbergh's flight across the Atlantic. No one mentioned the anti-semitic speech he had made in Des Moines. The children were taught about 'the critical importance of genetic inheritance'. When each of them came of age there would be numerous lectures from their father about natural selection. His sons he warned against becoming entrapped by a girl, his daughters against becoming ruled by their emotions. He was misogynistic in the casual way that many men of his generation were, in the casual way that he had also been anti-semitic. He believed women should not get involved in mechanical problems, though he also said that his wife was a better pilot than he was. When he returned from trips he shook his sons' hands, as a king might.

Their father was always prepared for any emergency: food and water, of course, but also, Reeve wrote, 'flashlights, extra blankets, medical equipment, rubber rafts, snakebite kits, and even a vial of morphine, which was probably illegal. We could have withstood famine, flood, plague, siege, and possibly even nuclear war.' There was a rope ladder at the upper windows in the event of fire. That they were so well prepared for any eventuality was all thanks to their father, and he rarely let them forget it. He told them always to be on the alert for danger, but the danger went unspecified. 'It's the unforeseen...' he would warn them, 'it's always the unforeseen.'

He railed against what he called punk design. Why were flashlights cylindrical and not hexagonal? If they were hexagonal they would not roll off ledges. He told them that machines were never forgiving; that though they, as their parents, would always forgive them, machines never would. He embarrassed the children by refusing to drive faster than 55 mph; he had determined that that was the speed at which a car was most efficient. He held the steering wheel with his hands permanently at the ten-to-two position. He was not interested in speed for its own sake, nor competition, never had been. He was forever pulling over to let other drivers pass. By the 1950s he preferred to travel by car or train if at all possible, not by plane. It meant that family trips often took a very long time. The only cars he praised unconditionally were the first car his parents had bought, a Model T Ford, and the Volkswagen Beetle: 'To watch him maneuver his six-foot-two-inch frame, like one of his own folding rulers, into the driver's seat of a VW Bug was to witness an engineering miracle in itself.' He liked everything to be tight-fitting, as if he were climbing into the cockpit of *The Spirit of St Louis*. He chose to sit in the smallest armchair.

He never checked in luggage, got everything he needed into his carry-on bag, even on long trips abroad. He was an ascetic: 'I had concluded that numerous possessions became formidable obstacles to my awareness and accomplishment, and that every unneeded article was best gotten out of the way – like clearing the decks of a battleship for action.' He never wasted anything: toothpaste, toilet paper, soap, time and space. He made endless lists. List-making had got him safely across the Atlantic. He had invented the safety checklist, and it had saved his life many times. Now he made lists of his children's misdemeanours. He would call them in one by one to talk about whatever was on it. Usually it was something about reading comics or chewing gum, but sometimes there was a more encompassing entry: 'Freedom and Responsibility', for example. If the entry read 'Instinct and

Intellect', they were in for a lecture 'about appreciating nature, using common sense, and not getting carried away with contemporary trends'. Nor was his wife exempt. Anne was expected to keep an account of every item of expenditure. Not because he was cheap, but because he was controlling.

With the publication in 1955 of her memoir *Gift from the Sea* Anne Morrow Lindbergh had one of her greatest successes. In hardback it was number one on the bestseller lists for a year. It sold over 2 million copies in paperback and put Pantheon on the map as a publishing house. But Charles was abroad and not there to witness her success. Anne said that her husband was restless and always sad because he had never properly mourned the murder of their firstborn. As a consequence he seemed to have condemned himself to wander the Earth endlessly like a Flying Dutchman of the air.

In 1968 Lindbergh visited Hana in Hawaii and fell in love with the place. 'What a romantic C is!', Anne wrote in a letter: 'Imagine buying a vacation home without even trying out the climate and locale for one season!' Charles told his wife that he intended to spend more time with her there. When they moved in he was almost immediately called away by Pan Am to an emergency meeting in New York. While Charles was away it rained unceasingly. Mud flowed into the house, and then an invasion of insects. Anne was furious to be left alone 'in this place in this state... I must harden my heart,' she wrote in her diary, 'not because I don't love him, but because I do.' After a year she had resigned herself to the place and Charles began to visit with increasing frequency, which brought its own problems.

Anne once wrote of her husband that his body had 'never said "no" to him'. But now his body did begin to say no. He looked thin and tired. She began to think that she had traded her 'strong companion' for yet another child. Anne felt cut off from her friends, isolated again, but now because Charles was with her nearly all the time, not because he was away so often. 'It is impossible to maintain other relationships outside the core one at this juncture,' she wrote in a letter, 'and one *needs others* – to live, to breathe, to grow, or to bring some kind of new life and air into the central relationship.'

In 1972 Lindbergh was diagnosed with cancer of the lymph nodes. Once he knew that it was cancer that was the cause of his physical decline, Lindbergh embraced and explored his final illness as if it was an adventure. Once again, sensing danger, he needed to get closer in and take a good look at it. He assessed his odds of surviving just as he had once assessed his odds of making it across the Atlantic alone. He thought there was a small chance he might pull through, but he also wanted to be prepared in case he did not. 'I'm not afraid,' he told his son Land, 'but my body is afraid.'

Lindbergh, being Lindbergh, orchestrated and managed his own death. He decided he wanted to be buried in Hawaii, even though he was currently dying in a bed in New York City. He persuaded his doctor to sign the death certificate in advance, so that he could be flown – one last flight – to die on the spot on the planet he had come to love best. He was told that there was a good chance he would die in the air. Anne said that that last flight was like his flight to Paris: 'No one believed he could do either and survive.' Charles had specified how the coffin was to be constructed: native wood cut by handsaw, a biodegradable lining. His sons dug their father's grave: 'It might seem that helping to build your father's grave even before he died would be very strange. But in actuality it felt an intimate, very loving family project.' There was to be a granite headstone inscribed with words from Psalm 139: 'If I take the wings of the morning, and dwell in the uttermost parts of the sea'. There was to be a short funeral service and a memorial service two days later. No eulogy. Instead, he chose a series of readings to illustrate his belief that 'no one culture or religion had a monopoly on truth'. Everyone from the local town was welcome but otherwise the services were to be private. Local men were to act as pall bearers dressed in their work clothes. The *Maui News* was to be informed of his death and no one else. He was to be buried within hours of his death, in order to thwart the press, who were closing in on him one last time. The instructions were written down to the last detail. He asked (instructed?) Anne to kiss him on the forehead the moment after he had died.

Anne asked Charles to describe what he was feeling, because, she said, 'You're going through an experience we all have to go through.' He had not realized it before, but 'death is so close all the time – it's right there next to you'. He felt completely relaxed. He thought it might be harder for them than it was for him. He said that he had been close to death three times that week, and he didn't feel as if he was confronting anything: 'It's not terrible. It's very easy and natural. I don't think it's the end. I think I'll go on, in a more generalized way, perhaps. And I may not be so far away, either.'

During a hospital visit made towards the end, Reeve noticed a sentence he had written on the notepad by his bed: 'I know there is an infinity outside ourselves. I wonder now if there is an infinity within, as well.' He had once wondered if the relationship between soul and body was like that between a pilot and his plane: the plane no more than an outer shell, 'a convenient tool for material accomplishments'.

Anne wrote that for those who were there 'it will always be one of the richest and deepest experiences of our lives'. She minded that she could 'not give him more sympathy, more extravagant expression of love,' but they were both barely hanging on to their self-control. Charles let his frustration out on Anne. But then, 'choking with tears', he said, 'I feel so awful to have hurt you.'

Anne was reminded of a moment from years before, when her son Scott as a little boy had said to her: 'You don't have to say "I'm sorry" to me Mother. You don't ever have to say "I'm sorry". I love you so much, you don't have to say "I'm sorry."'

Charles Lindbergh died on 26 August 1974. Anne Morrow Lindbergh wrote later that she remembered how much she had minded that her husband's body had been rushed out before she could say a prayer: 'I could only kiss his temple and go out, because he himself had wanted to be taken swiftly from his bed to his grave before the press could know of it and be there.' The funeral was held on the afternoon of 27 August. The service finished at 3pm. The first TV crew to arrive on the island was only half a mile away. But they were too late.

'There are portions of everyone's life,' Lindbergh wrote in his *Wartime Journals*, 'that could be improved if they could be lived again in the light of later experience; but Anne and I are not ashamed of the way we have lived our lives, and there is nothing in our record that we fear to have known. I wonder how many of our accusers would be willing to turn their complete files and records over for study in the future.'

'I believed,' Reeve wrote, 'he was unjust, demanding, and difficult, just as often as I believed he was the strongest, most exciting, most intelligent, and most truthful person in the world.' With his death 'he left behind a vast hole in our universe, as great as the death of a star'. But there was relief too, as there always had been when their father went away.

When Anne was given the will to read she was shocked to discover that Charles's fortune had been divided between her and two other unnamed parties.

The relationship with Brigitte Hesshaimer had lasted until Lindbergh's death, though they could not have seen much of each other during the two years of his final illness. Brigitte's children only spoke out after their mother's death in 2001, the same year

Anne died. Charles Lindbergh's other lovers and their children have remained silent. None of his three mistresses ever married. It is not known how many other, if any, illegitimate children Lindbergh fathered, but there was a rumour that he had fathered at least one child during one of the periods he lived with an indigenous tribe.

Sometime after Lindbergh had died a Swiss auto-dealer wrote to Anne to ask if the Charles Lindbergh who had regularly rented the same car from him was the famous aviator. Anne confirmed that it had been and asked that he destroy the car.

Lindbergh's immediate family were at first dubious of the claims that there was another family, perhaps more than one; they had, after all, seen off many a pretender. But DNA tests determined the Hesshaimer claim conclusively. Perhaps Lindbergh would have approved that it was in the end DNA that found him out. In 2004, Reeve flew to Germany to meet Brigitte's children, her half-sister and brothers, to welcome them to the family.

Many years after Charles Lindbergh had died, Gore Vidal wrote that Lindbergh had been 'the best that we are apt to produce in the hero line, American style'.

CHAPTER THREE

Von Braun had always believed that America must maintain its scientific lead in order to keep the world safe, but for decades he had also wrestled with the ethical problem: 'Is everything that is scientifically possible, also permissible?' On 2 September 1946, a couple of days after the *New Yorker* had devoted its entire issue to John Hershey's essay on Hiroshima, von Braun wrote: 'Today all of civilization is already in play . . . Will man's intelligence keep up with his technology? If not, it will be the end of the human race and then maybe people such as we technicians will be at fault.' By asking if the human species could keep up with its technology, he had identified one of the key questions of our age. What was needed, he believed, was wisdom, and statesmanship: 'ironclad international treaties' to protect mankind from 'its newly-born scientific capabilities'.

He had also come to see religion as a way of addressing the problems of technology: 'It is as difficult for me to understand a scientist who does not acknowledge the presence of a superior rationality behind the existence of the universe, as it is to comprehend a theologian who would deny the advances of science.' Humans need something outside themselves otherwise they have no perspective, he said. Science and religion, he believed, must walk hand in hand.

Von Braun had been inspired by Teilhard de Chardin's idea of planetization: a high-flown theological philosophy that attempted to unite the world religions and God's creation with evolution. We need to think of ourselves as not French or American or Chinese, Chardin wrote, but as terrestrials. Chardin's notion of a lovable and loving universe appealed to von Braun, as did Chardin's emphasis on 'the overriding importance of love and charity' in Christ's teachings. Von Braun also saw Christ as a revolutionary: no more so than in his injunction not only to love your neighbours but to love your enemies. Von Braun wrote that it had been a hard blow to Catholic scientists when Chardin's works were condemned by the Roman Church. In recent times the church has changed its position, just as Chardin's influence among scientists has dwindled to nothing. Popes John Paul II, Benedict XVI and Francis have all written positively about Chardin's contribution to Catholic theology.

Von Braun kept his religious beliefs largely to himself, yet, as his biographer Michael Neufeld says, even the small number of religious articles he wrote during his years in America – including one titled 'Why I believe in immortality' – were sufficient to make him 'a force for the promotion of deistic belief in the United States'. The last major paper von Braun wrote was at the invitation of the Lutheran Church of America, an 82-page testament titled 'Responsible Scientific Investigation and Application'. The paper was to have been presented at a conference to be held in late 1976, but by then von Braun was too ill to read the paper himself. He had probably had kidney cancer since the early 1970s, but he had ignored the symptoms – a terrible irony after a lifetime of hypochondria.

Shortly after the Apollo programme was cancelled in 1973, NASA's administrator, Thomas Paine, persuaded Wernher von Braun to move to Washington to take up a position as Deputy

Associate Administrator for Planning, his main role being to fight for the Mars project. During one speech, he drew a gasp from his audience when he told them that two spaceships would depart for Mars on 12 November 1981.

President Nixon talked up the Mars project in public but behind the scenes he refused to fund it. Paine saw the writing on the wall and left NASA for a job at General Electric Company, leaving von Braun isolated just a few months after he had taken up the post. Paine's deputy George Low got the top slot. It soon became apparent that von Braun did not have Low's support. Once more, von Braun was cut out of the picture. He had difficulty even getting an appointment to see Low, and was no longer invited to key planning meetings.

Low told journalists, 'The first time I met von Braun I was prepared not to like him, but we became the best of friends.' Jay Foster, however, who was on von Braun's planning team, said, 'I don't think Low ever forgave Wernher for the Nazi connection.' Which was hardly surprising given that Low had been born into a prosperous Jewish family in Vienna and that the family had been forced to flee their homeland when the Nazis came to power. Low may also have been jealous of von Braun's fame: whenever he testified at a congressional hearing only a few people would be present 'and some of those would fall asleep'; when von Braun spoke, the hearing room was packed. And yet, it seems that Low and von Braun were friends, and remained friends. It was simply that von Braun's position had become untenable. He quickly came to realize that now was not the time to push for Mars. He knew that there was no real public appetite for a manned Mars expedition. He said that he had 'always felt that it would be a good idea to read the signs of the times and respond to what the country really wants, rather than trying to cram a bill of goodies down somebody's throat'. Environmentalism was in the ascendant. The world had moved away from the grand gesture of space to more local concerns. E. F. Schumacher's best-selling *Small is Beautiful* was published the year Apollo was cancelled. In Washington von Braun was 'just another guy with a funny accent'.

Von Braun left NASA for a job at the engineering firm Uhl Fairchild, where he worked on the ATS-6 satellite. The company's stock value went up on the news of his appointment. For the first time in his life he took an industry-scale salary. At NASA he had never earned more than $35,000 a year. He loved the ATS project, and foresaw the satellite's power to reach the poorest regions of the world. He was dismayed by the lack of interest shown in the ATS-6 project by the American government. He thought there were disquieting signs that the country was turning in on itself, and hoped that such inwardness would be temporary.

Von Braun continued to believe that many of today's greatest problems might be solvable if the world could be managed as an entity. He isolated a number of challenges to civilization: how to provide energy, food, clothes and houses for all, how to protect the environment, the need for personal freedom (which includes the 'freedom to be wrong ... one of the most precious freedoms of all'), and the threat of nuclear holocaust. He believed that satellites and space research could help bring about that transformation. By continuously monitoring the land and ocean surface and its atmosphere, satellites could help us acquire a new perspective on ourselves. Satellites in Earth orbit could show us the harm humans were causing the planet, and help us to work out how to repair that damage. During his Apollo flights, Jim Lovell had been struck by how, from space, there was no evidence of human activity on Earth of any kind. The writer Kurt Vonnegut made a similar observation: 'Earth is such a pretty blue planet in the pictures NASA sent me. It looks so *clean*. You can't see all the hungry, angry earthlings down here – and the smoke and the sewage and trash and sophisticated weaponry.' By the 1980s Richard Underwood thought that it was now possible to see from space the damage humans had wreaked on the planet. Compared to the 'brilliant, clear photographs' of the Earth that had returned from the Gemini missions, the colours of the Earth looked drabber. The reason was air pollution.

Von Braun was already writing about the future possibility of climate change in 1972. 'We have no idea,' he wrote, 'how stable [the Earth's] present climate is.' He wondered how much more the Earth could take 'in the way of atmospheric and water pollution to upset the balance and perhaps alter the climate dramatically'. Clearly we had to do whatever we could to protect the Earth from further deterioration, but more than that he realized, as Lindbergh had, that the human species itself had to change fundamentally. We needed, he believed, to acquire *conscious* awareness.

Von Braun died on 16 June 1977. The paper he had written the year before at the invitation of the Lutheran Church of America covered subjects as wide-ranging as science and technology, environmental pollution, and the relationship between religion and science: 'In this reaching of the new millennium through faith in the words of Jesus Christ, science can be a valuable tool rather than an impediment. The universe revealed through scientific inquiry is the living witness that God has indeed been at work.' He wrote that it was 'of no purpose to discuss the issue of knowledge. Man wants to know and when he ceases to do so, he is no longer man.' From his hospital bed in the last days of his life, he repeatedly asked friends if they thought he had done the right thing in persuading the country to spend so much on space. He died, as perhaps we all should, full of doubt.

Curiously paralleling Lindbergh's, the news of von Braun's death was delayed so that he could be buried without the presence of news reporters, photographers and television crews, fans or protestors. The man who had shaken the hands of Göring,

Goebbels, Hitler, Eisenhower, Kennedy and Nixon was dead. His burial plot is marked by a simple headstone with the inscription: 'Wernher von Braun, 1912–1977, Psalms: 19:1'. The relevant verse reads: 'The heavens declare the glory of God; and the firmament sheweth his handiwork.' A week later, a memorial service was held in Huntsville. The Reverend George B. Wood officiated. He had been a US Army airborne combat chaplain during the Normandy landings. He described von Braun as a lover of freedom: 'If you read between the lines of his life's work and words, you sense a smoldering resentment and anger against oppression and tyranny.' Mike Collins gave one of the three eulogies:

Ten Saturn I, nine Saturn IB, and thirteen Saturn V rockets were launched, a total of thirty-two flights, all successful, all without loss of life, all on peaceful flights flown without weapons. Saturns sent nine astronauts up to Skylab, which was itself a converted Saturn V upper stage, and kept three of them in space for eighty-four days. And finally, a last Saturn sent an American crew up to join a Russian spacecraft in Earth orbit … A study in contrasts … a visionary and a pragmatist, a technologist and a humanist … He was a master of the intricacies of his machines, with their innumerable pipes, valves, pumps, tanks, and other vital innards, yet he realized that his rockets could only be as successful as the people who made them. And he assembled an extraordinarily talented team, people who worked well with each other, and who were totally devoted to Wernher.

In short, he was a leader, with the versatility that leaders of genius must possess. Because he worked with rockets, I would call him a rocket man, but that is a cold term, and he was anything but cold. He was a warm and friendly man, interested in everyone around him, no matter who they were. He had a marvellous knack for explaining his machines in simple, understandable, human language. And he never seemed too busy to share his ideas – and he was full of them – with others.

Wernher von Braun believed that the desire to explore is a fundamental part of mankind's nature.

Von Braun had once predicted that a Saturn V would take humans to Mars. Instead, not long after the moon race was over, the rocket that Ed Mitchell described as being 'like a living extension of everyone involved in the project', came to the end of its useful life. Even the original engineering drawings have been lost.

When Andrew Smith interviewed all the Apollo astronauts who had walked on the moon for his book *Moon Dust*, he was surprised 'by the grim determination' with which Apollo 14 astronaut Ed Mitchell defended 'the ex-Nazi'. 'I have great respect,' Mitchell said, 'and great caring for the man.' Al Bean, too, praised von Braun, particularly 'his ability to talk to anyone on their own level, in a way that they could connect with, without being condescending or patronizing'. Bean got the feeling in meetings that von Braun already knew what the best course was but encouraged debate because he wanted everyone to come to the same understanding through discussion. But we're left, Bean also acknowledged, 'with the problem that he did a lot of bad things'. He must have known what was going on in the camps, Bean said. 'Of course he did. He was a genius. We knew, so I'm sure he did ... But I imagine Braun thinking, "They're unfortunate humans just like me." I don't know that. But if it is the case, what does he do? He's not going to stop it. What does it all mean? Thank God we're not in that position, you and me, because we could have been born in it, and we might not know what to do any more than von Braun did.'

By his own account, von Braun admitted that even during the war he was ashamed of what was happening in Germany, but he claimed never to have seen a dead prisoner, a prisoner hanged or otherwise killed, nor had he ever participated in any violence or mistreatment of prisoners. It was the usual perfunctory denial of every functionary of the Nazi state. He said it was only natural, as a German citizen, that he had wanted his country to win the war. He added that he had wanted to go on living, and to carry on doing his work. 'The man living under dictatorship,' he wrote, 'adjusts himself to business-as-usual, whether he likes it or not, because he must, in order to survive.' He said that whenever he was summoned to Berlin he didn't know if he was going to be given a medal or be shot. In retrospect it seems highly unlikely that he would have been shot. It is hard to think of any figure of comparable importance to the Nazis who was executed; a different matter in the USSR, however.

In a 1971 interview von Braun was asked to defend his role as designer of the V-2. 'This is a grim problem every engineer or scientist working for the military has learned to live with,' he said. 'People have come to different conclusions as to what their ties are. My own belief is that when your country is at war and calls on you or drafts you, whether as an infantryman or as an engineer, you have to do your duty.' But what if the leaders of your country are wrong? 'Many people ask themselves that question today in this country. I think in the last analysis everybody has to live with his own conscience. All I can say is that if I were a young man today in America, disagreeing as I do with the continuation of the war in Vietnam, I would still serve in the Army. I would see this as my duty and I would not consider it morally right to defect. If other people come to other conclusions, I will still respect them. But they, too, will have to live with their consciences.' It was a defence that showed the strength of his obeisance to the strong state.

Could von Braun really not have known what was going on? He admitted himself that he had visited Mittelwerk 15 to 20 times. He said that he knew what was happening in the concentration camps. He suspected it, and in his position he could have found out, but he didn't and he said he despised himself for it. In 1984 his records were declassified. They show that on at least one occasion he had visited Buchenwald looking for specialist labour.

Several prominent public figures came to the conclusion that von Braun should have done more, Carl Sagan among them. Sagan found von Braun's smooth compliance with the militarism and racist ideology of Nazi Germany 'deeply disturbing ... It is the responsibility of the scientist or engineer to hold back and even, if necessary, to refuse to participate in technological development – no matter how "sweet" – when the auspices or objectives are sufficiently sinister.' Nor was Sagan impressed by von Braun's lifelong support of his Peenemünde team. When he was challenged about his loyalty to the German core of his team, von Braun turned on the charm: 'We're just old-timers who have been working on these things so long, we've had ... years to make mistakes and learn from them.' Even at the end of the Apollo programme, half of the original Peenemünde team was still in place. Michael Neufeld writes that you can search in vain for any admission of 'guilt over the Nazi period or the fate of concentration camp prisoners ... Like many Germans, he seems to have accepted the maxim of not looking back.' Was it his aristocratic background? Never apologize, never explain. Or did he realize that no apology would or could be sufficient, would just draw attention to its inadequacy? Or perhaps mundanely he kept quiet because during his lifetime his connection to the Dora camps was not public knowledge. Maybe after a while silence had become a habit. Neufeld calls von Braun 'a lesser war criminal'.

In an interview on NBC's *Today* show after von Braun's death, Tom Wolfe said that America had 'never had a philosophy of space exploration', that von Braun was 'the only philosopher NASA ever really had', but because von Braun had been 'a member of the German Wehrmacht, and had a Teutonic accent – had everything but a dueling scar on his cheek' – it hadn't been possible for NASA to push him forward as NASA's philosopher. 'But he *was* their philosopher.' And yet his subtle philosopher's nature hadn't always translated well into a public arena. His brilliance and apparent lack of doubt, together with his 'funny accent' and exuberance, made him look like a caricature. The eponymous mad scientist in Kubrick's *Dr Strangelove* (1964) – a German with a line to the President – was based in part on von Braun (others have added Edward Teller and even Henry Kissinger as models for the sinister doctor). Von Braun claimed never to have seen the film. In Jean-Luc Godard's dystopian science-fiction film *Alphaville* (1965), the creator of Alphaville's dictatorial computer is called Professor von Braun.

When the journalist Orianna Fallaci interviewed von Braun she was predisposed to dislike him. Her family had been involved in resisting German occupation in Italy. She wrote that when she first met him she noticed that von Braun carried about him a 'slight smell of lemon'. It reminded her of the smell of a particular soap that German soldiers had used. It was, to her, 'a visceral reminder of evil and terror'. She wrote that it was 'almost like a gas that penetrates through your nostrils right into your heart and brain'. And yet within half an hour she realized that in fact she liked him.

In Alabama the German engineers had once again found themselves in a brutal and segregated society. In Huntsville in the 1950s there was, von Braun said, a widespread belief among the mainly white population that black children were 'either not ready for first-class education, or that a general raise in their educational standards would only lead to trouble with job placements'. He said that over the years he had often been asked what he had done when Jews were disappearing in Nazi Germany. He wasn't going to 'sit quiet on a major issue like integration' and do nothing a second time. At one point, NASA had pushed to move von Braun's operation out of Alabama because of segregation but von Braun insisted that they stay and fight. From the early 1960s he encouraged the employment of black engineers at the Marshall Center. He lobbied local business leaders and attended black community events in an attempt to further integration. Von Braun is said to have helped ensure that the education system in Huntsville became 'highly integrated'. He and fellow Peenemünders ignored segregation laws; played tennis with black Americans, visited all-black jazz clubs: 'If the cops questioned them about it, they would just pretend they didn't understand, or speak German ... It was just kind of known that you didn't mess with them.' Von Braun became a significant voice in Alabama arguing for racial integration and tolerance. On at least one occasion he met with Martin Luther King privately.

Soon after they had arrived in the town, the von Brauns helped to found a Lutheran church, the Huntsville Symphony Orchestra and Huntsville Community Chorus, as well as the Broadway Theatre League, the German Club for International Travel and a Civilian Advisory Committee. At the first meeting of the Advisory Committee, von Braun told those assembled that 'good neighbors live and work together in harmony, and with understanding, each in his or her own way, working not only for the good of the family, but for the good of the community'. In 1961, at von Braun's recommendation, the University of Alabama in Huntsville was founded.

Huntsville became known locally as 'Rocket City' and thrived after the 'krauts' arrived. The population of some 17,000 grew over the next two decades to around 200,000. 'That damned Nazi', as the director of the Manned Space Station had once called him, became a beloved figure not just at the Marshall Space Flight Center but in Huntsville as a whole.

CHAPTER FOUR

In her diary for 1972 Madalyn O'Hair wrote of her intention to run for governor of Texas, and then President of the United States. Her chances were slim. During the 1950s the population of America grew by 19 per cent whereas church attendance grew by 30 per cent. In the 1960s most opinion polls indicated that 'all but a small percentage of Americans classified themselves as religious'. A survey at the end of 1972 showed that the most popular personality in America after President Nixon was Billy Graham. During that same year, Loretta Lee Frye's petition in support of religious readings in space reached 10 million signatures. Even in 1973, NASA administrator James C. Fields said that the Manned Space Center was 'sorely taxed' because of the volume of letters still flooding into NASA in support of religious observance in space. It has been estimated that by 1975 NASA received more than 8 million letters and petition signatures in support of the readings. In *To Touch the Face of God*, an account of religion and the space race, Kendrick Oliver writes that 'the magnitude, the duration, and even the existence of the correspondence campaign in defense of free religious expression in space has generally evaded the attention of historians'. The letters were warehoused, and then destroyed.

The day before the first moon landing, NASA scientist and evangelist Rodney W. Johnson told *Christians Today* that 'if the

space program can be faulted for anything, it is that it has ignored man's spiritual yearnings'. Rusty Schweickart said of the
Genesis reading that its intention had been 'to sacramentalize
the experience and to transmit what they were experiencing to
everyone back on Earth', as had Buzz Aldrin's attempt to celebrate
Communion publicly. But by choosing a Biblical text and a Christian rite, the attempts had been bound to fail. Kendrick Oliver said
that 'without realizing it, O'Hair had executed her most perfect
provocation'; and yet, by being against any kind of spiritual observance, O'Hair had also ensured that the opportunity that Apollo
had offered to find a new paradigm that included both science and
religion also failed.

By opposing religion and mocking it, O'Hair ended up creating a poisoned version of what she was so against; her version
of atheism was just another kind of fundamentalism. Two years
earlier she had founded her own atheist church: Poor Richard's
Universal Life Church of Austin, Texas. She went to court to
establish the church's tax status. In order to comply with the legal
requirements, she named herself bishop, and her husband, Richard, pastor and prophet. She said she was the Virgin Mary in her
fourteenth reincarnation. At a court hearing she wore black and a
clerical collar. She won her action. Like all churches in America,
Poor Richard's was now exempt from tax, and its members could
deduct tax from their donations. She had made her point, but
by always being in opposition to religion she had nothing to say
that was positive and outward-looking. When the Supreme Court

Madalyn O'Hair

ruled that businesses should accommodate those who wished to observe a day other than Sunday as the Sabbath, O'Hair encouraged atheists to take Thursday off: 'the day I led the children of atheism out of the wilderness of religion'. It was an amusing stunt that the newspapers covered but it did little to advance her vision of a secular America.

The legal actions continued, some of them important, some trivial. If a judgment went against her, she just issued a few more writs. It was as if they streamed from her head like snakes. She sued the President in an attempt to end religious services at the White House: *Murray v. Nixon,* 1970. Her action against NASA had been filed in 1971: *O'Hair v. Paine.* She filed an action against Austin city council for beginning their council meetings with an opening prayer: *O'Hair v. Cooke,* 1977. She challenged the inclusion of the phrase 'In God We Trust' on US currency: *O'Hair v. Blumenthal,* 1978. After seven years, often putting in 16-hour days, she won an action against the state of Texas overturning the requirement that persons holding offices of public trust swear to a belief in God: *O'Hair v. Hill,* 1978. The ruling was a disappointment to her; she had wanted the action to go all the way to the Supreme Court. She had succeeded in making a belief in God no longer a requirement of state office in Texas, but if it had gone to the Supreme Court the requirement might have become a national one. She was unsuccessful in her attempt to prevent John Paul II from celebrating Mass on the Washington Mall: *O'Hair v. Andrus,* 1979. She was again unsuccessful when she tried to have the nativity scene removed from the rotunda of the capitol building in Austin: *O'Hair v. Clements,* 1980.

Just as the flow of letters to NASA in support of religious observance in space was drying up, O'Hair provoked another protest; this time unwittingly. A rumour began to circulate among the Christian Right that the Federal Communications Commission (the FCC, a body that oversees radio and TV coverage in all 50 states) was about to ban religious broadcasting across the board. *Christian Crusade Weekly* ran the headline: 'Yes, she's at it again.' The FCC received 700,000 letters of protest in 1975. Though it was not in fact true – the FCC had no such plans – the rumour persisted. By early 1976 the FCC had received 3 million letters, and 25 million by the mid-1980s. Without uttering a word O'Hair had mobilized the new Christian Right. What O'Hair represented had been enough in itself.

By the end of the 1970s O'Hair had her own TV show, *American Atheist Forum*. It was carried on more than 140 cable TV channels. She was as famous as she had ever been.

Madalyn's relationship with her son Bill had been difficult for years. In 1980, on Mothers' Day, Bill converted to Christianity. When she heard the news, Madalyn disowned him. She said, as if casting a curse on him: 'One could call this a postnatal abortion on the part of a mother, I guess; I repudiate him entirely and completely for now and all times … he is beyond human forgiveness.' Bill's daughter, Robin, also disowned him, and continued to live, as she had done since she was a child, with Madalyn and Madalyn's other son, Garth. Richard had died of cancer in 1978. Her brother Irv, who never entirely escaped her gravitational pull, had died a virgin at the age of 71.

During the early 1980s Madalyn formed a bizarre partnership, bizarre even for her, with the self-proclaimed Reverend Bob Harrington, Chaplain of Bourbon Street, an ex-alcoholic, now evangelist pastor. Together they toured Texas in a bus provided by the pornographer Larry Flynt. Harrington and O'Hair were friends off stage, pretend enemies on stage. Harrington – handsome and charming – would warm up the audience, and then Madalyn, his stooge, appeared. They would engage in verbal battles. He would call her a bad smell that needed to be eradicated. The crowd would shout out 'Satan's Whore!', or sometimes, worse, 'Communist!' The show regularly attracted crowds of thousands.

Back-stage she was given 45 per cent of the collection: bags stuffed with dollar bills, typically taking home several thousand dollars a night. On occasion she was dragged off stage by armed police officers hired for the night. One time, the crowd got so riled up she had to flee, fearing for her life. The events got O'Hair and Harrington national attention and a spot together on the *Phil Donahue Show*. Donahue said it was one of his all-time favourites. They were in *Newsweek*. O'Hair was profiled on *60 Minutes*. They appeared together on *Good Morning America*.

O'Hair got to know the *Hustler* publisher Larry Flynt personally. By that time he was paraplegic and confined to a wheelchair, having been shot by a sniper outside a courthouse in Georgia while on trial for publishing pornography. For a time he had converted to Christianity, encouraged by the evangelist Ruth Carter Stapleton, Jimmy Carter's sister. Under O'Hair's influence he was persuaded to embrace atheism. She told him that he had turned to Christianity because of an iodine deficiency. Flynt described O'Hair as the most brilliant person he had ever met. When he decided to stand against Ronald Reagan in the 1984 election, he employed her as one of his speechwriters. She wrote polemics in support of his anti-war stance, his campaign against cigarettes and his belief in free speech.

When Flynt turned up in court during the trial of the sports-car manufacturer John DeLorean wearing diapers and swathed in the Stars and Stripes, he was given a six-month prison sentence for desecration of the flag. During his time in jail he gave O'Hair power of attorney over his $300 million publishing empire. Her attempt to transfer the corporation's assets to American Atheists was blocked by Flynt's brother. Nevertheless, by the end of the 1980s, O'Hair was wealthy. She bought a Mercedes for Garth and one for herself, for Robin a Porsche and a diamond necklace. They all went on holiday together to China for a month. In her diary she wrote, 'Jesus Christ it is wonderful to be rich.' But the chaos that had served her well was beginning to rebound on its author. Her life was out of control. She had developed diabetes and it was getting worse. Her mood swings were greater than they'd ever been. She and Garth fought all the time. He would call her a fat stupid bitch, she would call him a retard, and then pick up the phone to some potential donor and put on her sweetest voice. She described her supporters as 'longhair freaks, filthy bodies, alcoholics, Jesus freaks'. She was lonely and depressed. At one point she had launched a dating service called Lonely Atheists. She confided to her diary: 'I hope I have lived my life in such a manner that when I die, someone cares ... I want some human being, somewhere, to weep for me.' In 1991, at an American Atheists conference in Arizona, she turned her attention to the environment, proving that she was still in tune with the zeitgeist. But by now her speeches were becoming wilder. The passion remained but she was often incoherent and inconsistent. 'We simply have to change the system,' she told a journalist. 'It's crying out to be changed ... I love my country. I love the world. I love the people ... We are taking our whole ecosystem down with us ... the human community needs to be wiped out. What a pity we can't have a nuclear war.' In the same year, a set of commemorative stamps had been issued to celebrate the achievements of American Atheists, a first for an atheist organization. It was too little, too late.

On 31 January 1993 O'Hair took on a new employee, a typesetter named David Waters. She knew he had a criminal record but not how extensive it was. By then the O'Hairs – Madalyn, Robin and Garth – planned to flee to New Zealand, along with American Atheists' millions. Garth had already begun transferring funds into a New Zealand account. Like O'Hair, Waters was self-taught. He was a keen reader of history and biography. He listened to NPR, was eager to learn about anything. He was also good-looking; just the sort of man O'Hair always complained was lacking in her life. For almost a year all went well, but by Christmas Waters had come to loathe his employers. While the O'Hairs were on holiday, Waters fired all the staff and absconded with $54,000 worth of cheques. At the subsequent trial, details of his past emerged.

As a young boy he had been sent to reform school. By the time they were teenagers, he and his brother had become hustlers. Aged 17 Waters murdered a boy of 16, driving over his legs then beating him to death. He had dug out the boy's eyeballs with a can opener while he was still alive. He narrowly avoided being sent to the electric chair, and instead served 12 years. A year after his release he had been arrested again, for beating up his mother and urinating on her face. He got another year in jail. In 1982, the bullet-riddled body of a 36-year-old ex-con named Billy King was discovered. The last person to have seen him alive was David Waters, but the case against him was never proved. A year or so later, Waters was back in jail for beating up two friends, in an apparent road rage attack. Waters got bored easily and violence was his outlet. His girlfriend Carolyn said that he hated women and that all men feared him. O'Hair wrote up what had come out about Waters in an issue of the American Atheists newsletter. It was to be a fatal mistake.

Garth, Madalyn and Robin

On Monday 28 August 1995 the staff of American Atheists arrived to find a note pinned to the door. It said that the O'Hairs had been called out of town on an emergency. Inside were half-eaten dinners abandoned on the table. They seemed to have left in a great hurry. Over the next month the office was able to reach Garth on his cell phone. He sounded cool, but when Robin came on the line she sounded anxious. And then after a month there was silence. The missing funds suggested that the O'Hairs had fled the country, as indeed they had planned to do. The press reported their disappearance. The Austin Police Department did almost nothing about it. A spokesman was reported as saying, 'It's not against the law in Texas to be missing.'

A year later, the case was taken up by the *San Antonio Express-News*. The editor suggested to one of his longtime reporters, John MacCormack, that the story might be worth further investigation. MacCormack soon discovered that $612,000 was missing from American Atheists' bank account. No one at American Atheists had reported the loss. One of the staff told MacCormack that during the month of phone contact with the O'Hairs Garth had asked for $600,000. She said that she had seen no reason why Garth shouldn't have the money. MacCormack also discovered that Garth had subsequently commissioned a jeweller, for a fee of $25,000, to use the money to buy gold coins: a mixture of Canadian Maple Leafs, Krugerrands and American Eagles, each worth around $300. CNN, *Time* and NBC began to cover the story. The most likely explanation seemed to be that the O'Hairs had fled to New Zealand. The jeweller recognized Garth from photographs. He had turned up to collect the coins in person and had not appeared to be particularly anxious. By now the CIA and IRS were involved. Attention soon turned to Waters, who was arrested. But it was hard to bring a case against him. Where were the O'Hairs? And if they had been kidnapped, why had Garth apparently been so calm? One of Waters' defence witnesses, a preacher, said he had seen Madalyn eating pasta in Romania in 1997. In a later plea bargain, one of Waters' accomplices, Paul Karr, revealed what had actually happened. Waters and Karr, along with another accomplice, Danny Fry, had indeed kidnapped the O'Hairs. And then, in a curious case of Stockholm Syndrome, Garth had bonded with Waters. They even went shopping together to buy clothes for Madalyn and Robin. Sometimes they went out drinking or to play video games. One night, returning to the apartment where the O'Hairs were being kept captive, Waters knew immediately that something was wrong. It turned out that Karr had lured Robin away from Madalyn and had raped her. She had suffocated while he was holding a pillow over her face to prevent her crying out. Madalyn and Garth were unaware that Robin had been killed. Now Waters knew that he had to get Garth to collect the gold immediately even though not all the coins had been delivered. The following day, once the gold was in the kidnappers' possession, Waters and Karr killed Garth and Madalyn, most likely by strangling them. Fry panicked. Waters and Karr drove him to a riverbank near Dallas where he, too, was murdered. Karr cut off his head and hands and left the rest of the body. The O'Hairs' bodies were cut up and put – along with Fry's head and hands – into 50-gallon drums.

Two important question remained. Where was the money? And where were the bodies? In 1999, six years after the O'Hairs had first disappeared, a tip-off led MacCormack to one Joey Cortez and two accomplices. In exchange for immunity they admitted that they had stolen the coins. They claimed that they had found the gold in a storage unit. They had managed to get hold of the storage company's master key, which opened lockers at a number of different locations. Inside the first locker they opened they found a TV, but when they couldn't pawn it, they went to a different location and tried again. On just their second attempt they had come across the hoard of gold coins.

At the trial they said they had nothing left to show for it except for a three-year-old child, one of them said, which drew laughter from the courtroom. The money had gone on girls, big-screen TVs, black-leather couches, trips to Vegas, cars, guns, and on giving parties about which they had no memories. No connection was ever found between the gold thieves and the O'Hairs, nor between them and the O'Hairs' murderers. The FBI investigation concluded that the coins had been found by accident as Cortez and his accomplices claimed. But there were some puzzling features. They had driven 100 miles to get to the second storage unit even though there were plenty of others nearby. And why had they only opened one unit in each location? Off the record, one of the accomplices told MacCormack that there were 'just some parts of this we cannot reveal'.

The police retrieved a single gold coin, a Canadian Maple Leaf that had been made into a pendant for the aunt of one of the robbers.

Waters agreed to say where the bodies were buried in exchange for incarceration in a state jail rather than a federal one. What was left of Madalyn O'Hair was identified by the serial number on her hip-replacement joint.

Bill had to fight American Atheists to get possession of the bodies. A court ruling granted him custody of the bodies, but the ruling was being contested by the secretary of American Atheists, Ellen Johnson. She was flying in with a lawyer to try to put a stop to Bill's planned Christian burial. Bill got the bodies of his mother, daughter and brother into the ground before she touched down. They were buried in Austin, Texas on 23 March 2001.

In 2002, George W. Bush tried to bring back prayer in public schools. He enlisted Bill's support. His attempt failed.

Madalyn Murray O'Hair could be affectionate and funny, but she could also be withering, sarcastic and obscene. Like many charismatics, when she turned her gaze on you it was all-consuming. Phil Donahue, despite being a Catholic himself, greatly admired her. 'She was difficult to love,' he said. 'She was loud. She interrupted you. She would show you no deference. She was a zealot, but she was right. If you listened to her she'd make you stronger ... Madalyn was fighting against public servants who believed they had a pipeline to God, and once you've got somebody who talks to God and God talks back, you're in trouble. Madalyn was victimized by the piety of America. She wouldn't have received this abuse in Europe. They understand hypocrisy. In America, the piety rises to a messianic level. Those who would pray for Madalyn are condescending. They're saying, "I'm better than you. I'll take Jesus with me and I will win." There is an anger in them that is the beginning of war.'

Madalyn had been a powerful voice speaking out against the destructive forces of religious fundamentalism in America, but by defining herself entirely in opposition to what she had sought to destroy, in the end she represented only nihilism.

It was no surprise that NASA had been institutionally incapable of rising to the opportunity of acknowledging the metaphysics of space travel, but Madalyn almost single-handedly ensured that the conversation never took place.

CHAPTER FIVE

As early as in the mid-1960s, von Braun argued that if the trend of cutting back on the space programme were not reversed, we might as well put up a sign on the moon saying Kilroy was here. 'To make a one-night stand on the Moon and go there no more would be as senseless as building a railroad and then only making one trip from New York to Los Angeles.' Even America's space shuttle programme hung by a thread. When Caspar Weinberger argued in a memo written in August 1971 that not to support the shuttle programme would give the impression that 'our best years are behind us, and that we are turning inward', Nixon wrote, 'I agree with Caspar' on the memo, and that was that, the space shuttle was saved. But whether the project had been worth saving is a moot point. There was nowhere to shuttle to except back and forth between the Earth and the spaceship itself. The shuttle was certainly not the weigh-station to Mars that von Braun had once envisaged.

With the end of the Apollo programme came the end, too, of Apollo's distinctive management style. 'The bureaucrats moved in for good,' Apollo 15 astronaut Al Worden wrote. 'Many of them felt for years that we astronauts had far too much power, prestige, and responsibility. Things were never as informal as they had been before. NASA changed and lost some of its original pioneering and engineering spirit. All of the rules were now laid out in black and white, and every decision passed through multiple layers of middle and upper managers.'

Von Braun often made the point that Apollo wasn't about the moon landing, it was about everything that had made the landing possible. 'In reaching the moon,' he said, 'we will have proved the feasibility of space flight. All the hardware; the technical and scientific manpower; the launching, tracking and communication ... the test, management, and logistics support will remain with us as much more of a permanent asset than a handful of lunar dust.' He said the space programme had become 'America's greatest generator of new ideas in science and technology'. A random list might have on it scratch-resistant lenses, cordless power tools, memory-foam mattresses, household water-filters, ear thermometers, shoe insoles, safety grooving on pavements, advances in cryogenics, a better understanding of how radiation affects cell division, and of course Teflon: all things that have made life that little bit better. And yet such list-making remains trivial compared to the simple fact of the space programme as a human achievement. 'Man does not fit a common mold, either physically or spiritually,' von Braun wrote in criticism of Soviet collectivism. 'His soul is unique ... A man would be a robot without a heart that can feel joy, love, grief, compassion, or devotion.' He said that no matter how impressive our machines and instruments, 'by far the most decisive factor in the exploration of space is the human element'. We have not returned to the moon for almost 50 years. Perhaps we have concluded that space is for artificial intelligences – machines – not feeling human beings. But the eye of a machine is not the eye of a human. The returning Apollo astronauts struggled to tell us in words what they had experienced, but the struggle is precisely what makes the attempt human. Machines do not hold conversations. Machines do not write poetry.

'I do hope,' von Braun wrote, 'history will record that we were *aware* ... of the enormous implications of the lunar journey, that . . . we are reaching out in the name of peace, and that, while we take pride in this American achievement, we share it in genuine brotherhood with all nations and with all people.' In our cynical age it is hard to embrace such idealism, and hard, too, to accept the achievement of the space race from a purely political perspective. Konrad Lorenz said that it had absorbed mankind's aggressive and competitive instincts, that the space race stayed Presidential hands. Carl Sagan argued that by exploring space, 'the creative, the aggressive, the exploitive urges of human beings can be channelled into long-term possibilities and benefits'. They are surely romantic assessments given that the space race coincided with terrible proxy wars in Vietnam, Laos, Cambodia and a number of African countries. It might just as well be argued that the space race accelerated the arms race.

'We must open our vision to the unknown,' von Braun said in a speech he gave during the early years of the moon race. 'We must expect the unpredictable; we must value knowledge for its own worth; and we must cease to measure the new in terms of usefulness alone.' Arguably, we have learned none of these lessons.

In 1977, within two weeks, NASA launched two space probes – Voyager 2 and then Voyager 1 (presumably in that order because Voyager 1 would eventually overtake Voyager 2). Almost two years later, Voyager 2 flew past Jupiter. Back on Earth we saw detailed images of the Great Red Spot, a swirling storm the size of the Earth in Jupiter's atmosphere. On Io – one of the largest of Jupiter's many moons – the first active, non-terrestrial volcano was discovered.

The launch of the Voyagers had been timed to take advantage of a future favourable alignment of the outer planets. Jupiter's powerful gravitational field was used to sling Voyager 2 further out and headed on a course towards Saturn. A little more than two years later the probe sent back close-ups of the planets' rings. By then, NASA had decided to include Uranus and then Neptune in the Grand Tour. Voyager 2 reached Uranus in January 1986. Three years or so later it flew past Neptune, sending back images of its largest moon, Triton, an ice world that had only been visible as a speck on the Earth's most powerful telescopes.

Voyager 1 had set off on a more direct route out of the solar system. By 1990 it was almost 4 billion miles away from Earth (Voyager 2 was a mere 3 billion miles away). Voyager 1 had just enough energy left to take a final photograph.

Carl Sagan readily admitted that a photograph of the Earth would have no scientific value, but it would have human value. In our imaginations we can place ourselves aboard Voyager 1: we are leaving the solar system, turning our heads for one last look at the Earth before it fades from view forever. What would the Earth look like seen from 4 billion miles away? Would it be discernible at all? Sagan petitioned NASA to turn the craft towards the Earth one last time before it continued its journey into interstellar space. What we saw back on Earth was an image of the Earth less than the size of a single pixel. Remarkably, the pixel looks unlike the other millions of pixels that make up the image: it is a pale-blue colour, the last remnant of our specialness. The image – travelling at the speed of light – took five and a half minutes to make the journey back to Earth.

'Look again at that dot,' Sagan wrote in a famous essay, 'That's here. That's home. That's us. On it everyone you love, everyone you know, everyone you ever heard of, every human being who ever was, lived out their lives. The aggregate of our joy and suffering, thousands of confident religions, ideologies, and economic doctrines, every hunter and forager, every hero and coward, every creator and destroyer of civilization, every king and peasant, every young couple in love, every mother and father, hopeful child, inventor and explorer, every teacher of morals, every corrupt politician, every "superstar", every "supreme leader", every saint and sinner in the history of our species lived there – on a mote of dust suspended in a sunbeam.'

We might see the Pale Blue Dot as Earthrise's counterpart. Earthrise is an image of our home hoving back into view. 'It was the most beautiful, heart-catching sight of my life,' said Apollo 8 astronaut Frank Borman. 'It sent a torrent of nostalgia, of sheer homesickness, surging through me.' Earthrise is an image of greeting, of return. The Pale Blue Dot is an image of farewell, hinting at our future as space travellers. Further out, and even the blue pixel fades away.

'I believe that the long-term future of the human race must be in space,' Stephen Hawking once wrote: 'It's time to free ourselves from Mother Earth.' He follows in a now long tradition. Jules Verne, in his novel *From the Earth to the Moon* (1865), wrote of 'certain narrow-minded people, who would shut up the human race upon this globe, as within some magic circle which it must never outstrip'. But they will not prevail, he said. One day humans will 'travel to the moon, the planets, and the stars', and with the same ease, he predicted, as nineteenth-century travellers were then moving between Liverpool and New York. 'Man has no value save for that part of himself which passes into the universe,' wrote Teilhard de Chardin. 'Earth is our cradle, but one cannot live in a cradle forever,' Konstantin Tsiolkovsky wrote in a letter in 1911, the most quoted of all utopian declarations in support of space travel. In an interview with the journalist Oriani Fallaci, the science-fiction writer Ray Bradbury said, in language that sounds almost Biblical, 'Let us prepare ourselves to escape, to continue life and rebuild our cities on other planets: we shall not be long of this Earth!' Fallaci said that Bradbury spoke 'in a low voice, his eyes half shut . . . like a priest who recites the Pater Noster'. Fellow science-fiction writer Arthur C. Clark made a similar remark posed as a question: 'Can you believe that man is to spend all his days cooped and crawling on the surface of this tiny Earth – this moist pebble with its clinging film of air? Or do you, on the other hand, believe that his destiny is indeed among the stars, and that one day our descendants will bridge the seas of space?' In *A Man on the Moon* (1994), Andrew Chaikin argues that the possibility of a comet strike is reason enough for us to become 'a multi-planet species'. In the early 1990s, detailed images taken from the Hubble telescope showed a comet slamming into Jupiter. The Earth is a smaller target but a target nevertheless.

Whether or not the reasons are utopian or practical, the question remains: where are we to find another home? Today Voyager 1 is around 12 billion miles away; Voyager 2 about 10 billion. As far out as Voyager 1 is, it would take another 3,000 years – travelling at its current speed of 36,000 miles per hour and assuming it was pointing in the right direction (which it isn't) – before the craft reached our nearest neighbouring star, three light years distant. We can certainly speed up our spacecraft, but we are ultimately limited by the speed of light. The laws of physics as we currently understand them restrict us to regions nearby our own solar system. The rest is fiction.

Lindbergh predicted that the future of travel lay in dreaming and imagination. After his famous flight across the Atlantic, he had wondered what would come after conventional aircraft and had come to the conclusion that the future lay with rockets. And then when there were rockets he had wondered what would come after rockets. 'What lies for man beyond solar system travel?' he asked. 'What vehicle can be conceived beyond the rocket?' Lindbergh's answer was informed by his experience in the cockpit of *The Spirit of St Louis* during those hours of utter exhaustion when unearthly visitors had joined him there. Lindbergh's Atlantic crossing had transformed the age of physical travel, but gradually, throughout his life, Lindbergh came to understand that the secret those visitors vouchsafed in words he could never remember was that of travel itself. When the visitors arrived, everything 'unessential to [his] existence' fled, and for a time his consciousness grew independent of his ordinary senses. He saw, without using his eyes, that he was 'spirit masquerading in matter's form'. He felt himself departing from his body as he imagined a spirit might depart its body, to be reformed as 'awareness far distant from the human form' he had left behind. He was 'awareness spreading out through space, unhampered by time or substance'. Weightless, he felt he could travel freely across the universe. His understanding was not so different perhaps from Russian cosmism of the 1920s. He wrote that the experience in the cockpit made him 'cherish the illusion of being substance', yet in the knowledge that he was also the spatial emptiness inside the atom and the space between stars. He called it matterless awareness. Ed Mitchell had come to a similar conclusion. Mitchell saw that there was a connection between spirit and matter and that that connection was consciousness. 'To me,' he said, 'divinity is the intelligence existing in the universe.' After meetings with Native Americans and others living in non-material societies, he understood that 'visions perceived in trance and dreams are exactly what they appear to be: assistance obtained through either benevolence or supplication, from a spirit world to aid and guide humans'.

Of course Lindbergh's experience – the only time in his life, he said, that he conversed with ghosts – can be explained away in a material way as an eidetic or hypnogogic vision, a hallucination. But whatever the label, for Lindbergh the experience was as real as anything else was real for him. He said that the longer he lived the more limited he believed rationality to be: 'I have found that the irrational gives man insights he cannot otherwise attain.'

'Is it remotely possible,' Lindbergh wondered in his last years, 'that we are approaching a stage in evolution when we can discover how to separate ourselves entirely from earthly life, to abandon our physical frameworks in order to extend both inwardly and outwardly through limitless dimensions of awareness?' It was typical of Lindbergh that he should have found a way to hold fast both to his belief in Social Darwinism and to move onto ground few other materialists would dared to have trodden. He thought the future of travel might have no need for vehicles or matter: 'As Goddard's dreams resulted in the spacecraft . . . advancing man may discover that thought and reality transpose like energy and matter.' Because humans could now determine to a large degree their 'physical, mental, spiritual, and environmental evolution, limitless courses' were open to them.

I think the great adventures of the future lie – in voyages inconceivable by our twentieth-century rationality – beyond the solar system, through distant galaxies, possibly through peripheries untouched by time and space.

I believe early entrance to this era can be attained by the application of our scientific knowledge, not to life's mechanical vehicles, but to the essence of life itself: to the infinite and infinitely evolving qualities that have resulted in the awareness, shape, and character of man. I believe this application is necessary to the very survival of mankind. That is why I have turned my attention from technological progress to life, from the civilized to the wild. We will then find life to be only a stage, though an essential one, in a cosmic evolution of which our developing consciousness is beginning to become aware. Will we discover that only *without* spaceships can we reach the galaxies; that only *without* cyclotrons can we know the interior of the atom? To venture beyond the fantastic accomplishments of this physically fantastic age, sensory perception must combine with the extrasensory, and I suspect that the two will prove to be different faces of each other. I believe that it is through the sensing and thinking about such concepts that the great adventures of the future will be found.

Humans will reach the furthest regions of the universe, Lindbergh suggests – across spaces infinitely vast, and infinitesimal – as spirit not flesh, and fuelled by imagination. At the end of his life Lindbergh was as 'way out' as any devotee of the counter-culture:

> When I see a rocket rising from its pad, I think of how the most fantastic dreams come true, of how dreams have formed into matter and matter into dreams. Then I sense Goddard standing at my side, his human physical substance now ethereal, his dreams substantive. When I watched the fantastic launching of Apollo 8, carrying its three astronauts on man's first voyage to the moon, I thought about how the launching of a dream can be more fantastic still, for the material products of dreams are limited in a way dreams themselves are not. What sunbound astronaut's experience can equal that of Robert Goddard, whose body stayed on earth, while he voyaged through galaxies?
>
> Now that Goddard is dead, what difference does it make that his earthly individuality never left the ground in rocket flight? He thought of the stars; he became part of the stars. What physical fantasy of man can compare with my living memory of him – to my time-escaping vision? There is no better proof of immortality than this; dream is life and life is dream transposing.

Science extends our physical reach, but imagination comes first. We have long imagined what it might be like to fly, just as we long imagined what the Earth might look like seen from a great distance. We imagine what it might be like to live on other planets, and even in other universes. In our modern cosmology, we can travel beyond time and space into infinitely inflating quantum landscapes. But how do our dreams measure up to experience? In the light of facts and proofs old dreams quickly fade and disappear, to be replaced by new ones. Imagination only takes us so far, and we have often been wrong in our imaginings. Science and technology bring the power of our creative imagination into focus, so that the beam might be directed further out.

The target of human evolution long ago moved from the gene to tools and technology. Skill became more powerful than brute force. Over millennia our tools became more and more sophisticated. It's what we mean by progress. The tool might once have been an axe, now it might be a violin or an aircraft. A few rare human beings are so well integrated into their tools that it is as if the tools have become extensions of themselves. Lindbergh wore his plane as if it were a suit. Collectively, human beings have traded in direct apprehension of reality via their senses for detecting instruments. We do not have the hearing of a deer, or the sense of smell of a fox, and yet we can detect emanations from deepest space, even back to the beginning of the universe itself. The day will come – the revolution has already begun – when our machines will become so sophisticated that we will all be integrated into them as if they were extensions of ourselves. Today, we see a new generation, for the moment heads down, inseparable from their mobile devices. These are toddler steps. Virtual reality is still in its infancy. We have hardly begun to plug our machines into our human frames, but the day is coming when our bodies and minds will be integrated into our instruments: machine made flesh, flesh machine. Lindbergh might yet prove himself prescient. Even if we discover that we must leave our bodies behind, the machines we send deep into space may yet take with them the spirit that animates flesh. The reach of our minds might reach out into space along with our instruments.

There is something utopian about all our visions of space exploration, and something both dispiriting and fantastical about the motivation for space travel that tells us that we must find another home because we will at some point have to give up on this one, perhaps because we have trashed it, or because our sun will at some point run out of energy.

In the 1920s the rocket scientist Willy Ley said that it was a mathematical certainty that there was life elsewhere in the universe. 'The universe itself is our larger home. And without doubt we will meet other species along the way, if we haven't already,' Ed Mitchell wrote, decades later. 'To declare the Earth must be the only planet in the universe with life,' says astronomer Neil deGrasse Tyson, 'would be inexcusably big-headed of us'. It is hard to believe otherwise, and yet the question of whether or not there is life elsewhere cannot be decided by statistics; the numbers involved are too various and too unstable to be wrestled into some measure of probability. The more refined our measurements become, the closer we seem to get to finding other 'Earths' out there, and yet all those other possible 'Earths' – proto-Earths – have so far proved to be too hostile to host life. So, for the time being at least, what we have found is, not evidence of life elsewhere, but further evidence that life here exists within a very narrow band of possibility. The search for proto-Earths helps us understand our own Earth; it gives us something 'like' the Earth with which to compare our Earth. We humans understand by comparison. We are the only species, as far as we can tell, that tries to imagine what it is like to be something else. For now the question of whether or not there is life elsewhere cannot be answered. Richard Dawkins believes that there is life elsewhere in the universe, as I imagine most scientists do, but he readily admits that it is a belief, not a fact. He says that if life on Earth proves to be unique in the universe, then we will be forced to accept its existence as a miracle. Science, of course, does not do miracles.

Science is driven in part by the Copernican ideal that human beings are without privilege. Copernicus wondered if the sun, not the Earth, was at the gravitational centre of the solar system. We now know that 'we' are not at the gravitational centre of anything, whether we take that 'we' to be the Earth, Sun or even Milky Way. After Darwin, the same principle insists that neither is there anything special about being human from an evolutionary perspective. In short, the Copernican principle tells us that there is no universal privilege attached to being human. No wonder scientists seem to be obsessed with aliens. The Copernican principle makes their existence essential. Science is ultimately in search of a universal perspective (as religion is), a perspective that would be undermined if the only scientists in the universe turned out to be human beings. Science needs aliens in order to confirm that its universal laws really are universal: that aliens have reached the same understanding of the nature of reality.

It is possible that life is indeed profligate across the universe but so spread out that we will never make alien contact, constrained as we are by the laws of nature as we currently know them. Those other wildernesses where other life forms live may be forever beyond our reach. For us then, space would be not a place in need of taming, but beyond wildness, barely more than stuff and motion. Maybe there just isn't much out there of interest compared to the complexity that is here on Earth. Maybe space is as boring as a wall of white tiles.

If we ever have to leave Earth we will need to know how to fabricate wilderness. Out in space we will have to begin again, make from scratch what we can in imitation of what we have left behind. And if we are to succeed in that, then naturally we will have to know what wilderness is. Here, then, might be one of the most important reasons for space travel: not to relocate elsewhere, but to deepen our understanding of why we cannot leave.

Can we break the pattern of our species: move in, ravage, move on? Or is our narcissism – individual and collective – our doom? We might in principle move to other parts of the galaxy, other parts of the universe, and continue the same pattern forever; but what if, when the day comes, as it seems to be coming soon, when the Earth is exhausted and we must move on, we find that we do not have the technological wherewithal to do so; that to move on from the Earth is harder than we imagined? Or what if we discover, as we find out more about ourselves as seen from the outside, that we are more integrated into the Earth than we ever imagined, that human beings cannot be transplanted?

Have we humans got what it takes to see ourselves from without, and so save our Earth-home? The arrival of aliens would surely unite the world. We would fear *them* and not the *others* (whoever our others may be) for their difference. But what if aliens never come? What if there are no aliens? Can we learn, instead, to see ourselves as if from an alien's perspective? Might this be another reason to travel into space, in order to experience what an alien might experience: to see ourselves from the outside?

What would it take for human beings to change permanently? Are human beings even capable of change, either individually or collectively? The young Lindbergh thought that the experience of seeing the Earth from an aircraft would change humans fundamentally: 'I felt sure airplanes would bring peoples of the world together in peace and understanding.' But as he grew older he began to change his mind. Mike Collins believed that the view of the Earth from space had the power to change us, to make us realize that 'the planet we share unites us in a way far more basic and far more important than differences in skin color or religion or economic system'. But it had to be the experience itself; looking at a photograph is not the same thing, he said. Photographs 'deceive us . . . for they transfer the emphasis from the *one* Earth to the multiplicity of reproduced images'. He thought it a great pity that the view has so far 'been the exclusive property of a handful of test pilots, rather than the world leaders who need this perspective, or the poets who might communicate it to them'. Jim Lovell thought that perhaps even the experience itself might not be enough: 'But the mind easily forgets,' he wrote, 40 years after Apollo 8 had returned to Earth, 'and not too long after . . . people get back to the way they lived before – wars and disruption and human cruelty.' Rusty Schweickart said that after he had told the story a thousand times he grew sick of it, and even to wonder if he'd made the whole thing up.

If the experience of seeing the Earth from a plane or a rocket will not change us, if we cannot keep up with the speed of technological progress, if we cannot grow into our brains in time, if we cannot learn from history or art or science, if the Earth is indeed teetering on the edge of colossal environmental collapse, then the species that biologist Lynn Margulis once described as 'upright, mammalian weeds' will surely perish. Human life will have proved itself fragile and transitory. The robust Earth, however, will continue to plunge silently through black space.

Perhaps we need worry about the universe only when we do physics. 'As a philosophy,' Ed Mitchell wrote, 'science is terrible; as a method, it's superb.' We should use the method, he said, and forget the philosophy. Or – I would argue – we could improve the philosophy. Science is only bad philosophy when it is scientism; that is, the belief that science has the answers to all questions. If we could accept that the scientific method continually points to what is outside itself – which is always almost everything – we might see that the scientific method is actually even more powerful than certain reductionists would have us believe. Whether we are unique or not, we living complex forms – humans, flies and grasses – might live as if we were unique, as if this blue marble is never to be made again; never again this small damp spot. Perhaps life, whether profligate or not, is in balance with the universe: complexity in trade-off with size. If it is hubristic to rate life so highly, perhaps on the other side it is just as hubristic to rate the universe more highly than the Earth. Do we really want to risk the Earth itself for the sake of an idea: for a particular interpretation of the scientific method? Paradoxically, the human experience of space travel is to reject a scientistic obsession with infinity and eternity – to reject the eternal search for universal laws, as reflected in the exploration of infinite space and the search for eternal existence – for an understanding of the human species at a human scale, intimately connected, even embedded, in its home. The more distant our perspective, seemingly the more intimate. Perhaps a fragile tent of twigs really is more robust from a human perspective than a cluster of galaxies. Perhaps the imagination of a child does outweigh the destructive power of an approaching meteor, even though we die. Until the day comes when we might know otherwise, life is here, not there. And we do not know that the day will ever come to tell us otherwise. Perhaps the greatest achievement of the moon race lies hidden in Kennedy's original aim of 'landing a man on the moon and returning him safely to the Earth'. After centuries of wars in which countless human beings have slaughtered each other, here, in this simple statement, a nation committed itself to protecting the life of a single human being. Perhaps *we* are the miracle we refuse to acknowledge.

BIBLIOGRAPHY

A book like this, that is so wide-ranging, is bound to rely – even more than all books do – on the work of others who have come before. Below is a list of books out of which I constructed the skeleton of this book (and sometimes flesh and veins). I cannot overstate my indebtedness to each writer listed and the particular work referenced. I might have wished to show that indebtedness title by title, but there are only so many superlatives. My sincerity would soon have been called into question. Better perhaps to understate my gratitude and beg for understanding. My hope is that as well as being a bibliography, this list will serve as a reading list for those readers who wish to explore a particular area in more detail. There is no title listed here that I cannot recommend heartily.

I hope it will be clear from the titles alone which book provided me with the necessary expertise for which part of my own work. Where there may be some doubt I have appended a bald sentence or two of description.

This book would also be much the poorer were it not for the existence of NASA's website: nasa.gov

Magnificent Desolation: The Long Journey Home from the Moon, Buzz Aldrin and Ken Abraham, Harmony Books, 2009

Lindbergh, A. Scott Berg, G. P. Putnam's & Sons, 1998

> The authorized biography of Charles A. Lindbergh and his family.

The Man Who Ran the Moon: James E. Webb, NASA, and the Secret History of Project Apollo, Piers Bizony, Thunder's Mouth Press, 2006

Red Moon Rising: Sputnik and the Rivalries that Ignited the Space Age, Matthew Brzezinski, Times Books, 2007

Space Race: The Untold Story of Two Rivals and Their Struggle for the Moon, Deborah Cadbury, 4th Estate, 2005

> An account of the parallel lives and careers of von Braun and Korolev.

The Last Man on the Moon, Eugene Cernan and Don Davis, St Martin's Press, 1999

A Man on the Moon: The Voyages of the Apollo Astronauts, Andrew Chaikin, Foreword by Tom Hanks, Viking Penguin, 1994

> The definitive account of Project Apollo.

Rocket Man: Robert H. Goddard and the Birth of the Space Age, David A. Clary, Theia, 2003

Carrying the Fire: An Astronaut's Journeys, Michael Collins, Foreword by Charles A. Lindbergh, FSG, 1974, new edition 2009

Hitler's Scientists: Science, War and the Devil's Pact, John Cornwell, Penguin, 2003

Space Chronicles: Facing the Ultimate Frontier, Neil deGrasse Tyson, W. W. Norton, 2012

> A defence of space travel by the popular American astrophysicist.

Ungodly: The Passions, Torments, and Murder of Atheist Madalyn Murray O'Hair, Ted Dracos, Free Press, 2003

Lindbergh vs Roosevelt: The Rivalry that Divided America, James P. Duffy, Regnery, 2010

Into that Silent Sea: Trailblazers of the Space Era 1961–1965, Francis French and Colin Burgess, University of Nebraska Press, 2007

Lindbergh Alone, Brendan Gill, Harcourt Brace Jovanovich, 1977

A short illustrated account of Lindbergh's solo flight across the Atlantic in 1927.

Falling Upwards: How We Took to the Air, Richard Holmes, William Collins, 2013

A history of ballooning.

Atlantic Fever: Lindbergh, His Competitors and the Race to Cross the Atlantic, Joe Jackson, FSG, 2012

The Astronaut Wives Club, Lily Koppel, Grand Central, 2013

1968, The Year that Rocked the World, Mark Kurlansky, Ballantine Books, 2003

North to the Orient, Anne Morrow Lindbergh, Harcourt, Brace and Company, 1935

Listen! The Wind, Anne Morrow Lindbergh, Harcourt, Brace and Company, 1938

The Wave of the Future: A Confession of Faith, Anne Morrow Lindbergh, Harcourt, Brace and Company, 1940

Earthshine, Anne Morrow Lindbergh, Harcourt, Brace & World, 1969

Two essays, one of which describes the visit she and her husband made to Cape Kennedy in 1968.

Bring Me a Unicorn: Diaries and Letters of Anne Morrow Lindbergh, 1922–1928, Harcourt Brace Jovanovich, 1971

Hour of Gold, Hour of Lead: Diaries and Letters of Anne Morrow Lindbergh, 1929–1932, Harcourt Brace Jovanovich, 1973

Locked Rooms and Open Doors: Diaries and Letters of Anne Morrow Lindbergh, 1933–1935, Harcourt Brace Jovanovich, 1974

The Flower and the Nettle: Diaries and Letters of Anne Morrow Lindbergh, 1936–1939, Harcourt Brace Jovanovich, 1976

War Within and Without: Diaries and Letters of Anne Morrow Lindbergh, 1939–1944, Harcourt Brace Jovanovich, 1980

Against Wind and Tide: Letters and Journals, 1947–1986, Anne Morrow Lindbergh, Introduction by Reeve Lindbergh, Pantheon, 2013

> A last collection of Anne Morrow Lindbergh's writings, put together by her family and published posthumously.

We, Charles A. Lindbergh, Putnam, 1927

Of Flight and Life, Charles A. Lindbergh, Charles Scribner's, 1948

The Spirit of St Louis, Charles A. Lindbergh, Charles Scribner's, 1953

The Wartime Journals of Charles A. Lindbergh, Harcourt Brace Jovanovich, 1970

Autobiography of Values, Charles A. Lindbergh, Foreword by William Jovanovich and Judith A. Schiff, Harcourt Brace Jovanovich, 1976

> Lindbergh's last memoir, edited by his editor William Jovanovich and published posthumously.

Under a Wing: A Memoir, Reeve Lindbergh, Simon and Schuster, 1998

> An intimate memoir of Charles and Anne Morrow Lindbergh.

Gaia: A New Look at the Earth, James Lovelock, Oxford University Press, 1979

Dora: The Nazi Concentration Camp Where Space Technology Was Born and 30,000 Prisoners Died, Jean Michel, Holt Rinehart and Winston, 1980

Cosmonauts: Birth of the Space Age, edited by Doug Millard, 2014

> Catalogue of an exhibition held at the Science Museum.

The Way of the Explorer: An Apollo's Astronaut's Journey Through the Material and Mystical Worlds, Edgar Mitchell with Dwight Williams, Putnam, 1996

My Life Without God, William J. Murray, WND Books, 2012

Hermann Oberth: One of the Fathers of Rocketry, David Myhra, RCW Technology S&S, 2013

The Heroic Journey of Private Galione: The Holocaust Liberator Who Changed History, Mary Nahas, Mary's Designs, 2004; revised 2012

Von Braun: Dreamer of Space, Engineer of War, Michael J. Neufeld, Knopf, 2007

To Touch the Face of God: The Sacred, the Profane, and the American Space Program, 1957–1975, Kendrick Oliver, Johns Hopkins, 2013

> An investigation into the relationship between religion and the US space programme, including an account of the influence of Madalyn Murray O'Hair.

Moon Shot: The Inside Story of Mankind's Greatest Adventure, Dan Parry, Ebury Press, 2009

> An account of the Apollo 11 mission.

Earthrise: How Man First Saw the Earth, Robert Poole, Yale, 2008

> The story of the first colour photographs of the Earth, particularly Earthrise and the Blue Marble, and of their subsequent influence.

Soviet Russia's Space Program During the Space Race: The History and Legacy of the Competition that Pushed America to the Moon, Charles River Editors, Createspace, 2015

The Last Hero: Charles A. Lindbergh, Walter S. Ross, Harper and Row, 1964; revised 1976

> A biography of Charles A. Lindbergh.

Southern Mail, Antoine de Saint-Exupéry, translated by Stuart Gilbert, Harrison Smith and Robert Haas publishers, 1931

Night Flight, Antoine de Saint-Exupéry, translated by Stuart Gilbert, Crosby Continental Editions, 1933

Wind, Sand and Stars, Antoine de Saint-Exupéry, translated by Lewis Galantière, Harcourt, 1939

The Little Prince, Antoine de Saint-Exupéry, translated by Katherine Woods, Reynal and Hitchcock, 1943

Das Doppelleben Des Charles A Lindbergh: Der Berühmteste Flugpionier Aller Zeiten – Seine Wahre Geschichte, Rudolf Schröck, Heyne, 2005

> The German biography that uncovered the existence of Lindbergh's other families. Not published in an English translation.

America's Most Hated Woman: The Life and Gruesome Death of Madalyn Murray O'Hair, Ann Rowe Seaman, Continuum, 2005

Konstantin Eduardovich Tsiolkovsky: The Pioneering Rocket Scientist and His Cosmic Philosophy, Daniel H. Shubin, Algora, 2016

Moon Dust: In Search of the Men Who Fell to Earth, Andrew Smith, Bloomsbury, 2005

> A collective biography of the men who walked on the Moon.

Wernher von Braun: Crusader for Space, A Biographical Memoir, Ernst Stuhlinger and Frederick I. Ordway III, Krieger Publishing Company, 1994

Among Men and Beasts, Paul Tregman, Gazelle Book Services, 1979

Dr Space: The Life of Wernher von Braun, Bob Ward, Foreword by John Glenn, Naval Institute Press, 2005

The Overview Effect: Space Exploration and Human Evolution, Frank White, Foreword by Gerard O'Neill, Houghton Mifflin, 1987

> An account of the effect that seeing the Earth from outer space had on the astronauts and cosmonauts who experienced it.

The Voice of Dr Wernher von Braun, edited by Irene E. Powell-Willhite, Apogee Books, 2007

> A collection of von Braun's writings and talks.

Falling to Earth: An Apollo 15 Astronaut's Journey to the Moon, Al Worden with Francis French, Smithsonian Books, 2011

•

This is not an academic work; it is meant to be a narrative built on the expertise of others. In that spirit I have decided not to source every last quotation. There are, for example, many brief phrases put into the mouth of Charles Lindbergh. All of those words can be found in his books – or in the biographies of him – as itemized above. An Internet search will likely take the reader to a precise location. Elsewhere, the source of most quotations is obvious from the context. Where that is not the case, I list the sources below.

PART ONE

Chapter Two

'It appears completely incapable of flight ...' *The Spirit of St Louis*, Charles A. Lindbergh

Chapter Six

'I did not realize that we were ...' *Locked Rooms and Open Doors*, Anne Morrow Lindbergh

'the most healing and nourishing element ...' *Locked Rooms and Open Doors*, Anne Morrow Lindbergh

'held a great many more ...' from William Dodd's Diary

'Hitler is apparently more popular ...' Charles A. Lindbergh in a letter to Alexis Carrel

Chapter Seven

'A cry, a thud ...' *Dora*, Jean Michel

Chapter Eight

'For six days our train ...' *Among Men and Beasts*, Paul Tregman

Chapter Nine

'could barely walk or talk ...' *Red Moon Rising*, Matthew Brzezinski

'reverse-engineering puzzle ...' *Red Moon Rising*, Matthew Brzezinski

'We are living in a democracy ...' *Wernher von Braun: Crusader for Space*, Ernst Stuhlinger and Frederick I. Ordway III

'A vista into ...' Walt Disney's own words

Chapter Ten

'an elongated propellant tank ...' Wernher von Braun's own description, *The Voice of Dr Wernher von Braun*, edited by Irene E. Powell-Willhite

'I cannot understand why someone ...' 14 November 1957, *Against Wind and Tide*, Anne Morrow Lindbergh

'even scared the CIA ...' *Red Moon Rising*, Matthew Brzezinski

'the world's first nuclear detonation ...' *Red Moon Rising*, Matthew Brzezinski

'unflappable crew chief ...' *Red Moon Rising*, Matthew Brzezinski

'To hell with it ...' *Red Moon Rising*, Matthew Brzezinski

Chapter Eleven

'a delightful, bald-headed ...' *Carrying the Fire*, Michael Collins

'for space flight's sake ...' Von Braun's own words, *The Voice of Dr Wernher von Braun*, edited by Irene E. Powell-Willhite

Chapter Twelve

'forthright views on ...' *Moon Shot*, Dan Parry

PART TWO

Chapter One

'over-encouraged the development ...' *The Man Who Ran the Moon*, Piers Bizony

'Hey Dick, said Gilruth ...' from an interview with Richard Underwood for NASA

'the most spectacular ...' *Carrying the Fire*, Michael Collins

'mired in technical ...' *Red Moon Rising*, Matthew Brzezinski

Chapter Two

'a veil came down ...' *Carrying the Fire*, Michael Collins

'a jungle of wire ...' *Carrying the Fire*, Michael Collins

Chapter Three

'her cryogenic ...' *The Last Man on the Moon*, Eugene Cernan and Don Davis

'Instinctively I feel I am ...' actually a description by Mike Collins of a later flight

'was as sentimental as they come ...' *A Man on the Moon*, Andrew Chaikin

'a steel cone ...' *The Last Man on the Moon*, Eugene Cernan and Don Davis

Chapter Five

'a boldness of vision ...' *The Man Who Ran the Moon*, Piers Bizony

'At the time I could ...' *Magnificent Desolation*, Buzz Aldrin and Ken Abraham

'Is that all there is?' written by Jerry Lieber and Mike Stoller, and first performed by Peggy Lee

'his soul had been stirred ...' *A Man on the Moon*, Andrew Chaikin

Interlude

'One polar ice cap ...' writing in *Collier's* magazine in 1952

PART THREE

Chapter One

'The entire range of living matter ...' *Gaia*, James Lovelock

'deliver *samdhi*, a ...' *1968*, Mark Kurlansky

Chapter Two

'and before the astonished ...' the American physicist, Kenneth Bainbridge

'I don't know anyone better ...' Anne Morrow Lindbergh in a letter dated 3 April 1972, *Against Wind and Tide*

'I felt incredibly alone...' *War Within and Without*

'Too much of our life...' Anne Morrow Lindbergh diary entry dated 27 July 1953, *Against Wind and Tide*

'What a romantic C is...' Anne Morrow Lindbergh in a letter dated 1 February 1969, *Against Wind and Tide*

'had never said "no"...' Anne Morrow Lindbergh in a diary entry dated 29 March 1954, *Against Wind and Tide*

'It is impossible to maintain...' Anne Morrow Lindbergh in a letter dated 27 July 1973, *Against Wind and Tide*

Chapter Three

'By his own account, von Braun...' in a letter von Braun wrote to *Paris Match* in 1966

'a visceral reminder...' in *To Touch the Face of God*, Kendrick Oliver

'If the cops questioned them about it...' Diane McWhorter, Pulitzer-prize winning journalist, on NPR ('How a Nazi Rocket Scientist Fought for Civil Rights') discussing her forthcoming book, *Moon Over Alabama* about the German rocket engineers in Huntsville

Chapter Four

'all but a small percentage...' *To Touch the Face of God*, Kendrick Oliver

Chapter Five

'scratch-resistant lenses, cordless power tools...' extended from a list in *Space Chronicles: Facing the Ultimate Frontier*, Neil deGrasse Tyson

'As Goddard's dreams...' Charles Lindbergh in the Foreword to *Carrying the Fire*, Michael Collins

'I think the great adventure...' Charles Lindbergh in *Life* magazine, 4 July 1969

'When I see a rocket...' *Autobiography of Values*, Charles A. Lindbergh

IMAGE CREDITS

Plate section 1: Photos by NASA.

p.12 Image by Svenskt porträttgalleri, Stockholm 1880. Original author unknown.

p.15 (left) Photo by Bettmann / Contributor / Getty Images.

p.15 (right) Photo by Bettmann / Contributor / Getty Images.

p.35 Photo by Time Life Pictures/ Mansell/The LIFE Picture Collection/ Getty Images.

p.36 Photo by the United States Library of Congress's Prints and Photographs division.

p.51 Photo by chrisdorney/ Shutterstock.com.

p.55 Photo by Kwonghwa Studio, Nanking.

p.62 Photo by SSPL/Getty Images.

p.66 Photo provided by Wikimedia Commons. Author unknown.

p.85 Photo by Lindbergh Picture Collection (MS 325B). Manuscripts and Archives, Yale University Library.

p.89 Photo provided by Wikimedia Commons. Author unknown.

p.115 Photo by Stephenson / Topical Press Agency / Getty Images.

p.123 (top) Photo provided by Wikimedia Commons. Author unknown.

p.123 (bottom) Photo by United States Holocaust Memorial Museum, courtesy of Joseph Mendelsohn.

p.128 Photo by NASA/Marshall Space Flight Center.

p.134 Photo by SPUTNIK / Alamy Stock Photo.

p.139 (left) Photo by Horyzonty Techniki 11/1966.

p.139 (right) Photo provided by Wikimedia Commons. Author unknown.

p.143 Photo by Clyde T. Holliday.

p.163 Photo by US Air Force.

p.165 Photo by Chrysler Corporation and US Army.

p. 174 Photo by San Diego Air & Space Museum Archives.

p. 178 Photo by NASA.

p.180 Photo by Chrysler.

p.181 Photo by NASA.

p.183 Photo by Bettmann / Contributor / Getty Images.

p.184 (top) Photo provided by Wikimedia Commons. Author unknown.

p.184 (bottom) Photo by NASA.

p.191 (top) Photo by Kennedy Space Center Photo Archive.

p.191 (middle) Photo by NASA.

p.191 (bottom) Photo by NASA.

p.192 (top) Photo by NASA.

p.192 (bottom) Photo by Ralph Morse / Life Magazine / The LIFE Picture Collection / Getty Images.

p.194 (top) Photo by: Sovfoto / UIG via Getty Images.

p.194 (bottom)

p.218 Photo by The Realist.

p.219 Photo by Alan Light.

p.221 Photo by Bettmann / Contributor / Getty Images.

p.234 Photo by NASA.

p.235 Photo by NASA.

p.236 Photo by NASA.

p.237 Photo by SPUTNIK / Alamy Stock Photo.

p.238 Photo by NASA.

p.241 Photo by NASA.

p.242 Photo by NASA.

p.244 Photo by Photo by Time Life Pictures / NASA / The LIFE Picture Collection / Getty Images.

p.245 Photo by NASA.

p.248 Photo by NG Images / Alamy Stock Photo.

p.254 Photo by NASA.

p.256 Photo by NASA.

P.258 Photo by Royal Canadian Air Force.

p.259 Photo by NASA.

p.264 Photo by NASA.

p.276 Photo by NASA.

p.279 Photo by NASA.

p.280 Photo by NASA.

p.281 Photo by INTERFOTO / Alamy Stock Photo.

p.282 Photo by Granger Historical Picture Archive / Alamy Stock Photo.

p.304 Photo by NASA.

p.305 Photo by Photo by SSPL / Getty Images.

p.306 Photo by Photo by Central Press / Getty Images.

p.314 Photo by NASA.

p.329 Photo by NASA.

p.330 Photo by Photo by Ralph Morse / The LIFE Picture Collection / Getty Images.

p.333 Photo by Space Frontiers / Getty Images.

p.335 Photo by NASA.

p.340 Photo by NASA.

p.346 Photo by NASA.

p.348 Photo by NASA.

p.350 Photo by National Geographic Creative / Alamy Stock Photo.

p.381 Photo by Mondadori Portfolio via Getty Images.

p.406 Photo provided by YouTube. Author unknown.

Plate section 2: First photo by DODGE, all other photos by NASA.

ACKNOWLEDGMENTS

Stacey D'Erasmo, Tim Hughes, Martin McInnes, Cy O'Neal, Peter Parker and Noni Pratt all read the entire manuscript at various stages of its development. For their invaluable editorial comments I am hugely indebted. Peter Parker heroically read the manuscript more times than I'd now like to recall.

For their encouragement, support and advice I would also like to thank Ed Bartlett, Mary Bartlett, Sheridan Bartlett, Thomas Bartlett, Michael Benson, David Cafiero, Bill Clegg, Hazel Coleman, Elizabeth Cook, Sonali Deraniyagala, Freddie Fabiano, Jonathon Fairhead, Jane Haynes, Courtney Hodell, David Hopson, Daniel Kaizer, James Lecesne, Ken Legins, Jim Lowe, Simon Lunn, Adam Moss, Sylvia Nasar, Shabir Pandor, Beth Povinelli, Noni Pratt, David Purcell, Seth Pybas, Lisa Randall, Sally Randolph, Joyce Ravid, Salley Vickers and Robert Westfield

Whenever I despaired, my agent, Georgina Capel, brought calmness and hope. My editor at Head of Zeus, Neil Belton, saved me from a number of embarrassments. Whatever errors remain – embarrassing or otherwise – are entirely my own. Also at Head of Zeus, Ellen Parnavelas and Clémence Jacquinet turned my manuscript into an elegant-looking book. Octavia Reeve was a diligent and sensitive copyeditor. Jessie Price designed a beautiful jacket. Many thanks too to my editor at Pegasus, Jessica Case.

INDEX

Page numbers in *italics* indicate an illustration

References to NASA and quotations from Mike Collins, Charles Lindbergh and Wernher von Braun are so numerous throughout the book that it has not been deemed helpful to index them

airmail, 18–27, 90
Alcock, John and Arthur Brown, 27–8, 31
aliens, 427–8
America First, 94, 96
anti-semitism
    see Ford, Henry; Kennedy, Joseph; Lindbergh, Charles
Apollo project, 180, 189, 190, 233, 266–77, 332–3, 357, 416–17
    Apollo 1 disaster/'The Fire', 266–9, 271, 272, 277, 306
    see also astronauts: Chaffee, Grissom, White
    Apollo 8, 279–97, 282, 298–303, 305, 307, 316, 323, 424
    Genesis quoted, 293–4, 299–300, 303, 316, 328–9, 335, 405–6
    see also astronauts: Anders, Borman, Lovell; stamps
    Apollo 11, 305–29
    Command Module/Columbia, 307, 308–9, 310, 319, 321, 322

'Giant Step' tour, 327
Lunar Module/The Eagle, 308–9, 309–10, 313–15, 318–19, 321, 322
see also astronauts: Aldrin, Armstrong, Collins
Apollo 13, 332–4
    emergency, 333–4
    Howard, Ron *Apollo 13*, 334
    Lunar Module, 333–4
    see also astronauts: Haise, Lovell, Swigert
cancelled, 350
Command Modules, 266–71, 281–2, 286, 287, 296, 304, 305, 330, 341, 342–3
configuration at launch, *281*, 281–2
cost, 241
first launch, 230
Lunar Landing Training Vehicle/ Flying Bedstead, 313, *314*
Lunar Modules, 270, 271, 275–6, 281, 304, 305, 330, 347–8
Lunar Roving Vehicle/Moon Buggy, 340, 346

moon garbage, 318–19, 321
moonwalks/rides, 237, 318–21,
323–4, 341, 345
orbits, 289–93, 295, 305
Service Modules, 267, 281, 287,
290, 296, 308
simulator, 308–9, 331
see also rockets: manned moon
shots
army pilots, 90–1
astronauts, 24, 32, 169, 170, 171, 199,
233–4, *234, 235,* 247, 357, 363
Aldrin, Edwin 'Buzz', 76, *234, 235,*
253, 260, 280, 305–29, *306,*
406
his father, Edwin Sr, 311, 318,
327
his wife, Joan, 317, 318, 327
quoted, 327
on the moon, 320–1
Anders, Bill, *234, 235,* 279,
279–97, *280,* 302
quoted, 300, 302
Armstrong, Neil, 76, 233, *234,* 247,
248, 251, 280, 305–29, *306*
his son, Mark, 268, 320
his wife, Janet, 328
on the moon, 247, 315, 317,
318–21
as pilot, 247–9, 251
quoted, 247, 268
Bassett, Charlie, *234, 235,* 251–2,
260
Bean, Al, *234, 235, 329,* 329–32,
330, 345
quoted, 313, 331, 332, 400
bitchiness, 251–2, 260, 309
Borman, Frank, 233, *234,* 243, 244,
246, 269, *279,* 279–97, *280,*
299–300
his wife, Susan, 284, 288, 293,
295
quoted, 295
quoted, 300, 301, 420
Carpenter, Scott, 174, 174–5, 192,
199, 200, 202

Cernan, Gene, 252, 253–5, *254,*
260, 270, 271, 305, *305,* 347–50,
348
his wife, Jan: quoted, 339
quoted, 237, 252–3, 253, 254,
346, 347, 348, 349, 349–50
Chaffee, Roger, *234, 235,* 267–8
Collins, Mike, *234, 235, 256,*
256–9, 268, 285, 287, 297,
305–29, *306*
and expertise, 238–9, 246,
266–7, 336
his wife, Pat, 258
on Saturn V, 273
searching for the words, 258,
299, 347, 365, 373
on space pictures, 257, 319
von Braun eulogy, 399–400
see also 'earthly ennui'
Conrad, Pete, 233, *234,* 242, *242,*
259, *259,* 286, *329,* 329–32, *330,*
370
quoted, 332
Cooper, Gordon 'Gordo', 174,
174–5, *193,* 202–3, 242, *242,*
245
Cunningham, Walt, *234, 235,* 276,
276–7
Duke, Charlie, 345–6, *346*
his wife, Dotty: quoted, 346
'earthly ennui'/epiphanies/
spiritual journeys, 324–9,
331–2, 333, 338, 344–5, 370–3,
428–9
Eisele, Donn, *234, 235,* 276,
276–7
Evans, Ron, 347–8, *348,* 350
Freeman, Theodore, *234, 235,* 252
Glenn, John, 174, *192,* 194–5, 199,
201, 334
his wife, Annie, 194, 196
quoted, 198–9, 203
see also astronauts: Shepard,
Alan
Gordon, Dick, *234, 235,* 259, *259,*
286, *329,* 329–31, *330*

Grissom, Virgil 'Gus', 174, 174–5,
    191, 237–39, 238, 267–8, 277
    his wife, Betty, 268
    quoted, 238, 267–8
Haise, Fred, 280, 282, 332–4, 333
    quoted, 334
Irwin, Jim, 339–45, 340
    quoted, 340, 342
Lovell, Jim, 233, 234, 243–6, 244,
    253, 260, 279, 279–97, 280,
    332–4, 333, 347, 397
    and Charles Lindbergh, 307
    his wife, Marilyn, 295
    quoted, 334, 429
Mattingley, Ken, 345–6, 346
McDivitt, Jim, 233, 234, 239–40,
    241, 304, 304
Mitchell, Edgar 'Ed', 335, 335–9,
    346, 371–4, 422
    quoted, 338, 372, 373–4, 400,
    422, 426, 429
perks, 196
Roosa, Stu, 335, 335–9
Schirra, Wally, 174, 174–5, 193,
    202, 243–4, 246, 276, 276–7
    his wife, Josephine: quoted, 238
    quoted, 276
Schmitt, Jack, 347–50, 348, 350,
    361
    quoted, 349, 349–50, 366
Schweickart, Rusty, 234, 235, 304,
    304, 369–71, 372, 429
    his wife, Clare, 370
    quoted, 370, 371, 374, 406
Scott, Dave, 234, 235, 247, 248, 251,
    304, 304, 339–45, 340, 370, 371
See, Elliot, 233, 234, 251–2, 260
    his wife, Marilyn, 252
Shepard, Alan, 76, 174, 182, 184,
    184, 199, 246, 252, 335, 335–37
    quoted, 175, 183, 184, 338
    rivalry with John Glenn, 174–5,
    181–2, 195–6, 198
Slayton, Deke, 174, 174–5, 270,
    279, 313, 314, 347–8
    grounded, 196–7

harsh judgments, 252, 277, 370
    and religious/mystical issues,
        253, 303, 316, 336
    and Richard Underwood,
        240–1
Stafford, Tom, 233, 234, 243, 244,
    252, 253, 254, 255, 305, 305
Swigert, Jack, 332–4, 333
White, Ed, 234, 234, 239, 240, 241,
    241, 267–8
    his wife, Pat, 268
    quoted, 268
Williams, Clifton, 234, 235, 252
women, 194
Worden, Al, 339–45, 340
    quoted, 340, 341, 342–3, 344,
        416–17
Young, John, 233–4, 234, 237,
    238, 256, 256–7, 305, 305, 309,
    345–7, 346
    quoted, 257, 347
    see also Mercury project:
        Mercury Seven
atom bomb/nuclear device, 132, 140,
    141, 150, 153, 159, 160, 161, 186,
    397
    Hiroshima, 394
    nuclear deterrence, 150–1
    nuclear war simulation, 160
    simulated missions, 375
    tests, 201, 371–2, 375
    see also cold war: heats up;
        Manhattan Project

balloons/ballooning, 29, 49, 57,
    76–7, 168
'beatnik', 200
Berg, Scott: quoted, 14
Berry, Charles 'Chuck', 306, 343
    quoted, 331
Bixby, Harold M., 29–30
Bleriot, Louis, 49
Boeing, 263–4
Bradbury, Ray: quoted, 420
Brand, Stewart, 362–5
    CoEvolution Quarterly, 365, 368

Full Earth, 364
  quoted, 363, 364, 367, 371
Whole Earth Catalog, 364–5
  quoted, 364
Brezhnev, Leonid, 236, 244–5,
  260–1
Brzezinski, Matthew: quoted, 157,
  161, 163–4, 164–5
Byrd, Commander Richard E., 34,
  38, 40

Cape Canaveral/Cape Kennedy,
  152–3, 165, 167, 201, 267, 306
Carrel, Alexis, 98–101, 102
  quoted, 101
Chaikin, Andrew *A Man on the Moon*,
  336
  quoted, 294–5, 345, 421
Chamberlin, Clarence, 34–5, 38
Christ, Jesus, 134–5, 395, 398
Churchill, Winston, 109, 132, 151
  quoted, 103
Clarke, Arthur C., 172, 308
  quoted, 421
climate change, 398
cold war, 160, 166, 203–4, 302
  heats up, 229
Collier's magazine, 146, 147–8, 153,
  339, 340
Coolidge, Calvin, 22, 51
cosmism, 133–4, 198, 366, 422
cosmonauts, 170, 181, 182, 236, 247,
  351
  Bykovsky, Valery, *194*
  deaths beyond Earth's atmosphere,
    351
  Gagarin, Yuri, 182–3, *183*, 184, 187,
    197, 198, 203, 213, 271, 351
    death, 272
    quoted, 182, 198
    tours the world, 183
  Komarov, Vladimir, 271–2
    his wife, Valentina, 271
  Leonov, Alexey, 237
  Nikolayev, Andriyan, *193*, 198
  Popovich, Pavel, *193*

Tereshkova, Valentina, *194*, 347
Titov, Gherman, *191*, 197, 199,
  239, 300
  quoted, 203
Cronkite, Walter, 172, 200, 323
Cummings, E. E., 94–5
  quoted, 371
Cuxhaven, 132, 139, 140

Dawkins, Richard, 368–9, 427
defecation
  see space evacuation
Degenkolb, Gerhard, 107, 109
diarrhea
  see space evacuation
Disney, Walt, 94–5
  see also von Braun, Wernher:
    collaborations with Disney
Dornberger, Walter, 77, 78, 82, 83,
  106–13, 127, 129–30
  quoted, 108
Dubos, René: quoted, 367
DuPont chemical company, 56, 66
Durant, Frank C., 151

'Earthgazing', 24, 255, 285, 327,
  355–7
Earthrise, 291–2, 305, 323, 341, 420
  see also space photography:
    Earthrise shot
ecology movement/
  environmentalism, 366–7, 396–7,
  410
Eiseley, Loren, 366–7
  quoted, 366
Eisenhower, Dwight D., 150, 152, 153,
  159, 204, 217, 230, 398–9
  army career, 132, 169, 210
  backing in space race, 165, 168
  misgivings about/indifference to
    space program, 151, 154, 158,
    159, 169, 170–1, 183, 187, 190
  quoted, 185–6, 204
  spending on defense, 160, 161
  tainted by space setbacks, 166
  U-2 crisis, 185, 188

eugenics, 100–1

Federal Communications
    Commission (FCC), 407–8 Fiske,
Edward, 299, 301, 302
Flynt, Larry, 408–9
Fonck, René, 28, 29, 33
Ford, Henry/his company, 102–3,
    116–17, 126
Model T, 388
Fuller, Buckminster, 363–5
    'spaceship earth', 363
futurism, 133–4
Fyodorov, Nikolai, 133, 134

Galione, John, 120–2, 124–5
Gas Dynamics Laboratory (GDL), 136
    see also Reactive Scientific
        Research Institute
Gemini project, 233, 236, 237–62,
    266–7, 273, 282, 286, 307, 308
geology, 335–6, 340–2, 345, 347–8,
    350, 351
    Genesis rock, 341–2, 345
Gilruth, Bob, 171, 181, 240, 250, 313
    quoted, 250
Glennan, T. Keith, 171, 187, 230
Glushko, Valentin, 136, 138–41, 139,
    161, 162, 164
    sent to the Gulag, 137
Godard, Jean-Luc: Alphaville, 403
Goddard, Esther, 65, 68, 176, 177,
    322, 324
Goddard, Robert, 62, 66, 68–74, 147,
    311
    anticipates space travel, 57–8, 59,
        60–4, 73, 151–2, 297, 322, 324,
        423, 424
    boyhood/adolescence, 57–8
    character, 65
    contacted by Lindbergh, 56, 65,
        68, 93
    death, 130
    funding, 60, 61, 63, 64, 65–6,
        67, 79
    his parents, 57

of interest to the Germans, 83,
    130–1
launches rockets, 66–7, 68, 131,
    372
overtaken, 72
quoted, 58, 61, 64, 73
relocates to New Mexico, 66
    see also Roswell
reputation, 64, 135–6, 175–7
secretiveness, 70–1, 176
Göring, Hermann, 87–8, 92, 398–9
Gruppa Izucheniya Reaktivnogo
    Dvizheniya (GIRD), 136, 137
    see also Reactive Scientific
        Research Institute
Guggenheim, Carol, 56, 64
Guggenheim, Harry, 40, 53, 56,
    65–6, 67, 69, 70, 130, 131, 175, 177
Guggenheim Aeronautical Laboratory
    at the California Institute of
    Technology (GALCIT), 69–70

Hamilton, Margaret, 195
Hasselblad, Victor, 202, 362
    his cameras, 202, 255, 257, 323
Hawking, Stephen: quoted, 420
Hearst, William Randolph, 42, 52,
    86
Herrick, Myron T., 50, 53
Himmler, Heinrich, 106–7, 108, 109,
    113
    quoted, 111–12
Hitler, Adolf, 88–9, 105–9, 112–13,
    115, 126, 128, 355, 377, 398–9
    and America/Americans, 87, 100,
        102, 104
    Berchtesgaden, 119
    death, 127
    Lindbergh on, 92
    at Peenemünde, 82
    rise of, 78–9, 80
    Wolf's Lair, 105, 108
Holliday, Clyde T., 142, 143
Hotz, Robert: quoted, 327–8
Hovhaness, Alan: quoted, 369–70
Hoyle, Fred, 58–9

quoted, 58, 362
Hugh, Fitzroy, 50, 52
Huntsville, Alabama (under
    changing titles), 155, 158, 166, 169,
    172, 178, 201, 340, 403–4
    led by John Medaris, 155, 187
    memorial service for von Braun,
        399
    von Braun sets up, 144, 145, 154

International Geophysical Year (IGY),
    153, 157

jet engines/planes, 70, 106, 131, 136,
    145, 174
jet trainers, 251–2
Johnson, Katherine, 194–5, 334
Johnson, Lyndon B., 187, 188–9, 196,
    262, 307, 324
    quoted, 159, 194, 307

Kammler, Hans, 111, 113, 125–6, 127,
    129–30
Karth, Joseph: quoted, 264
Kennedy, Jackie (Onassis), 324
    quoted, 235
Kennedy, John F., 205–6, 217, 276,
    398–9
    assassinated, 235
    commits to Apollo project,
        188–90, 195, 231, 232, 248, 319,
        329
    Cuba crises
        Bay of Pigs invasion, 187–8
        Missile Crisis, 229
    initial indifference to space race,
        187, 234–5
    quoted, 189, 205, 231, 233, 235, 430
    takes office, 185–6, 187
Kennedy, Joseph, 92–3, 103
Khrushchev, Nikita, 161–2, 163, 168,
    183, 185, 188, 203–4, 229, 236
    quoted, 261
King, Martin Luther, 275, 404
    quoted, 301
Korean War, 145, 311

Korolev, Ksenia, 137, 138, 140
    quoted, 262
Korolev, Sergei, 137–41, 139, 181, 231,
    277, 278
    building rockets, 161–2, 168, 186,
        350–1
    death and reputation, 261–2
    and Gagarin, 182
    and orbit, 173–4
    quoted, 262
    sent to the Gulag, 137–8, 261
    and Sputniks, 162–5, 167
Kraft, Chris, 250, 273, 276–7
    quoted, 200, 295, 319
Krantz, Eugene, 293, 294, 314
    quoted, 293
Kubrick, Stanley,
    2001: A Space Odyssey, 147, 286,
        371
    Dr Strangelove, 160, 403

Laika, 164
Lambert, Albert Bond/his field, 29, 37
Land, Charles Henry, 13, 14
Lasswitz, Kurd Auf zwei Planeten,
    75, 77
Life magazine, 165, 183, 195, 196,
    200, 213, 223, 320, 348, 362
Lindbergh, Anne (née Morrow), 55,
    283, 302, 375, 379, 381, 382–7
    and America First, 96
    on diplomacy, 158
    Gift from the Sea, 389
    on Goddard, 64–5, 68
    Listen! The Wind, 96, 382
    marriage (progress of), 54–5,
        84–6, 91, 96, 378, 379, 381–6,
        389–2
    on Nazis, 92
    as a pilot, 55, 387
    The Wave of the Future, 96–7, 384
    as writer, 86, 96–7, 382–3
Lindbergh, Ansy, 378, 386
    quoted, 385–6
Lindbergh, August (Ola Månsson),
    11–12, 12

Lindbergh, Charles, 75, 89, 311, 315, 374, 375, 381, 425
  aerobatics, 16–17, 18
  as an executive, 55, 170
  anticipates space travel, 56, 59, 69
  anti-semitic views, 87, 95–7, 387
  and astronauts, 297, 306, 307, 308, 312, 312–13, 323, 334, 378–9, 424
  automobile driver, 14–15, 239, 388
  birth/boyhood, 13–14, 15, 239
  bombing raids, 118
  brushes with death, 20–2
  character, 15, 16, 19, 23–5, 43–5, 47, 50, 54–5, 386–9
  courting danger, 16–17, 18–22, 25, 27, 37, 41, 43, 44, 47, 84, 117
  decline and death, 389–93
  diplomacy, 49–50
  emigrates, 86
  environmentalism, 377–81, 398
  farming, 15–16
  on Hitler, 92
  as husband and father, 84–5, 381–2, 384–9
  injuries, 65
  isolationist, 88, 91–2, 93–8, 103
  learns flying, 16, 17
  lend lease, 94
  listomania, 32–3, 386, 388–9
  mail pilot, 18–27, 90–1
  marksman, 14, 15
  meeting Göring, 88, 92
  meets/supports Robert Goddard, 65–6, 69, 98, 130, 131, 175–6, 177
  mysticism/religion, 23–4, 99–100, 376–7, 379, 380–1, 390–1, 423–4
  New York to Paris flight, 27–48, 36, 190, 387, 390, 421–2
    American reception, 51–2
    French reception, 49–50
    his speech, 49–50
  philosophy/overview, 375–9
  psychology/hauntings, 39–40, 45–6, 53, 386, 389, 422–3
  reading of international politics, 88, 91–2, 93, 97, 101, 150
  secret families, 381–2, 385, 392–3
  The Spirit of St Louis (the book), 384–5
  The Spirit of St Louis (the plane), 29–48, 35, 36, 50–1, 51, 52, 274, 312–13, 313, 388, 421–2
    construction, 31–2, 34, 35, 51
    equipment, 32–3
  on a stamp, 51, 51
  training fighter pilots, 117–18
  We (memoir), 53, 197
Lindbergh, Charles Jr, 85, 387
  kidnapped/murdered, 84–5, 382, 384, 389
Lindbergh, Charles Sr, 12–14, 15, 15, 17, 18, 28, 37
Lindbergh, Eva, 13, 14
Lindbergh, Evangeline Lodge Land, 13, 14, 15, 15, 16, 18, 28–9, 41–2, 53
  quoted, 28, 42
Lindbergh, Jon, 85–6, 382
Lindbergh, Land, 384, 390
Lindbergh, Reeve, 25, 97, 384, 386, 391, 393
  quoted, 25, 97–8, 386, 387, 392
Lindbergh, Scott: quoted, 392
Longfellow, Henry Wadsworth
  The Builders: quoted, 32
Lovelock, James, 367, 369
  Gaia hypothesis, 368
  quoted, 368
Low, George, 269, 271, 275, 280, 396
  quoted, 396
Luftwaffe, 79, 80, 87, 106
  doodlebug, 106
Lunar Orbiter program, 263–5, 318–19, 364
Lutheran Church, 11, 149, 395, 398, 404

MacCormack, John, 412, 413, 414
MacDonald, Charles, 117

quoted, 117, 118
MacLeish, Archibald, 300–1
   quoted, 301
Magee, John, 258, *258*, 345
   quoted, 258–9
Manhattan Project, 147, 150
Mansur, Charles, 67, 68–9
Mariner program, 263
Mars, is there life on?, 367
   see also rockets: to Mars; van Braun
McNamara, Robert, 229–30
Medaris, John B., 155, 156, 157, 165,
   166, 167
   quoted, 156, 157, 167
Mercury project, 170–1, 174, 178, 180,
   181, 247–8, 250, 253, 260, 282,
   307, 334
   capsule, 184, *184*, *191*, 199, 233
   manned flights, 184, 191, 192, 193,
     195, 199, 200, 202, 206
   Mercury Seven, *174*, 174–5, 181–2,
     195, 196, 197, 233, 234, 238,
     249, 339
     We Seven (collective memoir),
       197
   post-Mercury, 229, 233, 242, 267
   test flights, 180–1
Minnesota, 12, 14
Mittelwerk, 109–11, 119–24, *123*, 130,
   131, 140, 401
   camp crimes trial, 144
   Dora camp, 110, 111, 119–20, 122,
     124, 402
   Lindbergh visits, 119, 124–5
   Nordhausen camp, 110, 119–20,
     122–6, *123*, 138–9
moon trees, 338–9
Morrow, Dwight, 53–4
Morrow, Elizabeth (Lindbergh's
   mother-in-law), 54–5, 92
   quoted, 387
'mother Earth', 20, 301, 420
Mueller, George, 230, 231, 269, 270
Murray, Garth, 212–13, 214, 220,
   408, 409–10, 411, *412*
   disappearance and fate, 412–14

Murray, Robin, 408, 409, 410, *411*,
   412–13
Murray III, William 'Bill', 211–23, *221*,
   408, 414
   quoted, 212, 213, 215, 216, 222
Murray O'Hair, Madalyn, 206,
   209–226, *219*, *221*, 363, 405–15,
   406, *411*
   birth/youth, 209–10
   campaign against religion, 14–15,
     210–26, 299–300, 303, 316,
     322, 342, 405–9
   defection (attempted), 213
   disappearance and fate, 412–14
   flight from police, 221–5
   lawsuits, 215–21, 302, 328, 342,
     406–7
     Supreme Court rulings,
       219–20, 406–7
   quoted, 213, 214, 217, 221, 222,
     407, 408, 409, 410

Neufeld, Michael *Von Braun: Dreamer
   of Space*: quoted, 80, 149, 250, 395,
   402
New York Times, 100, 154, 164, 324
   on Apollo, 299, 300–1, 325
   on Goddard, 60–1, 324
   on Lindbergh, 50, 52, 53, 91, 116
   on Murray O'Hair, 220
   on von Braun, 159
Nicolson, Harold, 87, 91
Nixon, Richard, 185, 321–2, 323, 396,
   398–9, 405, 407
   public enthusiasm, private
     ambivalence to the space
     program, 306, 307, 324, 350,
     396
   quoted, 324, 416
Noordung, Hermann, 147
North American (engineering firm),
   266–70
Nungesser, Charles and François
   Coli, 34, 35–6, 38, 39, 49

Oberammergau, 126, 127

Oberth, Hermann, 75, 76, 136, 147, 151–2, 176, 308
   Die Rakete zu den Planetenräumen, 72–4, 75
   quoted, 261, 308
O'Hair, Madalyn
   see Murray O'Hair
O'Hara, Dee, 308
   quoted, 344
Orbiter project (aka Project Slug), 152, 153, 154, 155, 158, 159, 165, 166, 178
   see also satellites: Explorer
Orteig, Raymond/his prize, 27, 28, 30, 33, 34, 35, 38, 39
Our World (BBC satellite hook-up), 272–3

Paine, Thomas, 328, 332, 395–6, 407
Pan American Airways, 84, 85–6, 96, 116
Paul VI, Pope, 322
   quoted, 301–2
Pearl Harbor, 103–4, 210
   as metaphor, 158
   Peenemünde (Army Research Centre), 107, 110, 145, 146, 152, 253, 402, 404
   air attack, 109, 111
   dispersal of the staff, 130, 139, 144, 172
   evacuation, 125–6
   Himmler's visits, 106–8
   Hitler's visit, 82
   Werk Ost (army), 79–80, 82–3
      rocket launches, 81, 83, 108
   Werk West (Luftwaffe), 79–80
perfusion pump, 98–9
Petrone, Rocco: quoted, 270, 273
Phil Donahue Show, 226, 409
   Donahue quoted, 415
Picard, Auguste, 76, 77, 81
Poole, Robert *Earthrise*, 249
   quoted, 317, 369
Pravda, 163, 183, 262
press,

cooling towards Lindbergh, 116
enthusiasm for Lindbergh, 50, 52
interest in Goddard, 60–1, 63–4, 64–5, 70
interest in von Braun, 171
international, 34, 35, 51, 164, 325
Lindbergh's hatred of, 41–2, 64–5, 86, 391, 392
pictures from space, 273, 299
preoccupations/misreporting, 54, 99, 175, 318, 324
scorn at the US space program, 158, 166, 183
sensational stories, 84–5, 269, 398
proto-Earths, 426
Proust, Marcel: quoted, 25, 26
Psalms, 323, 342, 345, 390, 399

Reactive Scientific Research Institute (RNII), 137
The Realist magazine, 217, 218, 225, 363
recovery vehicles, 179
Rhineland, occupation of, 88–9
rockets/spaceships/capsules, 66–7, 68, 74, 106, 135, 148, 168, 177, 399–400
   Agena, 243, 251, 253, 256, 259, 260
   Atlas, 170–1, 180, 181, 184, 186, 243, 251, 253, 284
   dreams of, 56–60, 75, 79, 108, 135, 136, 189, 213, 346, 421, 424
   gravity defied, 163, 173, 174, 189, 307
   Jupiter/Juno, 153, 154–5, 165, 167, 174, 177, 178, 179, 180, 186, 188
   a launch and flight in detail, 279–97, 305–29
   liquid-fuelled, 58–77, 62, 66, 82, 128, 137, 176
      alcohol/liquid oxygen, 78, 113
      gasoline/liquid oxygen, 56
      kerosene/liquid oxygen, 283

liquid hydrogen/liquid oxygen, 61–3, 73, 281
with living creatures, 143, 164, 177–8, 180–1, *181*, *191*, 277–8, 351
Luna, 173, 179, 262, 351
manned moon shots, 59, 188–9, 230–2, 244, 246–8, 256, 260, 262–3, 307
  cost, 190
  envisaged, 180
  far side of the moon, 289–90, 291–2, 295, 321, 334, 348
  fly-by, 277
  moonwalks post-Apollo 11, 248, 330–1, 337, 341–2, 349–50
  see also Apollo
manned space flight, 76, 81, 168, 170, 180–1, 182–4
  see also Gemini; Mercury; Vokshod
to Mars, 57–8, 64, 79, 136, 141, 147, 231–2, 339, 395–6
  reached, 262
to the moon
  crash-landing, 179, 232
  missed, 173–4, 232
  orbited, 262, 263
  past the moon, 277–8
  public/press scepticism concerning, 146, 164
  soft-landing sought, 188, 232–3, 262
  speculation, 60–1, 63–4, 77, 108–9, 129, 135–6, 141, 144
  see also Apollo
multistage, 59, 135, 145, 151–2
N1, 261, 278, 350–1
to outer space, 72, 82, 106, 112, 125, 128, 135, 173
projected commerciality, 72–3
Proton, 256, 277, 351
Redstones, 145, 152–3, 154–5, 157, *165*, 180, *180*, *181*, 184, 201
re-entry, 143, 182–3, 191, 276, 277, 296, 329–30, 351

rocket clubs, 72
R-2/R-3, 140–1
R-5/R-7, 161–2, 186
R-12/R-14/R-16, 186, 188
R-11/SCUD missile, 262
Saturn, 170–1, 180, 230–1, 235, 250, 261, 267, 272–5, 281–4, 351, 399–400
  see also von Braun: 'his masterpiece'
solid-fuelled, 59–60, 152
Surveyor (lunar lander), 263, 318–19
Thor/Thor-Able, 154, 178, 179, 186
Titan, 211, 232, 250, 251, 253, 284
in the UK, 186
V-1, 106, 110, 113, 131
V-2 (aka A-4/R-1), 82, 105–14, *115*, 124–7, 130–2, 138, 139–45, 152, 161, 175–7, 372, 401
  attacks on London, 114–15
Venera, 262
Viking, 152, 153
with warheads, 78, 83, 93, 106, 113
  ballistic missiles, 108, 113, 152, 153, 154, 160, 161, 186
  cruise missiles, 113, 145
  guided missiles, 113, 114, 141, 144
  long-range missiles, 140, 141, 160, 161, 162
  SCUD missiles
    see rockets: R-11
  see also atom bomb
Roosevelt, Franklin D., 89–91, 95, 116, 150, 210, 307
  death, 119
  'fireside chats', 94, 165
  and Pearl Harbor, 103–4
  quoted, 91
  and war in Europe, 88–9, 93, 94, 95–6, 97
Roswell, New Mexico, 66, 67–8, 93
  Goddard Museum, 176
Ryan, Claude/his airline, 30–4

Sagan, Carl, 347, 368, 419
  quoted, 402, 418, 419–20
Saint-Exupéry, Antoine de, 117–18,
  320, 382–4
  quoted, 22, 23, 24, 25, 26, 383
  samdhi (expanded consciousness),
  370, 372
Sandys, Duncan, 109
  quoted, 114
satellites, 70, 147, 150–2, 153–4, 272,
  348, 397
  ATS, 273, 274, 320, 364, 397
  broadcast pictures, 272–3
  Department of Defense
    Gravitational Experiment
    (DODGE), 272, 273, 364
  Discoverer, 179
  dummy, 155
  Explorer, 166, 167, 168, 178–9, 230
  geosynchronicity, 272–3, 290–1
  Intelsat/Early Bird, 273, 289
  Sputnik, 156–9, 160, 161, 162–4,
    166, 168, 213, 310
    'Sputnik panic', 169–70
  Sputnik 2, 164–5, 168
  Sputnik 3, 168
  spy, 154, 179
    see also U-2 spy plane
  Vanguard 1, 168
  weather, 154, 179, 273
Schopenhauer, Arthur: quoted, 369
Sedov, Leonid, 164
  quoted, 164
Shea, Joe, 269–70
Sikorsky, Igor, 116
  quoted, 116–17
Smith, Truman, 87–8, 89, 89, 92
Smithsonian Institution, 60, 61, 63
    National Air and Space
      Museum, 178
Soviet Union
    Communism/Communist, 101,
      185, 220
    Cold War, 163–4, 166, 185, 203,
      220, 302
    eye on German talent, 127, 129

    potential fellow travellers, 213
    Soviet Embassy in Washington,
      157, 213
    space exploration, 133–41, 146
    US ease with Communism, 150
    US fear of Communism, 86, 149,
      159
    World War II, 104, 126, 127
    see also space race
Soyuz project, 261, 262, 271, 277–8
    Zond missions, 277, 351
space docking, 233, 247, 249, 251,
    253, 256, 259, 260, 277, 304, 309
space evacuation, 184, 194, 245–6,
    279–80, 288
space laboratory, 188
space photography/picture
    transmission, 199, 201, 240, 243,
    255, 257, 260, 333, 418–19
    Blue Marble shot, 351, 361–2, 367
    of Earth from space, 240–2, 247,
      249, 259, 263–5, 264, 274–5,
      304–5, 349
      partial/occluded, 178, 178–9
      as a whole sphere, 286, 351,
        361–8
    Earthrise shot, 298, 300, 301, 302,
      361, 362, 364, 367
    of the moon, 232, 241, 262–5, 264,
      319, 330, 343
    Pale Blue Dot shot, 419–20
    from V-2s/Aerobee, 142–3, 143
space planes, 249
    X-15, 248–9
space race, 146, 153–4, 156–9,
    168–70, 187, 188, 190, 229,
    244–5, 261–2
    arms race/missile gap, 160, 169,
      186
    domestic race, 165
    moon race, 231, 236, 249, 277, 351
    Russian 'threat', 150–1, 161, 174
space rendezvous, 232, 233, 239, 243
space shuttle, 232, 262, 416
space sickness, 288–9, 296, 304

space stations, 135, 147, 150, 180, 231–2, 335, 339
  Salyut 1, 351
space theology, 203–6
spacewalk/Extra-Vehicular Activity (EVA), 237, 240, 253–4, 257, 259, 260, 304, 343, 350, 370–1
Speer, Albert, 105–6, 107, 108, 111, 112–13
  quoted, 105
Stalin, Joseph, 137, 138, 139–40
stamps, 51, *51*, 302
Stapledon, Olaf: quoted, 368

Teilhard de Chardin, Pierre, 394–5
  quoted, 395, 420
Teller, Edward, 158, 169, 403
Thompson, William Irwin, 347, 369
Thoreau, Henry, 366
  quoted, 377
Time magazine, 99, 141, 153, 174, 320, 370
The Times (London), 102
Toftoy, Colonel Holger Nelson, 124, 129, 144
Tolstoy, Lev, 134
Transcontinental Air Transport (TAT), 55–6
Treaty of Versailles, 27, 88–9
Truman, Harry S., 130, 150
Tsiolkovsky, Konstantin, 64, 74, 133–6, *134*, 147, 151–2
  anticipates space travel, 134, 135
  arrest and imprisonment, 135
  quoted, 134, 135, 420

Underwood, Richard, 201–2, 240–2, 255–6, 263, 298, 323, 343, 349, 351, 361, 397–8
  quoted, 240–1, 241, 263, 298, 324
urination
  see space evacuation
U-2 spy plane, 185, 186, 188

Vanguard project, 152–9, 164, 166–7, 168, 171, 178
Verein für Raumschiffahrt (VfR), 72, 73, 74, 76, 77, 112, 147, 148
  funding, 77, 78
Vernadsky, Vladimir, 366
  quoted, 368
Verne, Jules, 74, 134, 136, 296, 310
  quoted, 420
Vidal, Gore: quoted, 393
Vietnam War, 196, 270, 307–8, 332, 401, 418
Vokshod project, 236, 260–5
  crew, 236, *236*
Volkswagen, 239, 310, 382, 388
vomiting
  see space sickness
von Brauchitsch, Field Marshal Walther, 107
von Braun, Baron Magnus, 74, 80, 144, 167
von Braun, Magnus Jr, 112, 127, *128*
von Braun, Maria (*née* von Quistorp), 144, 149
von Braun, Sigismund, 80–1
von Braun, Wernher, 74–6, 110, 126–8, *128*, 169, 275, 374, 399–400, 417
  to/in America, 125, 129–30, 141–9
  American citizenship, 148, 176
  anticipates manned space flight, 81, 82, 168, 339
  anticipates moon landings, 76, 77, 108–9, 129, 146
  anticipates satellites, 128–9, 272, 397
  anticipates space rockets, 112, 125, 128, 146
  anticipates value of space photography, 304–5, 357
  boyhood, 74–5
  character, 77, 80, 110, 111, 146–7, 172–3, 250
  collaborations with Disney, 148, 153, 166
  courting danger, 173

excluded, 153, 164, 168, 170–1, 250, 396
fame, 148, 153, 165–6, 167–8, 172
his family, 74–5, 75–6, 79, 80
illness and death, 395, 398
included, 151, 159, 165, 167–8, 172, 274, 324
injured, 126
launches rockets, 77, 78, 81, 83, 105–8, 141–2, 145, 155, 177, 181, 184, 261
  'his masterpiece': Saturn V, 273–5, 281–4, 351, 399–400
and Lyndon Johnson, 188
and Mars project, 395–6, 400, 416
The Mars Project, 147, 148, 232, 339
meets Hitler, 82, 105, 108
military promotions, 107, 108
and Moon Buggy, 340
and NASA, 171–2, 231
in 'protective custody' and release, 112–13
reading of international politics, 150, 162, 229, 397
religion/mysticism, 148–9, 336, 394–5
in the SS/on Nazi business, 106–7, 372, 396, 400–2
working for the German army, 78–9, 105
von Grosse, Aristid, 150, 151
von Kármán, Theodore, 70, 106
  Kármán line, 106, 248–9

quoted, 70, 176–7
Vonnegut, Kurt: quoted, 397
Vostok (Soviet manned rocket programme), 170, 182–3, 187, 197, 260
Voyager project, 418–20, 421

Wainwright II, Loudon, 195, 196
Warhol, Andy *Empire*, 364
Waters, David, 410–14
Weather Bureau, 38, 64
Webb, James, 187, 188, 189, 190, 231, 233, 241, 269–70, 275
  quoted, 230, 269
weightlessness/zero gravity, 20, 177, 197, 198, 246–7, 248, 254, 285, 287, 296, 343, 344, 422
Wendt, Guenther, 253, 283, 307
White Sands (rocket test site), 142, 143, 145, 372, 375
Wiesner, Jerome 'Jerry', 187, 189
  quoted, 233
Wilson, Defense Secretary Charles E., 154, 155, 158
Wolfe, Tom,
  quoted, 402–3
  The Right Stuff, 249
Woodruff, Dean, 315–17
  quoted, 316–17
World War II, 82–3, 93–8, 103–4, 105–15, 116–30, 131–2, 150, 151, 372

Yangel, Mikhail, 186
Yeager, Chuck, 249
  quoted, 249

21982031792785